AF323508

Accelerator and Radiation Physics

Accelerator and Radiation Physics

P.K. Sarkar
Samita Basu
Maitreyee Nandy

Narosa Publishing House
New Delhi   Chennai   Mumbai   Kolkata

Accelerator and Radiation Physics
358 pgs. | 164 figs. | 43 tbls.

P.K. Sarkar
Health Physics Division
Bhabha Atomic Research centre
Mumbai

Samita Basu
Maitreyee Nandy
Chemical Science Division
Saha Institute of Nuclear Physics
Kolkata

NAROSA PUBLISHING HOUSE PVT. LTD.

22, Delhi Medical Association Road, Daryaganj, New Delhi 110 002
35-36 Greams Road, Thousand Lights, Chennai 600 006
306 Shiv Centre, Sector 17, Vashi, Navi Mumbai 400 703
2F-2G Shivam Chambers, 53 Syed Amir Ali Avenue, Kolkata 700 019

www.narosa.com

ISBN 978-81-8487-182-1

Published by N.K. Mehra for Narosa Publishing House Pvt Ltd.,
22, Delhi Medical Association Road, Daryaganj, New Delhi 110 002

Printed in India

We gratefully acknowledge cooperation and support from

CBAUNP Project, Saha Institute of Nuclear Physics
Board of Research in Nuclear Sciences, Govt. of India
Department of Science and Technology, Govt. of India
Atomic Energy Regulatory Board, Govt. of India
Council of Scientific and Industrial Research,
Govt. of India

Foreword

The present volume on Accelerator and Radiation Physics is an endeavour to exchange information and share knowledge between experts and researchers in the field of basic radiation physics, accelerator physics, safety and related issues.

For the last several decades accelerators and other radiation facilities have become an integrated part of our society with its varied applications for benefit of mankind in various facets in spite of the harmful effects when handled inappropriately. Physics and technology of accelerator, reactor and the radiation environment around such facilities have emerged as important and independent research arena to deal with the physics and safety of its application and future development. The last few decades presented us with the state-of-art radiation technology with the development of synchrotron radiations, high energy particle storage rings, accelerator driven sub-critical systems, laser plasma, etc.

Efficient and safe use of all these facilities have called for improved modern radiation dosimetric techniques which are integral part of radiation physics research. But, changes in the ICRP defined fluence-to-dose conversion coefficients over time require determination of radiation energy distributions (rather than dose), which remains invariant for a given reaction system and can be converted to radiation dose. Evaluation of spectrum and dose needs to be done using both theoretical and experimental methods but practical measurements is always preferred over the computational results. To meet this requirement particle and gamma radiation detection systems should be continuously upgraded.

Accelerator science and technology has come a long way since the invention of the first accelerator. But accelerator radiation protection is very different from the conventional radiation protection modalities of reactors. Thus accelerator science has necessitated a completely different approach for physics and safety evaluation in accelerators. Radiation effects to the environment including human and non-human biota are too costly to be neglected. All these effects have called for in-depth studies and judicious practice of radiation safety protocols in the area. This subject is of prime importance to our Institute as we plan to take up development of a high-energy high-brilliance synchrotron source.

The present symposium, ISARP, is the first international symposium in the field of accelerator radiation physics in eastern India over the last several decades. The symposium has endeavored to address all the above mentioned areas of accelerators and radiation physics, to provide the practitioners in the field with a common platform for exchange of ideas and to generate new collaborations.

Professor Milan Kumar Sanyal
Director
Saha Institute of Nuclear Physics
Kolkata

Preface

This volume presents a world-wide scenario of research and development in the field of accelerators and radiation applications containing 58 contributions from scientists from India and abroad. The first accelerator of India was built in Saha Institute of Nuclear Physics, and the Institute is going to install another accelerator shortly and plans to take up development of a high energy high brilliance synchrotron source.

We are grateful to Professor Milan Kumar Sanyal, Director, Saha Institute of Nuclear Physics, Kolkata for his whole-hearted support and constant encouragement to make the Symposium a success. Our sincere thanks to our invited speakers Kamalesh Kar (India), S. Kailas (India), S. Rokni (USA), K. Pradeepkumar (India), K. B. Sainis (India), L. C. Tribedi (India), A. Datta (India), S. Augosteo (Italy) A. Esposito (Italy), S. Pomp (Sweden), K. P. N. Murthy (India), H. Nakashima (Japan), S. Markondeya (India), T. Bandyopadhyay (India), C. Sunil (India), S. P. Tripathy (India) and G. Haridas Nair (India) and chairpersons of the technical sessions P. K. Sarkar (India), N. P. Bhattacharyya (India), D. Dasgupta (India), P. Banerjee (India) and S. C. Roy (India) for their help and support. We also express our sincere thanks to the authors of all contributed papers.

We gratefully acknowledge the financial support received from CBAUNP Project, Saha Institute of Nuclear Physics. We also thank various Government of India organisations like Board of Research in Nuclear Sciences, Department of Science and Technology, Atomic Energy Regulatory Board, Council of Scientific and Industrial Research, for financial support.

We are thankful to all members of the Organising Committee for their keen interest and help in the management of the Symposium. We express our gratitude to all members of the International Advisory Committee for their support and guidance. We also thank all members of Chemical Sciences Division, SINP for their cooperation. Lastly we would like to express our sincere gratitude for all others who have helped us in various ways to make the Symposium a success.

P.K. Sarkar
Samita Basu
Maitreyee Nandy

Contents

Section C: Application of Accelerators in Advanced Nuclear Research and Technology

Section D: Recent Advances in Radiation Detection, Measurement and Applications

Section E: Application of Accelerators in Biomedical and Material Science Research

Section F: Enivironment Radiation Monitoring and Control

Radiological Safety and Prevention and Preparedness for Response to Radiological Emergencies

Pradeepkumar K.S.

Emergency Response System & Methods Section Radiation Safety Systems Division
Bhabha Atomic Research Centre Mumbai 400085, India
E-mail: *pradeep@barc.gov.in*

ABSTRACT

Large number of nuclear facilities including nuclear power Reactors are safely operating world over and usage of radiation and radioisotopes in medical, agriculture, industry are increasing. The major nuclear accident at Chernobyl and many radiological accidents associated with radioactive sources have led to serious concerns related to possible radioactive contamination of environment and radiation exposure to public. International agencies like IAEA and ICRP are concerned of the radioactive contamination in public domain and the capability of 'denial of area' and the fear factor which can be injected during any malicious acts using radioactive sources. A number of highly sensitive monitoring systems and methodologies have been developed to detect illegal movement of radioactive sources and to locate, identify and search stolen radioactive sources. As required for the emergency preparedness, the nation has to (i) develop large number of Emergency Response Centres (ERCs), trained first responders with Central and state governments, more state of the art radiation monitoring systems, (ii) strengthen the capacity for medical management of radiation emergencies, (iii) increase the public awareness programme etc. to prevent and reduce the consequences of radiological emergencies.

Keywords: Radiological, Emergency-response, RDD, First responder, ERC

1. INTRODUCTION

There is always high level of public anxiety in any 'nuclear/radiation' related news. There are several common misconceptions to the extent that any news on a minor incident in a nuclear facility is misinterpreted and equated to a Chernobyl-type major nuclear accident or it is compared with a nuclear explosion or for projected occurrence of high cancer incidence, deformed children etc. Hence, it is essential that any possible nuclear/radiation emergency including their preparedness for response is addressed in rational manner without any preconceived notion/bias. While any terrorist events can spread fear among the society, any suspected presence of radiation, which cannot be seen, smelt or felt can create severe psychological impact, if, not responded in time.

Large number of nuclear facilities including nuclear power reactors are safely operating world over and usage of radiation and radioisotopes in medical, agriculture, industry are increasing. The major nuclear accident at Chernobyl and many radiological accidents as Goiania, Yenango [1, 2] etc. associated with radioactive sources have led to serious concerns related to possible radioactive contamination of environment and radiation exposure to public. These have forced the world to prepare for responding to situations where those believing they are radiologically affected can be extremely large in proportion to small number of persons who actually may be affected. The radiological incident at Delhi made many believe to be affected by gamma radiation from the Cobalt 60 sources that had inadvertently reached Mayapuri scrap market from Delhi University. While this incident exposed the lack of awareness of radioactivity and radiation among the public, police and medical community, it also confirmed the fact that any news related to radiation can create more panic than any other emergency situation.

All nuclear facilities are designed and operated to ensure the safety of the occupational workers, members of the public and the environment. Though the nuclear power reactors are having very large quantity of radioactive isotopes inside the reactor core, during the normal operation the release of radioactivity to the environment is extremely low and exposure to the member of the public from this is negligible compared to the exposure from natural radiation. Due to engineering safety features incorporated into the design along with 'defence in depth' philosophy adopted, probability of a major nuclear accident having potential for release of significant quantity of radioactivity into the environment is extremely small. Inspite of this, emergency preparedness is in place for all Indian nuclear facilities to reduce the consequences, if at all any major release occurs to environment.

2. 'ORPHAN SOURCES' AND THE NEW CHALLENGES

Adequate security through approval of secured installation and storage of these sources and the regulatory, engineered safety and additional administrative controls ensure 'cradle-to-grave' control of these sources. A source becomes 'orphan' i.e., goes out of regulatory control due to: (a) lack of or breach of the provisions, (b) import of sources without licence deliberately or out of ignorance, (c) loss or misplacement, (d) theft, (e) unauthorised disposal, (f) mishandling of source consignments at entry and exit ports of the country, (g) lack of awareness in the public and the officials. It is reported that, within the last few years, many radioactive sources were lost or misplaced in certain countries/regions. These have posed radiological risk to the mankind the world over. Possibility of radioactive sources stolen or illegally brought into any country also cannot be ruled out.

Radiogical Emergency situations due to orphan sources are reported from many counties [1, 2]. The sources responsible were mainly ^{60}Co, ^{137}Cs and ^{192}Ir. The situations resulted in fatalities or severe injuries to people. In India, in medical institutions there were few cases of lost brachytherapy sources and those involving industrial radiography sources and nucleonic gauges, though there was no fatality reported in such incidents.

Malicious use of Nuclear and other radioactive material can be divided into three distinct types:

Radiation Exposure Device (RED)

An RED is a sealed radioactive source (partially or fully unshielded) that is intended to expose people in the vicinity of the device to radiation emitted from it. Radioactive source hidden on a subway or in a sports arena, where people would unknowingly receive radiation exposure can cause high radiation exposure to the victims who were near to it, but may not lead to spread of contamination.

Radiological Dispersal Device (RDD)

An RDD may be designed by the terrorists to spread radioactive contamination by exploding radioactive material integrated with explosive and can lead to spread of contamination depending on the source used as well as the quantity of explosive used.

Improvised Nuclear Device (IND)

An IND incorporates nuclear materials designed to produce a nuclear explosion
Radiological consequences of RED and RDD would depend on:
(a) Source: Type, isotope, activity
(b) Proximity of the person to the source/shielding in between
(c) Total time spent in proximity to the source
(d) Whether it is whole body or portion of the body received exposure

3. RDD AS A THREAT

World over millions of radiological sources are in use/available-majority of sources are relatively harmless. Many countries are reported to lack effective control over radioactive sources/nuclear materials which increases the chances of radiological terrorism. Usage of an RDD may trigger panic out of proportion of actual risk to human health and safety, and hence:

   (i) intelligent use of RDD is to be anticipated

  (ii) terrorist trends to be assessed

(iii) Effects of RDD explosion under various conditions are to be studied

Likely Targets for Dirty bombs

  (a) Crowded markets.

  (b) Railway stations.

  (c) Parliament

  (d) Crowded Religious places.

  (e) Stock exchange

  (f) Major Public Events

3.1 Why Assembling e RDD with a 'Dangerous Source' is Difficult

It will be very difficult to assemble RDD with strong radioactive sources that would deliver radiation doses high enough to cause immediate health effects or fatalities in a large number of people. Due to the high contact radiation field, large number of persons may have to share the job of assembling the RDD (to keep individual exposure within threshold). Before the execution of the RDD explosion, all persons involved in their assembly, transportation as well as placing at various locations may get very high radiation burn (unbearable pain) and also may get incapacitated with radiation acute syndrome. This may lead to terrorist's efforts getting leaked out to the security agencies. May be due to the lack of the required technical capabilities to overcome this challenge, till now there is no case of successful RDD explosion anywhere in the world, though, few cases of attempts of smuggling of sources along with explosives are reported. Still, understanding the impact of RDD explosion as well as the developing the capability for response to such emergencies is very essential.

The threat from RDD can be of many dimensions:

1. The blast and fragmentation effects from the conventional explosive

2. The radiation exposure from the radioactive material used

3. The fear and panic that its use would spread among the target group or population

4. Denying the use of area/locations for extended period of time

Blast and thermal effects following a RDD explosion will be the only cause for immediate deaths, since lethal levels of radiation exposure from the dispersed materials requires extremely large quantity of radioactive material to be integrated with the RDD. Panic caused by fear of radiation can result into disruption and the required cleanup of the radioactive material and any consequent avoidance of the location (if not cleared of contamination even after many attempts of cleanup) may cause economic losses.

3.2 Pathways of Eradiation Exposure After an RDD Attack

(a) Immersion in plume-direct on ground and above ground
(b) Deposition of activity on walls and ground (i.e., contamination of surfaces and areas)
(c) Re-suspension of particulates (i.e., settled particles
(d) Getting air borne under favorable weather condition)
(e) Inhalation of radioactive dust
(f) Intake through ingestion route (food/ water)
(g) Intake through grass, cow, milk pathway (for ^{131}I, ^{137}Cs etc.)

Keeping in view of the potential exposure conditions IAEA [7] has recommended the Inner Cordoned Area Radius (safe distances) for Radiological Emergencies [Table 1]

Table 1 *Inner Cordoned Area Radius (safe distances) for Radiological Emergencies (IAEA-2006)*

Situation	Initial radius of inner cordoned area (safe distance)
Damaged package Unshielded or unknown source (damaged or undamaged)	30 m radius or at: Ambient dose readings of 100 μSv/h 1000 Bq/cm^2 gamma/beta deposition 100 Bq/cm^2 alpha deposition
Spill	Spill area plus 30 m around
Major spill	Spill area plus 300 m around
Fire, suspected RDD, spent fuel, plutonium spill	400 m radius (or more to protect against effects of an explosion) or at: Ambient dose readings of 100 μSv/h 1000 Bq/cm^2 gamma/beta deposition 100 Bq/cm^2 alpha deposition
Explosion/fire involving nuclear weapons (no yield: only dispersion)	1000 m radius or at: Ambient dose 100 μSv/h 1000 Bq/cm^2 gamma/beta deposition 100 Bq/cm^2 alpha deposition

4. RESPONSE TO RADIOLOGICAL EMERGENCIES

Following a nuclear disaster/radiological emergencies, the response actions for reducing the consequences [4] are:

(a) Monitoring and quick assessment of the magnitude of damage
(b) Mobilization of resources at short notice
(c) Relief and rescue operation
(d) Distribution of iodine tablets at the earliest (wherever radioiodines are involved)
(e) Restriction of the consumption of potentially contaminated foodstuffs and substitution of food and water based on measurement of contamination
(f) Relocation of the affected population

(g) Sheltering

(h) Decontamination

(i) Access Control

(j) Evacuation

Taking into account of the likelihood of possibility of large number of people/area getting contaminated following a RDD explosion, the following are identified as the requirements for an effective response:

1. Monitoring large number of people (suspected to be contaminated)
2. Radiation Survey of large affected area
3. Décontamination facilites (personnel, houses, public places, vehicles, etc.)
4. Medical triage and treatment
5. Isolation and confinement of contaminated areas
6. Removal and disposal of contaminated soil
7. Teams ready to work in complex conditions and territories

While prevention of smuggling and illicit trafficking (Fig. 1) of nuclear and radioactive material are being achieved by the installation of large number of sensitive state of the art detection system at various strategic locations in the country, nuclear and radiological terrorism, though not successful till today, cannot be neglected. Hence development of 'First Responders'/Quick Response teams, preparedness for medical management of radiation injuries and public awareness on protective measures during radiation emergencies etc are being taken up at national level to reduce the consequences of any malicious usage of nuclear/radioactive material.

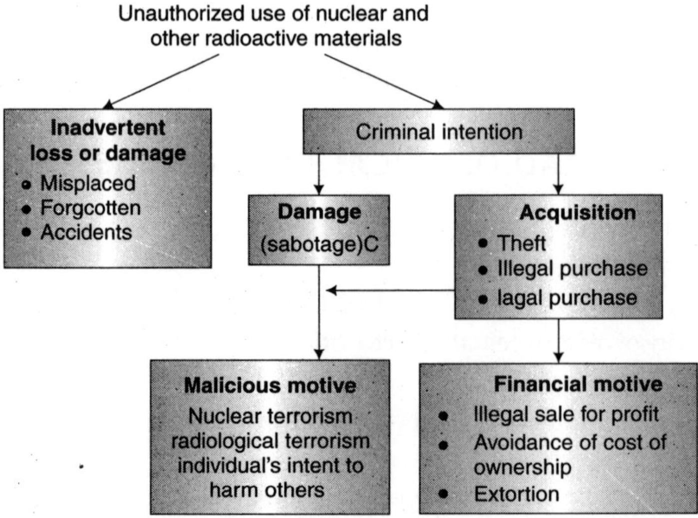

Fig. 1   *Routes for the unauthorized usage of nuclear and other radioactive material*

5. MONITORING SYSTEMS FOR PREVENTION AND RESPONSE TO RADIOLOGICAL EMERGENCIES

Various types of sensitive monitoring systems and methodologies have been developed for use at facilities handling/storing sources and for use in public domain to detect illegal movement of radioactive sources and to locate, identify and search stolen sources. Some of the important systems developed in BARC for prevention and response to nuclear and radiological emergencies are: (1) Aerial Gamma Spectrometry system (AGSS) for aerial radiation monitoring for detection and assessment [5], (2) Compact Aerial Radiation Monitoring System (CARMS) [6] for remote aerial monitoring using Unmanned Aerial Vehicles (UAVs), (3) Portal Monitor and Limb Monitor, (4) Vehicle scanning monitor (for inspecting goods/scrap carried by vehicles), (4) Installed Environmental radiation monitoring systems (IERMON) with data transfer facilities to ERCs. AGSS and CARMS are to be deployed on a mobile platform like airplane or helicopter, train or car/van, boat etc to perform quick environmental monitoring. Many aerial survey exercises and mobile radiation monitoring surveys/exercises are carried out to demonstrate the capability of these systems for searching of orphan sources and qualitative and quantitative estimation of radioactive contamination over large area on ground. Fig. 2 shows a typical display on AGSS screen showing the flight path and identification of 'source on ground' during an Aerial 'source search survey'.

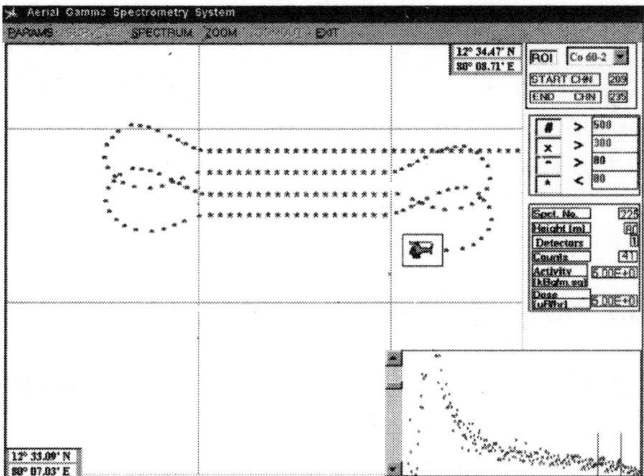

Fig. 2 *Typical display on AGSS screen showing the flight path and identification of 'source on ground' during an Aerial 'source search survey'.*

Portal, Limb and Vehicle monitors are being deployed as stationary but re-deployable monitors to detect unauthorised movement of radioactive material either by persons or in vehicles in public domain as well from nuclear facilities.

6. DAE-EMERGENCY RESPONSE CENTRES (DAE-ERCS) FOR NATIONAL LEVEL PREPAREDNESS

A network of Emergency Response Centres of Department of Atomic Energy (18 DAE-ERCs) spread over country has been planned with a nodal centre at BARC for responding effectively to any nuclear or radiological emergency anywhere in the country. ERC closest to the site of incident/accident will be activated by the centralised Emergency Communication Room (ECR) situated at Crisis Management Group (CMG), of Department of Atomic Energy (DAE), Mumbai on receipt of any relevant message. The CMG has been identified as the coordinator between the various agencies to facilitate a well-coordinated and effective response to nuclear or radiological emergencies. During the last few years, public functionaries like custom officials, police, fire brigade personnel and paramilitary forces are being trained in handling such radiological emergencies as 'First Responders'[7]. Expert Emergency Response Teams (ERTs) are being raised and trained at each DAE-ERC for developing an effective response to such emergencies.

The First Responders (Police, Fire Brigade is expected to reach the affected site immediately followed by the trained first responders from CPMFs after getting the information to carry out assessment of radiological status, rescue operation and to implement counter-measures. The First Responders will be supported by Emergency Response Team (ERT) of the DAE who may be called upon to reach the affected site at a later stage.

In India, Ministry of Home Affairs (MHA) and NDMA, in addition to raising many battalions of NDRF (National Disaster Response Force trained in radiological emergency response by DAE) for the nation, are equipping many police stations of major cities with radiation monitors which will also act as deterrents to radiological terrorism.

7. SUGGESTIONS FOR FUTURE STRATEGY

The following are suggestions for strengthening the emergency preparedness/response to radiological crimes:

(a) Radiation monitors/Systems to be kept at various locations.

(b) Have large number of Trainers with various agencies to develop trained Responders to radiation emergencies.

(c) Include "Disaster Management and knowledge on response to radiation emergencies/Nuclear disasters' in the syllabus of NDA, IAS etc.

(d) Forensic experts also to be trained on self protection and sampling of radiological crime scenes.

(e) Fire stations and Police stations to have radiation monitors with knowledge on monitoring.

(f) Create Deterrence for "smuggling of radioactive sources".

(g) Media to play an active role and to be well informed about the psychological aspects.

(h) National Disaster Management Authority (NDMA) has already developed guidelines for preparedness for response to nuclear and radiological emergencies. The responsibility of strengthening the preparedness is to be followed up by all the agencies including medical community and joint exercises with the involvement of these agencies are to be periodically conducted.

8. CONCLUSION

While preparedness for response to various radiological emergency scenario have many common factors, the main challenge associated with radiation emergencies is managing the fear factor which may be generated by the presence of radiation or radioactive contamination, even if it may be extremely small. Identifying the potential emergency scenario, nation has to develop large number of Emergency Response Centres (ERCs) spread over the country with adequate radiation safety experts, trained first responders and many state of the art systems for quick assessment of radiological impact as well as for the detection and prevention of smuggling of radioactive sources.

References

1. IAEA, Radiological accident in Goiania, February (1989).
2. Abel J. Gonzalez, Lost and found dangers-Orphan Radiation Sources Raise Global Concerns, Timely action, Strengthening Radiation Safety And Security, 2-17, IAEA Bulletin, Vol **41**, No.3 (1999).
3. Pedro Ortiz, Vilmos Friedrich, John Wheatley, Modupe Oresegun, Lost and found dangers-Orphan Radiation Sources Raise Global Concerns 18-21, IAEA Bulletin, Vol **41**, No.3 (1999).
4. IAEA, International Atomic Energy Agency, Method for Developing Arrangements for Response to a Nuclear or Radiological Emergency, Emergency Preparedness and Response Series EPR-METHOD, 2003, Vienna (2003).
5. Pradeepkumar K.S, Methodology for quick assessment of deposited radioactivity on ground in a radiation emergency, Radiation Protection and Environment, Sharma D.N, Krishnamachari G and Kale M.S., Vol. **21**, No.1 (1998).
6. Raman, N., Probal Chaudhury, Padmanabhan, N., Pradeepkumar, K.S. and Sharma, D.N., Design and development of a compact aerial radiation monitoring system (CARMS), Proceedings of DAE-BRNS Symposium on Compact Nuclear Instruments and Radiation Detectors, 2005, pp 233-238.
7. IAEA, EPR-2006, Manual for First Responders to a Radiological Emergency, (2006).

Radiation Shielding of Hadrontherapy Accelerators

Stefano Agosteo

Politecnico di Milano, Dipartimento di Energia, Sezione di Ingegneria Nucleare, via Ponzio 34/3, 20133 Milano, Italy and INFN, Sezione di Milano, via Celoria 16, 20133 Milano, Italy.
E-mail: *stefano.agosteo@polimi.it*

ABSTRACT

The shielding design for hadrontherapy facilities is ruled by the secondary radiation field generated by the interaction of the primary beam with the structural materials of the accelerator system and with the patient. This secondary radiation field is constituted mainly by neutrons covering a wide energy range (from thermal energies up to several hundreds MeV per nucleon). The attenuation curves in concrete of secondary radiation generated by 250 MeV protons on iron and by 400 MeV per nucleon carbon ions on copper and carbon were calculated with Monte Carlo simulations. The methods adopted for calculating these attenuation curves will be discussed in this work, together with some outlines for the design of the access maze to the treatment room.

Keywords: Radiation shielding, Hadrontherapy, Monte carlo simulations.

Pacs No.: 28.41.Qb, 87.55.N-, 87.10.Rt, 87.55.N-, 87.56.bd.

1. INTRODUCTION

This paper runs through previous works [1-5] discussing shielding data for hadrontherapy facilities (i.e. intermediate energy accelerators).

Radiation therapy with hadron beams is growing worldwide. The number of operating and proposed facilities is being updated constantly by the Particle Therapy Co-Operative Group (PTCOG) [6]. PTCOG reports the data collected in March 2011, showing thirty-four facilities in operation, out of which ten in America (nine in the US and one in Canada), thirteen in Europe (one in the UK, two in France, four in Germany, one in Italy, three in Russia, one in Sweden and one in Switzerland), ten in Asia (two in China, seven in Japan and one in South Korea) and one in South-Africa. Among these facilities, six have being treating patients with carbon ions and the remainder with protons. The number of patients treated at these operating facilities is 69233, out of which 6701 with carbon ions and 62532 with protons. The total number of patients treated with hadron beams up the end of 2009 is 78275 if the facilities out of operation are also considered (2054 with helium ions, 1100 with pions, 7151 with carbon ions, 873 with other ions and 67097 with protons). It should be mentioned that twenty-four hadrontherapy centres have been proposed, out of which nineteen are being constructed (five of which will deliver both proton and carbon ion beams).

Cyclotrons and synchrotrons are usually employed for particle acceleration for medical applications, although linear ion accelerators are also being studied [7, 8]. The maximum acceleration energy for treating deep seated tumours is 250 MeV and 400 MeV per nucleon, for protons and carbon ions respectively, while 70 MeV protons are necessary for irradiating eye tumours. These energies are obviously related to the particle range in tissue which is about 4, 37 and 27 cm, for 70 MeV protons, 250 MeV protons and 400 MeV per nucleon carbon ions, respectively.

2. RADIATION SOURCES

The main sources for shielding design are the secondary radiation fields generated through the interactions of the primary beam with the structural materials of the accelerator, the beam delivery system and the patient. The angular and energy distributions of these radiation fields are of primary importance for calculating the thickness of the shielding barriers.

A set of Monte Carlo simulations was performed [1] with the FLUKA [9, 10] code for calculating the double-differential distribution of secondary neutrons, photons, protons, electrons and positrons generated by 250 MeV protons striking thick iron and tissue targets. Iron was taken as representative for the structural materials of the acceleration system and the 4-element ICRU tissue (density 1 g cm^{-3}, mass composition: oxygen 76.2%, carbon 11.1%, hydrogen 10.1%, nitrogen 2.6%) for the patient. The targets were cylindrical and the particle source was a pencil beam of protons. The iron and the tissue targets were 5.8 cm in radius and 7.5 cm in thickness (the range of 250 MeV protons in iron is 6.9 cm) and 15 cm in radius and 40 cm in thickness (the range of 250 MeV protons in tissue is about 37 cm), respectively. The iron target dimensions were the same employed in refs. [3, 11] for calculating the attenuation curves in ordinary concrete and iron shields. Simulations were performed in that work to investigate the effect of the target dimensions on the yield and the energy distribution of neutrons. As a general rule, it was demonstrated that, at a given proton energy, the larger is the transverse dimension of the target, the higher is the yield and the lower is the average energy of the spectrum.

The energy distributions of secondary neutrons generated in iron and tissue in the angular intervals 0°-10°, 40°-50°, 80°-90°, 130°-140° and 170°-180° are shown in Fig. 1.

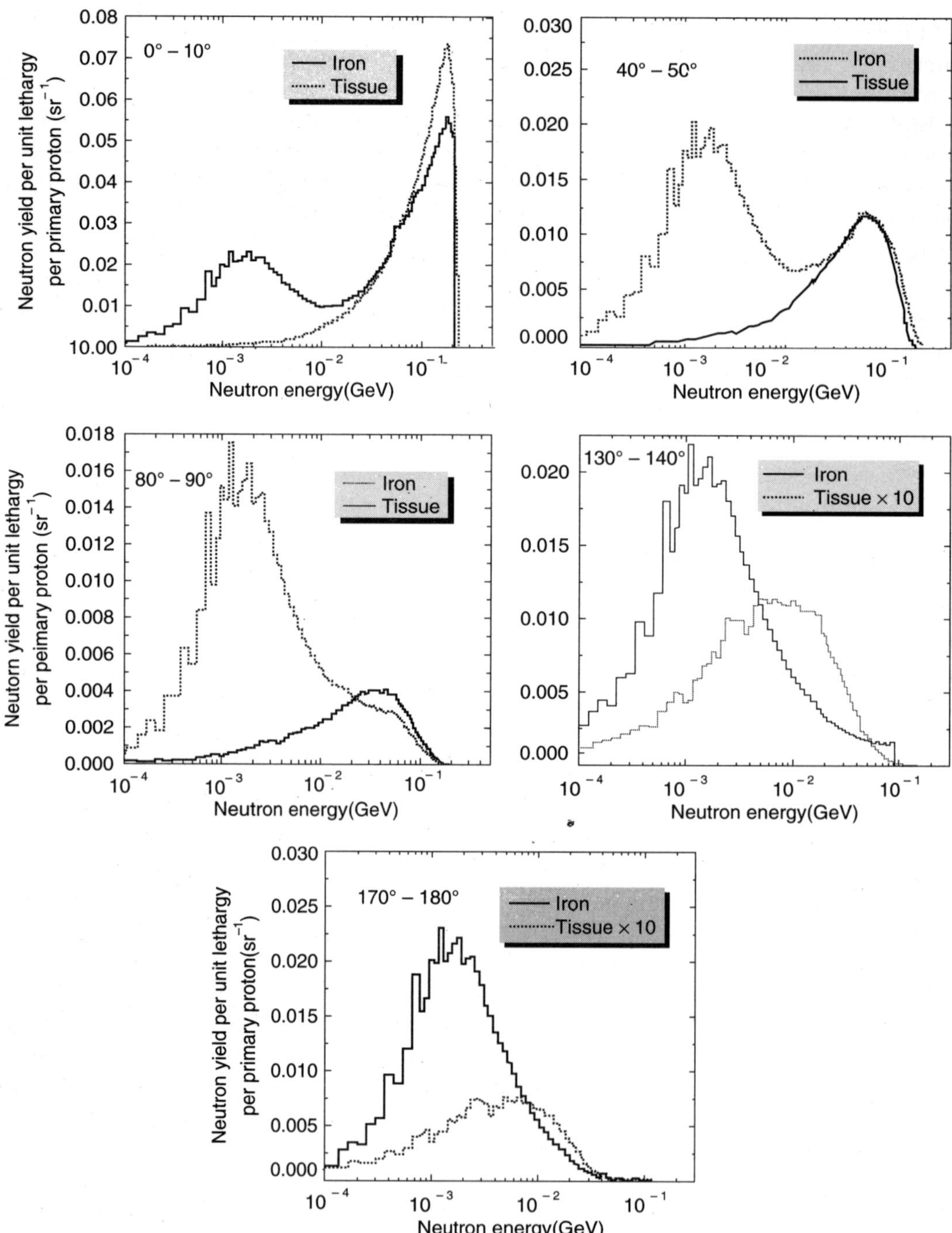

Fig. 1 *Energy distributions of secondary neutrons generated by 250 MeV protons on thick iron and tissue targets in the angular intervals 0°-10°, 40°-50°, 80°-90°, 130°-140° and 170°-180°. The tissue target spectra in the angular intervals 130°-140° and 170°-180° were multiplied by 10.*

At forward angles, the neutron spectrum from iron shows the typical double-peaked structure: a high-energy peak at about 100 MeV from direct and pre-equilibrium hadron-nucleon interactions and an evaporation peak in the MeV region. At larger angles the high energy peak decreases and tends to disappear in the backward directions, while the evaporation peak tends to dominate. For thin-targets, the evaporation peak is isotropic. The slight angular variation of the evaporation peak is due to neutron attenuation in the thick-target considered in the present set of simulations.

The neutron spectra from tissue do not show any evaporation peak in the MeV-region. It should be underlined that the nuclear evaporation model cannot be applied for light nuclei (typically $A \leq 16$), since their level density is not high enough to be approximated to a continuum. Moreover, the nuclear excitation energy can be a substantial fraction of the total binding energy of the target nucleus. Other de-excitation models can be adopted for light residual nuclei. The FLUKA code employs the Fermi break-up model, in which the nucleus is supposed to break in one step into two or more fragments [12]. For light elements, the sequential emission scheme of nucleons (mainly neutrons) and light fragments at the basis of nuclear evaporation is not applicable and the Maxwellian distribution of neutrons in the MeV-region loses its meaning.

A high-energy peak from direct and pre-equilibrium interactions is observable at forward angles for the tissue target. It should be noted that in the angular range 0°-10° this high-energy peak is more pronounced than that from the iron target. At larger angles, the average energy and the intensity of this energy distribution decreases sharply.

The energy distributions of secondary photons, protons, electrons and protons from iron and tissue targets bombarded by 250 MeV protons are discussed in details in ref. [1].

A set of simulations was also performed [5] to compare the double-differential distribution of neutrons generated from a thick carbon target bombarded by 400 MeV per nucleon carbon ions with the ones measured by Kurosawa et al. [13]. In that work, the neutron energy distribution was measured at several angles in the interval 0°-90° with the time-of-flight method for various combinations of target-projectile. The neutron detector was a NE213 liquid scintillator 12.7 cm in diameter. For 400 MeV per nucleon carbon ions on a carbon target, the following detection angles were selected: 0°, 7.5°, 15°, 30°, 60°, 90°. The detector was placed at different distances from the target, depending on the measurement angle (e.g. 5 m at 0° and 2 m at 90°). Therefore, the solid angle subtended by the detector surface varied with the measurement position. The target was a 20 cm thick parallelepiped. The area of the target surface perpendicular to the beam was 10×10 cm^2. The carbon density was 1.77 g cm^{-3}. The neutron spectra were measured above a few MeV depending on the angle (e.g. above 8 MeV at 0°, and above 3 MeV at 90°). The maximum energy of the spectra ranged from about 800 MeV at 0° to about 200 MeV at 90°.

The simulation geometry was described in order to reproduce strictly the experimental set up. A pure carbon target with a parallelepiped shape of the same dimensions and density of the target used in the experiment was simulated. A pencil beam of 400 MeV per nucleon carbon ions was directed to the target. The target was placed in vacuum, since the influence of air is negligible for most of the secondary high-energy neutrons. The neutron energy distributions were calculated at the same scoring angles of the experiment. The angular resolution of each scoring angle was set according to the solid angle subtended by the detector at each measurement position (by considering the target centre as the solid angle vertex). The simulated neutron energy distributions yields are shown in Fig. 2 together with the measured ones.

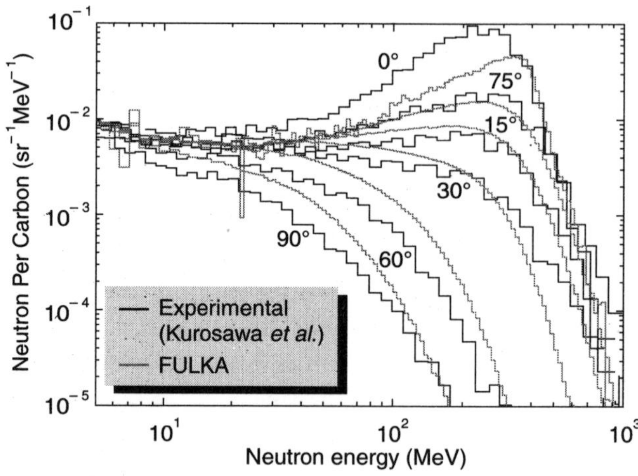

Fig. 2 *Simulated and experimental (from Kurosowa et al. [13]) energy distribution of neutrons produced at several angles by 400 MeV per nucleon carbon ions on a thick carbon target.*

The yield calculated at 0° is affected by the largest statistical uncertainties below about 100 MeV, since the scored angular bin is very narrow (about 0.7°) and the production of low energy neutrons is rather scarce at forward angles. The agreement of the simulated and experimental spectra is fairly satisfactory. The broad high-energy peak at 0° shows the largest deviation. A similar behaviour was observed in ref. [13] when comparing the experimental data with the predictions of a code based on Quantum Molecular Dynamics [14]. The simulated and experimental forward yields (2π) of neutrons above 3 MeV were 4.441 ± 0.001 and 4.750 neutrons per incident carbon ion, respectively.

3. ATTENUATION CURVES

Radiation shielding of hadrontherapy facilities can be performed through a set of attenuation curves which are usually calculated with radiation transport codes. Various data are available in the literature. A fairly extended list of references on this subject is given in refs. [3, 11]. It should be emphasised that after having calculated the shield thickness through a parametric formula, it is worth checking that the dose is below the prescribed limits with specific simulations accounting for the actual geometry of the shielded area. This is because the attenuation curves are usually calculated with simplified geometries which are far from representing the real ones. Moreover, these curves do not account for radiation scattering inside the hall/room where the acceleration system is installed. It should be noted that the influence of the scattered radiation field depends strictly on the geometry of the room which cannot be taken into account by a general purpose parametric formula.

Before describing the methods adopted for calculating the attenuation curves, a brief discussion about the particle spectrum equilibrium in an attenuating material is given. The particle spectrum equilibrium occurs when the shape of the energy distribution does not vary with depth in the attenuating material. It should be noted that the shape should be conserved, but the intensity of the spectrum lowers with depth (because of attenuation). If the spectrum equilibrium is satisfied, the attenuation curve shows an exponential behaviour. Deviations from spectrum equilibrium reflect in deviations from the exponential trend, thus showing, for example, a build-up region.

Simulations were performed for a 250 MeV proton beam bombarding a thick iron target, to emphasise the relation between spectrum equilibrium and the shape of the attenuation curve. Fig. 3 shows the spectral fluence at various depths in concrete for neutrons generated in the angular interval 40°-50° and the related attenuation curve. The attenuation curve shows a double-exponential trend. The first exponential extends from 10 cm up to about 40 cm in concrete and the shape of the related spectral fluences is fairly similar. The same shape is maintained at higher depths, but the attenuation length is higher. At 10-60 cm depth the high energy peak is broader with a maximum at around 50-60 MeV. At larger depths, the peak tends to get narrower and to be displaced towards higher energies (about 100 MeV). This trend is quite smooth below 80-100 cm and yields to a "quasi-equilibrium" situation. The spectral fluence then reaches its equilibrium. In other words, the lower energy components of the spectrum are attenuated mostly up to about 60 cm concrete depth with a short attenuation length, giving rise to a harder and more penetrating spectral distribution (although less intense), which is characterised by a larger attenuation length.

The simulations for calculating the attenuation curves in concrete for a proton beam striking a thick iron target are discussed in the following. This subject is described in details in ref. [3].

The attenuation through an ordinary concrete shield of the total dose equivalent produced by 100, 150, 200 and 250 MeV protons stopped in a thick iron target was calculated in ref. [3] with the FLUKA [9, 10] code (version 2006.03) at four emission angles: forward, 45°, transverse and 135°. The calculations were meant to reproduce the

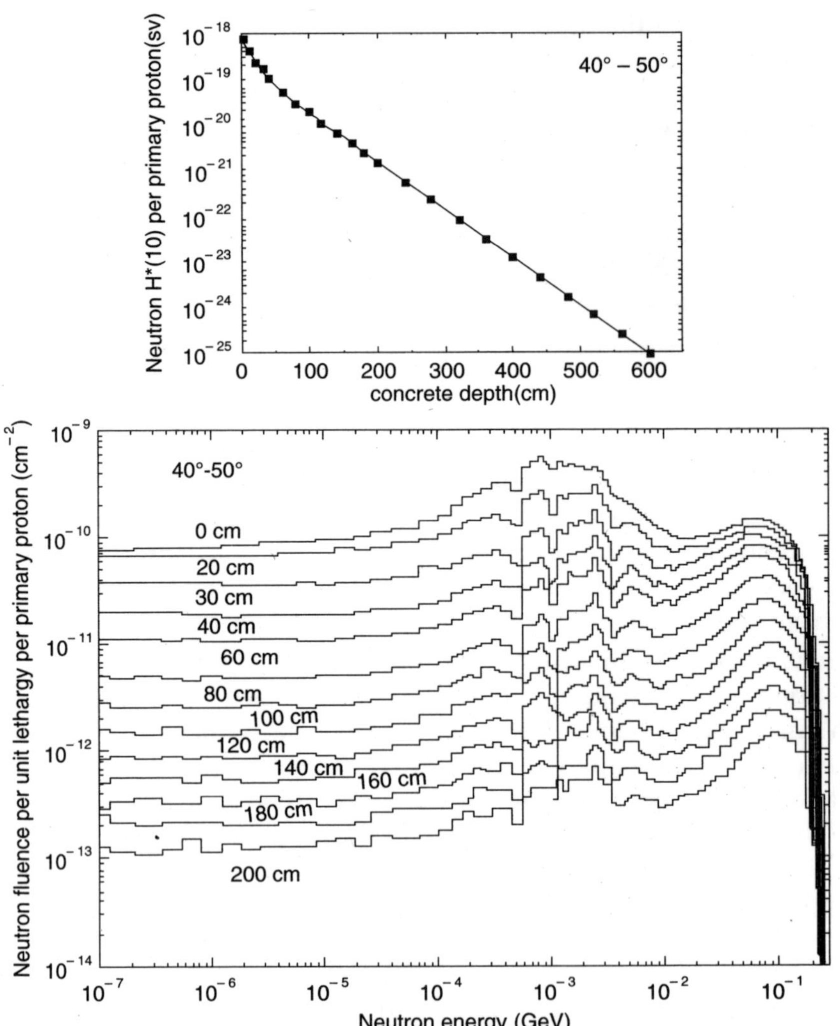

Fig. 3 *Attenuation of neutron dose equivalent in concrete for 250 MeV protons striking a thick iron target (top) together with the related spectral fluence of neutrons at various depth in concrete (bottom).*

dominant secondary radiation field created by a beam loss in a thick metallic target such as a magnet, a collimator or a vacuum chamber. Iron was chosen as representative of other materials of similar density and atomic number (such as copper and stainless steel), which are the main constituents of accelerator components.

The correct way to estimate the attenuation of radiation in a shield as a function of its thickness would be to perform individual sets of simulations for a series of slabs of different thicknesses placed at the given emission angle and to score the dose equivalent behind each slab. However, this process is very consuming in terms of computing time, as it requires a large number of simulations: $N_{total} = N \times N_{slab} \times N_{angle}$, where N is the number of simulations for each case, necessary to achieve statistical significance (typically N is in the range 5 to 10). In order to study different emission angles with a single geometry set-up, the calculations were performed in a spherical geometry.

Secondary neutrons, photons, protons and electrons produced by a monoenergetic and monodirectional proton beam ("pencil beam") impinging on an iron target located at the centre of a large spherical shield made of ordinary concrete (type TSF 5.5 [15]) were transported in four angular bins with respect to the incident proton beam direction: 0°-10° (forward), 40°-50°, 80°-90° (transverse) and 130°-140° (backwards). Transport of secondary radiation was limited to these four angular bins by assigning "blackhole" (a material with infinite absorbency) to the bins in-between. The volume of the cavity delimited by the inner surface of the shield was filled with vacuum. The dose equivalent behind the shield was calculated offline by folding the fluence with the appropriate fluence to ambient dose equivalent conversion coefficients [16, 17].

The target was cylindrical, with its axis coincident with the incoming beam direction, and it was slightly thicker than the proton range in iron at the given energy. The inner radius of the sphere was sufficiently large (90 m) to make curvature-related effects negligible. Simulations were run for 14 values of the shield thickness, up to 600 cm. The computed particle fluence was scored by means of inverse cosine-weighted boundary crossing estimators (i.e., fluence across a surface) at the interface between shielding and the outside blackhole, taking into account only particle fluences directed outwards ("one-way" scoring option). Variance reduction techniques were used, namely "geometry splitting" and "Russian roulette". Each region importance was adjusted so as to maintain the number of particles approximately constant throughout the shield. Each data point is the average of the results of at least 10 independent simulations. The total number of histories per data point was at least 2,000,000.

In order to better reproduce reality, the FLUKA geometry was made such as to minimize the following three effects:

(a) neutron exchange between contiguous angular bins during transport;

(b) neutron scattering back and forth across a boundary estimator during transport (typical of neutron scoring inside an homogeneous medium and not occurring while scoring outside a shield, that is at the interface between the shield material and air);

(c) neutron transport in one angular bin after being back-scattered (or backward emitted through inelastic processes in concrete) from another angular bin ("cross-talk" from the inner surface of the sphere).

The influence of the various effects listed above was studied with a set of preliminary simulations, which are described in details in ref. [3]. Depending on the energy, angle of emission of the secondary radiation and actual geometry, in a real case of assessing radiation shielding of a given facility, some of the effects which in the preliminary simulations mentioned above are excluded, may in fact be present. These specifically concern scattering within an extended barrier, room scattering and secondary radiation emitted backwards generated within the barrier. Therefore, as mentioned above, after a first estimate of the shield thickness through the data given in the present paper, a (partial) simulation of the facility in its real geometry may nonetheless be necessary.

In general, the attenuation through a thick shield of the total ambient dose equivalent is fitted with the classical two-parameter formula (see, for example, ref. [18]):

$$H\,(E_p,\,\theta,\,d/\lambda) = \frac{H_0\,(Ep,\,\theta)}{r^2}\,\exp\left[-\frac{d}{\lambda\,(\theta)\,g\,(\alpha)}\right] \tag{1}$$

where H is the ambient dose equivalent beyond the shield, E_p is the proton energy, r is the distance between the radiation source (the target stopping the protons) and the scoring position, θ is the angle between the direction $\vec{r}$ and the beam axis, H_0 is the source term, d is the shield thickness, $\lambda\,(\theta)$ is the attenuation length for the given shielding material at emission angle θ, and α is the angle between the direction $\vec{r}$ and the normal to the shield surface. The function $g\,(\alpha) = 1$ for the spherical geometry used in the present simulations and $g\,(\alpha) = \cos\alpha$ in all other cases. This is the expression used in the present work to fit the data in the forward direction (after the build-up region extending to a depth of about 100 cm, after which the attenuation becomes exponential).

As mentioned above, in certain cases the attenuation of the dose equivalent is better expressed by a double-exponential function [2]:

$$H\,(E_p,\,\theta,\,d/\lambda) = \frac{H_1\,(E_p,\,\theta)}{r^2}\,\exp\left[-\frac{d}{\lambda_1\,(\theta)\,g\,(\alpha)}\right] + \frac{H_2\,(E_p,\,\theta)}{r^2}\,\exp\left[-\frac{d}{\lambda_2\,(\theta)\,g\,(\alpha)}\right] \tag{2}$$

where H_1, $\lambda_1\,(\theta)$ and H_2, $\lambda_2\,(\theta)$ are the source terms and the attenuation lengths for the shallow-depth and large-depth (deep penetration) exponential functions, respectively. The second term of expression (2) describes the attenuation for a shield thicker than about 100 cm and obviously cannot be applied for a thinner shield, because it would lead to an underestimate of the ambient dose equivalent. In practice, expression (2) includes expression (1) by setting $H_0 = H_2$, $\lambda_1\,(\theta) = \lambda_2\,(\theta)$ and setting the first term to zero (i.e., $H_1 = \lambda_1\,(\theta) = 0$). As shown in the next section, a double-exponential attenuation is characteristic of the attenuation curves in the 40°-50°, 80°-90° and 130°-140° angular bins.

The attenuation of the total dose equivalent in ordinary concrete at the four emission angles are plotted in Fig. 4 for 250 MeV protons impinging on a thick iron target. The statistical uncertainties on the data points are less than 6%. For the sake of clarity only the fit on the second exponential in expression (2) is shown. The attenuation curves for 100, 150 and 200 MeV protons on iron are given in ref. [11]. The resulting source terms H_0(Svm^2 per proton) and attenuation lengths λ(g cm^{-2}) are listed in Table 1. It should be underlined that the source term is the dose equivalent per primary particle extrapolated from the attenuation curve at 1 m from the target (assumed to be a point source of secondary radiation). Therefore, H_0 is not the dose equivalent at a zero depth in concrete. It can be higher when the attenuation curve shows a build-up region at shallower depths.

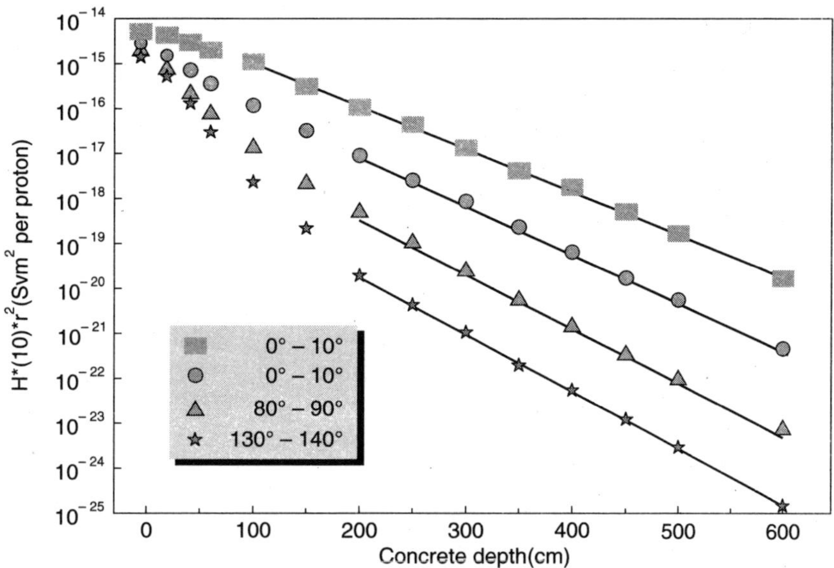

Fig. 4 *Attenuation of total dose equivalent in concrete for 250 MeV protons on a thick iron target. The statistical uncertainties are smaller than the size of symbols.*

Table 1 *Attenuation of the total dose equivalent in ordinary concrete for 100-250 MeV protons impinging on a thick iron target: source terms H_1 and H_2 (Sv m^2 per proton) and attenuation lengths λ and λ_2 (g cm^{-2}) resulting from the fits shown in ref. [3].*

		First Exponential		Second Exponential	
	Ang bin	$H_1(10)$ (Sv m^2 per proton)	$\lambda_1(\theta)$ (g cm^{-2})	$H_2(10)$ (Sv m^2 per proton)	$\lambda_2(\theta)$ (g cm^{-2})
	0°-10°	—	—	$(8.9 \pm 0.4) \times 10^{-16}$	59.7 ± 0.2
100 MeV	40°-50°	$(5.9 \pm 1.3) \times 10^{-16}$	47.5 ± 2.7	$(1.5 \pm 0.1) \times 10^{-16}$	57.2 ± 0.3
	80°-90°	$(5.3 \pm 0.8) \times 10^{-16}$	33.7 ± 1.2	$(1.1 \pm 0.3) \times 10^{-17}$	52.6 ± 0.7
	130°-140°	$(4.7 \pm 0.4) \times 10^{-16}$	30.7 ± 0.5	$(8.0 \pm 5.1) \times 10^{-18}$	46.1 ± 2.8
	0°-10°	—	—	$(3.0 \pm 0.2) \times 10^{-15}$	80.4 ± 0.5
150 MeV	40°-50°	$(1.2 \pm 0.2) \times 10^{-15}$	57.8 ± 3.4	$(3.3 \pm 0.8) \times 10^{-16}$	74.3 ± 1.4
	80°-90°	$(10.0 \pm 2.2) \times 10^{-16}$	37.4 ± 2.7	$(1.2 \pm 0.3) \times 10^{-17}$	70.8 ± 1.3
	130°-140°	$(7.8 \pm 2.0) \times 10^{-16}$	32.7 ± 1.5	$(2.1 \pm 0.6) \times 10^{-18}$	61.8 ± 1.1

Contd...

Contd...

	0°-10°	–	–	$(5.6 \pm 0.4) \times 10^{-15}$	96.6 ± 0.8
200 MeV	40°-50°	$(1.9 \pm 0.3) \times 10^{-15}$	68.3 ± 5.9	$(6.8 \pm 0.5) \times 10^{-16}$	86.4 ± 0.5
	80°-90°	$(1.3 \pm 0.4) \times 10^{-15}$	43.8 ± 4.4	$(3.7 \pm 0.8) \times 10^{-17}$	78.3 ± 1.3
	130°-140°	$(1.3 \pm 0.3) \times 10^{-15}$	32.8 ± 1.6	$(2.8 \pm 2.4) \times 10^{-18}$	70.0 ± 4.1
	0°-10°	–	–	$(9.8 \pm 1.0) \times 10^{-15}$	105.4 ± 1.4
250 MeV	40°-50°	$(2.3 \pm 0.5) \times 10^{-15}$	77.0 ± 7.9	$(1.2 \pm 0.1) \times 10^{-15}$	93.5 ± 0.5
	80°-90°	$(1.4 \pm 0.4) \times 10^{-15}$	49.7 ± 5.7	$(9.0 \pm 2.5) \times 10^{-17}$	83.7 ± 2.0
	130°-140°	$(1.9 \pm 0.6) \times 10^{-15}$	34.4 ± 3.4	$(6.5 \pm 2.6) \times 10^{-18}$	79.1 ± 3.4

At small depths and at the forward angle a dose build-up region is visible, particularly at the highest energies. Moving towards larger angles this situation is reversed, as the low energy component starts to dominate and a fast attenuation region appears at small thicknesses. As already mentioned, this is because the neutron energy distributions vary with depth in concrete. The spectrum hardens until a depth of about 100 cm, after which it reaches equilibrium. The lower energy component of the spectrum is attenuated mostly up to a depth in the shield of about 100 cm with a short attenuation length–the first term in expression (2)–giving rise to a less intense but harder and more penetrating spectral distribution, which is characterised by a longer attenuation length–the second term in expression (2). For the forward direction expression (1) was used, taking into account data points between 100 cm and 600 cm. For the other directions the data were fitted with two independent exponentials (rather than with the single 4-parameter fit of expression (2) as done in a previous work [2]), taking into account data points between 20 cm and 150 cm for the first exponential and data points between 200 cm and 600 cm for the second exponential. For the forward direction the associated uncertainties are less than 10% on both the source terms and the attenuation lengths, whereas for all other angular bins the uncertainties on the attenuation length are below 15%, whilst for the source term they can reach 85%. This is because the dose is proportional to the source term and depends exponentially on the attenuation length. Thus for a given dose past the shield, a small variation of the attenuation length leads to a large variation on the source term. Including in the fit for the second exponential data at shallower depths (between 200 and 350 cm), generally λ tends to decrease and H_0 to increase.

A fairly different simulation geometry was used for calculating the source terms and the attenuation lengths of neutrons from 400 MeV per nucleon carbon ions on thick carbon and copper targets [2]. These simulations were performed a few years before those described above for proton beams [3]. At that time, ion transport with secondary particle generation was not treated by the FLUKA code. Therefore, the spectra of secondary neutrons from 400 MeV per nucleon carbon ions on thick carbon and copper targets, measured by Kurosawa et al. [13], were used as input data for the FLUKA simulations.

A point neutron source was placed at the centre of a spherical shell, 6 m thick, made up of concrete TSF-5.5 [15], with inner radius large enough (90 m) to make effects related to curvature negligible. Effects due to neutron scattering are also negligible, since the fluence of neutrons diffusing inside the concrete shell is inversely dependent on its inner surface area [19]. The fluence of outward-directed particles was scored in boundary crossings placed at various depths inside the concrete shell. The latter was also subdivided into polar sectors to account for the angular distribution of the fluence. At projectile energies of 400 MeV per nucleon, where the energy distribution of the secondary neutrons exceeds the threshold energy for pion production (about 280 MeV [12]), the fluence of pions generated in concrete was scored in addition to that of neutrons, photons (from neutron absorption and residual nucleus de-excitation) and secondary protons. The ambient dose equivalent was estimated with the conversion coefficients of refs. [20, 17]. Geometry splitting and Russian roulette were used as variance reduction techniques for neutron transport inside the concrete shield.

As mentioned above, fluence scoring in each boundary crossing inside the concrete shells accounted only for outward-directed particles. This should minimize the effect of reflections (especially for neutrons) from the outer concrete shells, which leads to overestimate the fluence and consequently $H^*(10)$. However, reflections are not eliminated completely because, as a second order effect, neutrons can be backscattered more than once inside the shield. Each time a multi-reflected neutron crosses a boundary outwards, it is counted in this one-way fluence scoring. Moreover, since neutrons are slowed down in scattering events, the prompt gamma-ray component from low-energy neutrons may also be overestimated. As already mentioned, the exact evaluation would have required a set of different simulations, each one with the correct shielding thickness, and a much longer computing time. The effect of this approximation was investigated with separate simulations considering shells with different thickness, for the case of 400 MeV per nucleon carbon ions on copper [1]. Data were fitted with the classical two-parameter formula (single-exponential function), given by expression (1). As expected, the source terms resulted slightly lower (in the range 2%-13%, excluding the angular bins 30°-40° and 80°-90°) than those calculated with fictitious shells. The difference in the attenuation lengths was found to be comparatively small (a few percent at maximum). It can therefore be concluded that the data obtained with the fictitious shell model are sufficiently correct and representative of the real situation, if the overall non-statistical uncertainties (i.e. cross sections, nuclear models, concrete composition, etc.) are also taken into account.

The double-exponential function (2) was used for fitting the attenuation curves of 400 MeV per nucleon projectiles for angles above 40° to 60°. The resulting source terms and attenuation lengths are listed in Tables 2-3.

Table 2 *Source terms and attenuation lengths in concrete for neutrons generated by 400 MeV per nucleon carbon ions on C, resulting from the fits shown in ref. [2]. The data were fitted using expression (2).*

Angular bin	H_1 (Sv m^2 per ion)	λ_1 (g cm^{-2})	H_2 (Svm^2 per ion)	λ_2 (g cm^{-2})
0-10°	–	–	$(1.93 \pm 0.02) \times 10^{-12}$	120.98 ± 0.21
10-20°	–	–	$(4.37 \pm 0.02) \times 10^{-13}$	120.21 ± 0.14
20-30°	–	–	$(1.50 \pm 0.01) \times 10^{-13}$	122.15 ± 0.20
30-40°	–	–	$(5.75 \pm 0.03) \times 10^{-14}$	122.18 ± 0.21
40-50°	–	–	$(2.28 \pm 0.02) \times 10^{-14}$	117.03 ± 0.21
50-60°	$(1.03 \pm 0.05) \times 10^{-14}$	49.28 ± 2.27	$(7.53 \pm 0.15) \times 10^{-15}$	111.74 ± 0.34
60-70°	$(8.98 \pm 0.33) \times 10^{-15}$	50.07 ± 1.59	$(3.19 \pm 0.14) \times 10^{-15}$	103.86 ± 0.65
70-80°	$(7.62 \pm 0.28) \times 10^{-15}$	48.43 ± 1.40	$(1.49 \pm 0.07) \times 10^{-15}$	102.23 ± 0.64
80-90°	$(6.11 \pm 0.33) \times 10^{-15}$	39.88 ± 1.07	$(9.54 \pm 0.26) \times 10^{-16}$	95.87 ± 0.36

Table 3 *Source terms and attenuation lengths in concrete for neutrons generated by 400 MeV per nucleon carbon ions on Cu, resulting from the fits shown in ref. [2]. The data were fitted using expression (2).*

Angular bin	H_1 (Sv m^2 per ion)	λ_1 (g cm^{-2})	H_2 (Svm^2 per ion)	λ_2 (g cm^{-2})
0-10°	–	–	$(1.93 \pm 0.02) \times 10^{-12}$	120.98 ± 0.21
0-10°	–	–	$(8.79 \pm 0.12) \times 10^{-13}$	122.98 ± 0.43
10-20°	–	–	$(2.13 \pm 0.01) \times 10^{-13}$	121.62 ± 0.14
20-30°	–	–	$(8.75 \pm 0.06) \times 10^{-14}$	121.21 ± 0.19
30-40°	–	–	$(3.58 \pm 0.01) \times 10^{-14}$	122.47 ± 0.15
40-50°	–	–	$(1.93 \pm 0.02) \times 10^{-14}$	119.16 ± 0.19
50-60°	$(1.11 \pm 0.13) \times 10^{-14}$	30.93 ± 2.28	$(8.10 \pm 0.09) \times 10^{-15}$	120.91 ± 0.24
60-70°	$(7.83 \pm 0.56) \times 10^{-15}$	47.47 ± 2.63	$(2.91 \pm 0.10) \times 10^{-15}$	116.03 ± 2.53
70-80°	$(6.78 \pm 0.51) \times 10^{-15}$	45.61 ± 2.05	$(1.88 \pm 0.06) \times 10^{-15}$	102.46 ± 0.39
80-90°	$(7.67 \pm 0.29) \times 10^{-15}$	35.88 ± 1.42	$(1.30 \pm 0.04) \times 10^{-15}$	97.42 ± 0.32

3. CONCLUSION

The Monte Carlo simulations for calculating the attenuation curves in concréte of intermediate energy accelerators, such as the ones devoted to hadrontherapy, were discussed herein. The simulation data were fitted with parametric expressions. These expressions allow calculating the shielding barrier thickness fairly easily, but the results should be checked through numerical simulations of the actual geometry of the environment to be shielded.

References

1. S Agosteo, *Radiation Protection Dosimetry* **96**, 393 (2001)

2. S Agosteo, T Nakamura, M Silari, Z Zajacova *Nuclear Instruments and Methods* **B 217** 221 (2004)

3. S Agosteo, M Magistris, A Mereghetti, M Silari, Z Zajacova *Nucl. Instrum. Meth.* **B 265**, 581 (2007)

4. S Agosteo *Radiation Protection Dosimetry* **137** 167 (2009)

5. S Rollet, S Agosteo, G Feherenbacher, C Hranitzky, T Radon, M Wind *Radiation Measurements* **44** 649 (2009)

6. Particle Therapy Cooperative Group, ptcog.web.psi.ch

7. U Amaldi and M Silari, Editors. *The TERA project and the centre for oncological hadrontherapy* (INFN-LNF Frascati, Italy) ISBN 88-86409-09-5 (1995)

8. U Amaldi, P Berra, K Crandall, D Toet, M Weiss, R Zennaro, E Rosso, B Szeless, M Vretenar, C Cicardi, C De Martinis, D Giove, D Davino, M R Masullo, V Vaccaio *Nuclear Instruments and Methods* **A 521** 512 (2004).

9. G Battistoni, S Muraro, P R Sala, F Cerutti, A Ferrari, S Roesler, A Fassò, J Ranft *AIP Conference Proceedings* **896** 31 (2007).

10. A Ferrari, P R Sala, A Fassò,. J Ranft *FLUKA: A Multi-Particle Transport Code* (CERN-2005-10, INFN/TC_05/11, SLAC-R-773) (2005)

11. S Agosteo, M Magistris, A Mereghetti, M Silari Z Zajacova *Nucl. Instrum. Meth.* **B 266**, 3406 (2008)

12. A Ferrari and P R Sala *Proceedings of the Workshop on Nuclear Reaction Data and Nuclear Reactors Physics, Design and Safety, International Centre for Theoretical Physics, Miramare-Trieste (Italy) 15 April-17 May 1996, A Gandini and G Reffo, Eds.* (World Scientific) 424 (1998)

13. T Kurosawa, N Nakao, T Nakamura, Y Uwamino, T Shibata, N Nakanishi, A Fukumura and K Murakami *Nuclear Science and Engineering* **132** 30 (1999)

14. K Niita, S Chiba, T Maruyama, H Takada, H Fukahori, Y Nakahara *Phys. Rev.* **C 52** 2620 (1995)

15. National Council of Radiation Protection *Radiation protection design guidelines for 0.1-100 MeV Particle Accelerators Facilities*, NCRP Report 51 (1977)

16. International Commission on Radiological *Protection Conversion coefficients for use in radiological protection against external radiation* ICRP Publication n. 74 (Pergamon) (1996)

17. M Pelliccioni *Radiation Protection Dosimetry* **88** 279 (2000)

18. R H Thomas and G R Stevenson *Radiological safety aspects of the operation of proton accelerators* International Atomic Energy Agency Technical report series no. 283 (IAEA, Vienna) (1988)

19. R C McCall T M Jenkins R A Shore *IEEE Trans. Nucl. Sci.* **NS-26** 1593 (1979)

20. A Ferrari and M Pelliccioni *Radiation Protection Dosimetry* **76** 215 (1998)

Radiation Protection for Laser based Accelerators: The LNF FLAME Project

Adolfo Esposito

INFN-LNF, Via Enrico Fermi 40 00044 Frascati Roma Italia
E-mail: *Adolfo.Esposito@lnf.infn.it*

ABSTRACT

Radiation particles are widely applied in all fields of science. Up to today the radiations were produced by radiation sources like accelerators, X-ray tubes, radioactive sources with the well known problems of costs, parameters and safety. Since few years, following the development of laser able to focus ultra-short high intensity pulses onto targets, became possible the generation of ionizing radiation by intense lasers. The paper will describe the actual status of FLAME, the 300 TW laser in commissioning at LNF, focusing on some radiological protection aspects.

Keywords: Radiation protection, Laser based accelerators

Pacs No.:

1. INTRODUCTION

Since few years, it has been demonstrated both theoretically and experimentally [1, 2, 3, 4] that ionizing particles, mainly electrons and protons but also X-rays and ions can be produced and accelerated by laser, opening a new era in a different field of science. From then on, all practices concerning the use of such lasers have been regarded as practices with radiation risks. The aim of this paper is to highlight some radiological protection issues in designing the FLAME (Frascati Laser for Acceleration and Multidisciplinary Experiments) facility [5].

Particles Production and Acceleration

Laser-plasma acceleration is based on different physical principles compared to particle accelerators. In fact, it allows the achievement of energies up to GeV levels in accelerators long a few millimeters instead of hundreds meters (Fig. 1).

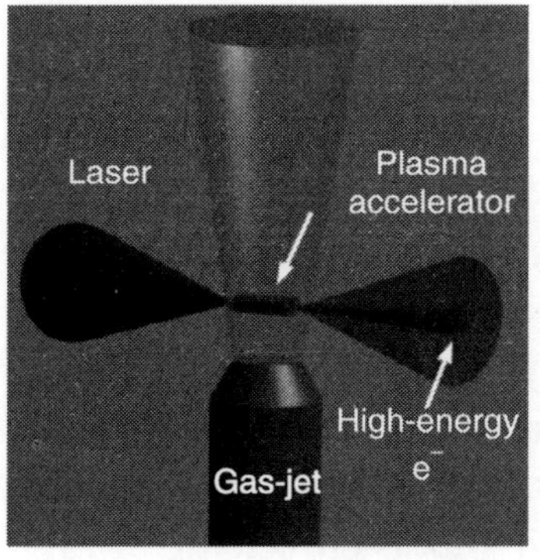

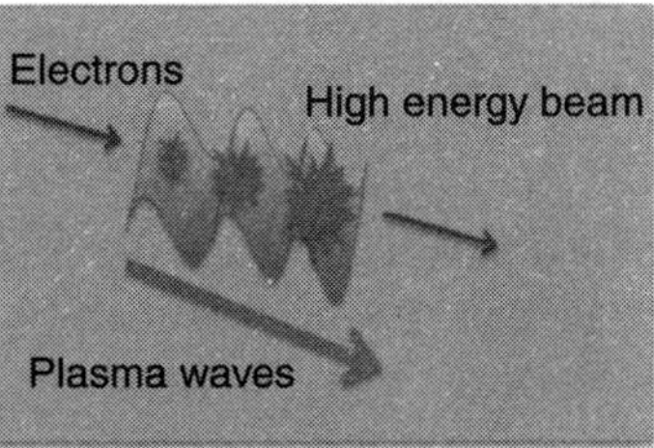

Fig. 1 *Production mechanism of high energy electrons*

The laser wake-field acceleration (lwfa) in low-density plasma provides the most promising approach to high performance compact accelerators because it takes advantage of very high gradients (100 GeV/m) [6].

At National Laboratories of Frascati (LNF) is under commissioning the FLAME Laser (Frascati Laser for Acceleration and Multidisciplinary Experiments) whose main parameters are:

- Peak power = 300 TW;
- Pulse length $\geq$ 20 fs;
- Repetition rate = 10 Hz
- Output energy = 8 J
- Laser intensity up to 10^{20} Wcm^{-2}.

The laser will be installed in a new building (Fig. 2) while the experimental chamber is foreseen in a pit five meter underground.

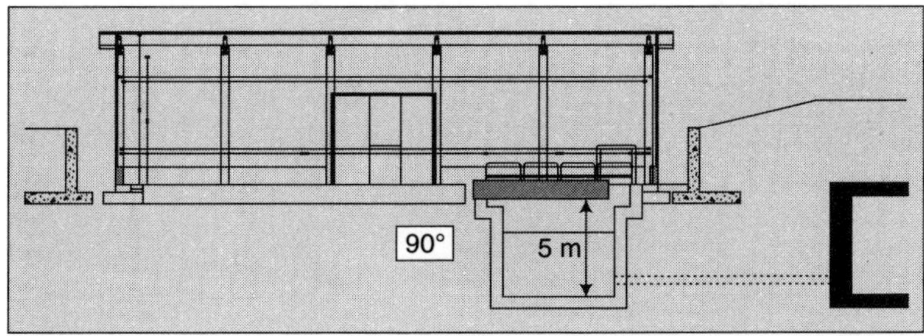

Fig. 2 *FLAME new building and target area pit*

The laser beam will be used for experiments of focalization on solid or gaseous target to produce and characterize charged particles, mainly electrons but also protons.

Studies and experiments or plasma acceleration and monochromatic x-ray production are foreseen combining the high brightness LINAC accelerator of the LNF-Sparc project [7] with the ultra-short, high energy, 300 TW FLAME laser. The most important approved experiments are

- self-injection of electrons in plasma waves driven into the bubble regime by FLAME pulses into supersonic gas-jets;
- external injection of ultra-short SPARC electron bunches into plasma waves driven by FLAME pulses;
- ions/protons production by FLAME pulses onto metallic foils;
- development of a monochromatic and tuneable X-ray source in the 20-1000 keV range, based upon Thomson scattering of laser pulses by relativistic electrons
- advanced radio-logical imaging with mono-chromatic X-rays from Thomson source.

At intensities greater than 10^{17} Wcm^{-2} a significant part of the laser energy is converted into the generation of high energy particles. In Fig. 3 are reported the focused intensities reachable by FLAME laser (Fig. 3).

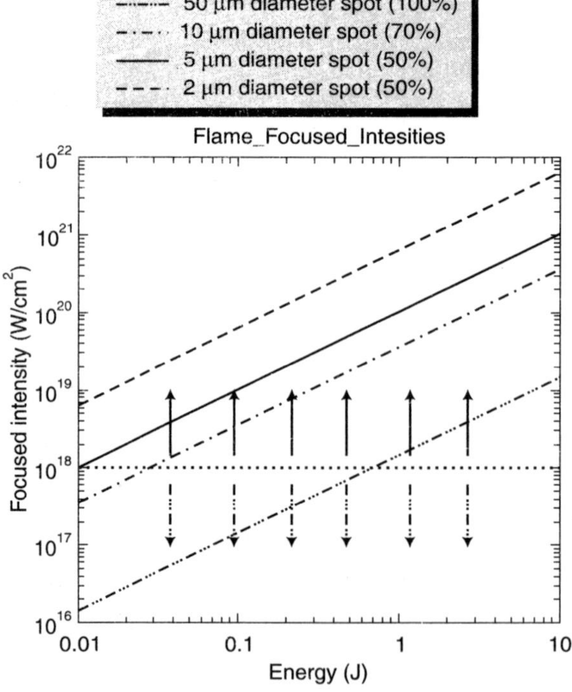

Fig. 3 *FLAME focused intensities versus energy*

We obtained, during some preliminary radiation protection test at very low power (laser energy before compression 550 mJ, 40 fs, gas-jet pressure 1700 kPa, only one over ten laser pump used), the first self injection of electrons (Fig. 4)

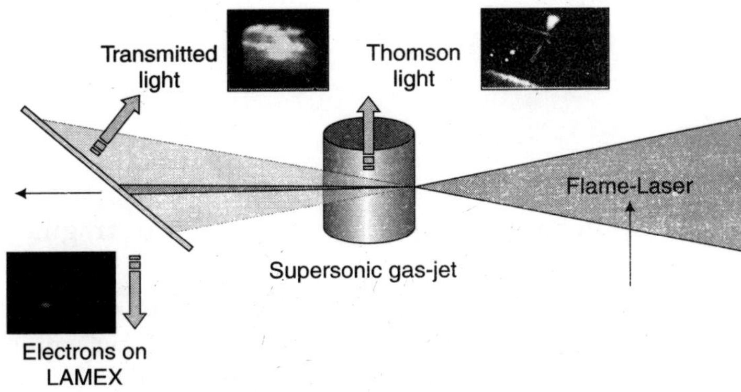

Fig. 4 *Experimental setup*

The Flame Shielding

The aim of an efficient accelerator shielding design is to attenuate the prompt radiation produced to levels that are acceptable to humans outside the shield, at a reasonable cost and without compromising the utility of the apparatus for its design purposes.

The steps needed to arrive at a preliminary understanding of radiological impact of the facility are below listed:

Specifications of design parameters
- type of particles accelerated;
- beam characteristic;
- maximum energy and current;
- duty cycle;

Specifications of assumptions on expected operation
- operation time per year;
- occupational factor for different building and location;

Determination of applicable radiation protection goals
- assessment radiological risk for workers and general public;
- under normal working condition;
- under accident condition;
- estimation of radiation source strength;

  particle yields have to be reported in terms of physical distribution such type, energy, fluence and angle of emission.

The higher the energy of particle accelerated, the more complex the characteristic of the prompt radiation field, that exist only while the accelerator is in operation.
- primary particles production (electrons, protons, ions);
- prompt radiation production;

- bremsstrahlung;
- neutron;
- muons;
- pions;
- kaons;
- any other particle (charged particles, ions, nuclear fragments and delayed radiation);

At large distance the radiation field due to the accelerator operation comprises two components direct and scattered radiation. The terms "skyshine" refers to all radiation whether scattered by the ground, air or neighboring buildings

- neutron skyshine;
- photon skyshine.

In considering the environmental impact when particle accelerators are operated two sources of radiation have to be taken into account. The over mentioned prompt radiation and induced radioactivity:

- radionuclides produced in solid material;
- accelerator structures and its ancillary components shielding materials;
- radionuclides produced in cooling water;
- radionuclides produced in air;
- radionuclides produced directly in air;
- radionuclides produced in dust;
- radionuclides produced in earth shielding and groundwater.

The thickness of the shielding depends from the attenuation of particles produced and from radiation protection policy chosen.

According to the recommendations of ICRP, to the European Directives as well as the laws in force in such matter, the recommended effective dose limits in planned exposure situations are 20 mSv/year for occupational, averaged over a period of 5 years, and 1 mSv/year for the members of the public.

The radiation protection policy adopted by LNF was to maintain the ambient dose equivalent, within the areas outside the shield frequented by the staff, at values much less than the limits for public in normal working conditions.

Electrons are produced when a very powerful laser interacts with a gas jet. Such radiation, after the interaction with the experimental chambers walls and/or the shielding materials, will generate, via electromagnetic cascade, the so-called prompt radiation.

The operating conservative parameters used in shield calculation of FLAME laboratory are

- electron energy = 200 MeV;
- electron charge = 1nC/pulse;

- repetition rate = 10 Hz
- operation at 10 Hz for 250 hour/year

For shielding evaluation purpose, three components of radiation field which are produced when an electron beam, with energy of hundreds MeV, hits a thick target or a thin walls of the vacuum chamber under a small angle of incidence have to be considered, namely bremsstrahlung X-rays, low energy neutron (giant resonance neutron), produced through photons interaction with the electromagnetic field of the nucleus, and high energy neutron produced through photons interaction with neutron-proton pair (quasi deuteron model) or nucleon interaction via pions production.

For each of these components, whose characteristic are known enough, a source term has been considered according with semi-empirical models for radiation source terms and attenuation through matter. Such models, represented by analytical formulae and values of relevant parameters, are available in the literature.

For an heterogeneous shield the attenuation formula used for shielding evaluation is

$$\Sigma \dot{H}^*(10)_i = \Sigma_i \frac{S_i}{r^2} e^{-d/\lambda i} * f_i$$

Where

$\dot{H}^*(10)_i$ = ambient dose equivalent rate

S_i = source term

d = thickness interposed

λ_ι = attenuation coefficient

Φ_i = conversion coefficients for use in radiological protection against external radiation

r = distance of interest

The values of attenuation length used for calculations are listed in the following Table 1 according to values found in literature.

Table 1 *Attenuation lengths for different radiation components*

Material	Radiation components	Attenuation length λ(cm)
Concrete	Bremsstrahlung	20.4 (0°)
Concrete		18.7 (90°)
Lead		2.2
High Density Polyethylene		69.3
Concrete	Giant resonance neutron	17.4
Lead		18.3
High Density Polyethylene		6.36
Concrete	High energy neutron	8.9
Lead		16.8
High Density Polyethylene		61.4

The ambient dose equivalent rate was evaluated only at 90°, due to the position of target area underground surrounded by not less 8 m of earth, and for the most conservative case from the point of radiation protection view (200 MeV, 1nC/pulse, 10 Hz).

The neutron fluence to ambient dose equivalent conversion coefficients used for this calculation are equal to 420 and 600 pSv cm^2 respectively for giant resonance neutrons and high energy neutrons choosing the most conservative values found in literature [8.9].

In Fig. 5 the annual ambient dose equivalent is reported versus concrete thickness.

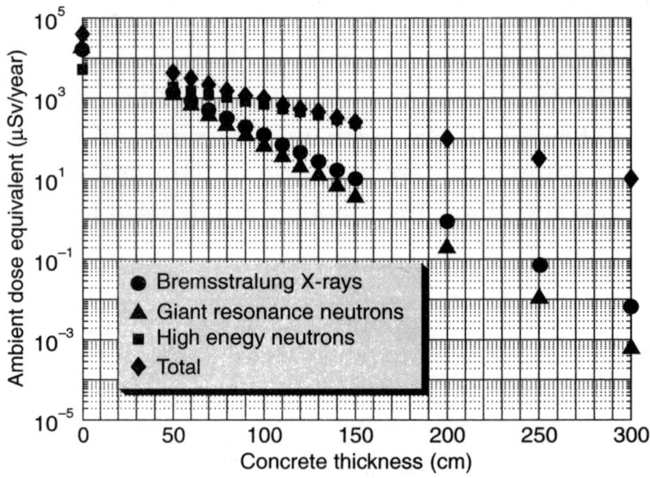

Fig. 5 *Ambient dose equivalent for the three components of radiation field, in the 90° direction at 5 m from the source, as function of concrete thickness plus an additional shield of 10 cm of lead and 10 cm of polyethylene*

According to results shown in Fig. 6 a shielding configuration of 10 cm of lead and 10 cm of polyethylene, installed just after the experimental chamber top, followed by 1 meter of ordinary concrete, was chosen in order to comply with an effective annual dose limits of about 1 mSv/year just behind the shield, obtaining values lower in the FLAME building and trivial outside it (Fig. 6)

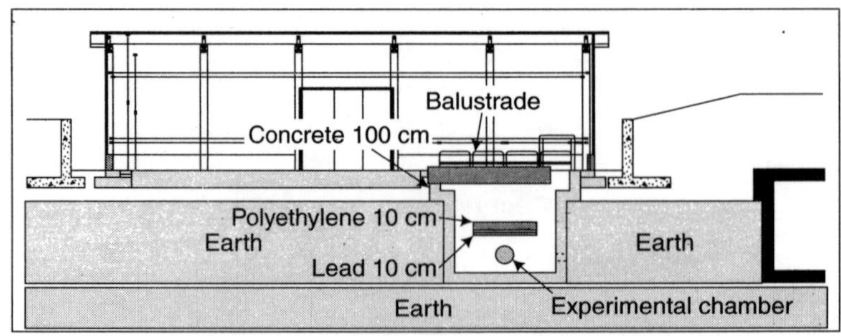

Fig. 6 *The FLAME shielding*

Radiological Risk for the Workers and the Members of Public from the Prompt Radiation

In normal working condition the ambient dose equivalent rates in the various areas of FLAME laboratory as well as in the external areas will be quite negligible and however difficult to measure with the radiation protection instruments used in routine monitoring.

In accident condition consisting in the irradiation of person close to the target area during the FLAME operation, the evaluation of the effective dose is not an easy job because of the impossibility to take into account the distance from the source, the condition of the exposition, the numbers of shot and so on.

One shot can in principle give at 1m from the target an ambient dose equivalent of 3-4 mSv in the worst conditions. This event is quite unlikely taking into account of the redundancy of the radiation safety system.

Radiological Risk from Induced Activity

The main radiological risk for the workers can only arise from radioactivity induced in solid components, air and water.

The three photonuclear reactions previously described are responsible for most of the produced activity in the "electrons machine" components . Furthermore, neutrons resulting from these reactions can activate surrounding materials (e.g. soil and air and cooling water). Taking into account the operation parameters of FLAME the activation of materials surrounding the target will be negligible.

2. CONCLUSION

The commissioning of multi-terawatt laser, used for studies in ultra-high intensity laser interaction with solid, gases and plasmas, as well as for high energy gradient acceleration technique, is in continuous increase in the world, posing new problems for the radiation protection communities. The paper focused some radiological protection aspects around the FLAME, 300 TW laser in commissioning at LNF.

References

1. T. Tajima, and J.M. Dawson, Laser electron accelerator Phys Rev. Letter 43,267-270 (1979)
2. C.D.R. Geddes, High quality electron beams from a laser wakefield accelerator using plasma-channel guiding Nature Vol. 432, p. 538-541, (2004)
3. S. Mangles *et al.* Mono-energetic beams of relativistic electrons from intense laser-plasma interactions Nature Vol. 432, p. 535-538, (2004)
4. J. Faure *et al.* A laser plasma accelerator producing mono-energetic electron beams Nature Vol. 432, p. 541-544, (2004)

5. L.A. Gizzi *et al.* An integrated approach to ultrai tense laser sciences: the PLASMON-X project accepted for publication on *EPJ - Special Topics* Vol. **175** (2009)

6. Lu W. *el al.*, Generating multi-GeV bunches using single stage laser wakefield acceleration in a 3D non linear regime Phys.Rev. S.T. Accelerators and beams 10 (20070 061301)

7. Alesini D, *et al.* Status of the SPARC project Nucl. Inst. Meth. A, Vol. 528 Issue: 1-2 Pages: 586-590

8. ICRP Publication 74 Conversion Coefficients for use in Radiological Protection against External Radiation Pergamon Press 1996

9. M. Pelliccioni Overview of Fluence-to-Effective Dose and Fluence-to-Ambient Dose Equivalent Conversion Coefficients for High Energy Radiation Calculated Using the FLUKA Code, Radiat. Prot. Dosim., 88, 279 (2000).

Neutron Facilities for Nuclear Data Measurements

Stephan Pomp

Uppsala University, Department of Physics and Astronomy,
Box 516, 751 20 Uppsala, Sweden
E-mail: *Stephan.pomp@physics.uu.se*

ABSTRACT

The continued demand for accurate nuclear data for neutron-induced reactions is connected to the progress made by a wide variety of applications. To meet the needs, new high-energy neutron sources have to be developed and maintained. An overview on existing quasi-monoenergetic neutron sources offering energies above 20 MeV is given. In addition, two new projects currently under development in Europe are presented; the Neutrons For Science (NFS) facility at GANIL in France, and nIGISOL in Jyväskylä, Finland. NFS, will offer white and quasi-monoenergetic neutron beams with energies up to 40 MeV that will be available from 2013. The IGISOL (Ion Guide Isotope Separator On-Line) facility is a high-resolution mass separator at the University of Jyväskylä. A neutron source allowing for measurements of independent fission yields is being designed. This source will produce a fast neutron spectrum that can optionally be moderated to thermal energies. First experiments are expected for 2012.

Keywords: Quasi-monoenergetic neutron beam, Fast neutron spectrum

Pacs No.: 29.25.Dz, 25.40.Fq, 25.85.Ec

1. INTRODUCTION

Neutrons play the key role in many important applications. These applications range from radiation treatment in cancer therapy and neutron dosimetry, to single event effects in micro-electronics, to nuclear energy production and waste management. The highest demand comes certainly from the latter and the efforts towards a closed fuel cycle.

To illustrate the challenges for nuclear energy production one may take a look at India's plans for the next decades. In February 2011, India had 20 nuclear power plants in operation, producing 4385 MWe or 2.5% of the total electricity production [1]. An annual growth of electricity consumption of 6.3% is expected. Therefore, India's ambitious plan to have 20,000 MWe from nuclear fission power plants on-line by 2020 and 63,000 MWe by 2032 will still only lead to a slow increase to 10% electricity from nuclear power stations. The long term plan is to supply 25% of electricity from nuclear power by 2050. From these sample numbers it seems clear that the global role of electricity production from nuclear fission power plants will increase dramatically over the next decades. As of January 2011, over 60 power stations are under construction world-wide, another 155 are planned and 320 are being proposed [1].

Efforts towards new types of nuclear power plants are also being made. In Europe this is driven by the Europeans Commissions SET-plan, a "strategic plan to accelerate the development and deployment of cost-effective low carbon technologies" [2]. Details concerning future systems for nuclear energy production are developed within the Sustainable Nuclear Energy Technology Platform, SNETP [3] and its European Sustainable Nuclear Industrial Initiative, ESNII. The Strategic Research Agenda of SNETP states that "Availability of accurate nuclear data is the basis for precise reactor calculations both for current and new generation reactors. Additional experimental measurements and their detailed analysis and interpretation are required in a broad range of neutron energies and materials. This is particularly true for fuels containing minor actinides for their transmutation in fast spectra."[4]

Continued progress in all other applications mentioned above also demand for additional and more accurate nuclear data for neutron-induced reactions, leading to better reaction models, and improved evaluated data libraries. In order to meet these needs neutron sources for a wide energy range, from thermal up to a few hundred MeV, have to be developed and maintained.

Producing neutrons is relatively easy. Any beam of particles hitting some material will result in neutron emission. It is the objective of a measurement that sets constraints on the choice of beam and target. In the high-energy domain, neutron beams are usually produced using protons impinging on either a thin lithium target leading to quasi-monoenergetic neutron (QMN) beams, or, *e.g.*, a thick tungsten target that fully stops the protons and leads to a so-called white neutron-energy spectrum.

I will give an overview about existing quasi-monoenergetic neutron facilities around the world offering beams with energies above 20 MeV and then turn to two new projects currently under development in Europe; the Neutrons For Science (NFS) facility at GANIL in France, part of the Spiral2 project, which will offer both QMN and white beams, and nIGISOL in Jyväskylä, Finland, which will make use of a fast reactor like neutron energy spectrum.

2. QUASI-MONOENERGETIC NEUTRON BEAMS AROUND THE WORLD

Currently, as of March 2011, there are just five quasi-monoenergetic neutron (QMN) sources in the world that can deliver beams above 20 MeV. These are located at The

Svedberg Laboratory in Sweden, iThemba in South Africa, and the Japan Atomic Energy Agency (JAEA), Tohoku University and Osaka University in Japan. All these sources produce neutrons using Li targets. The mono-energetic peak comes from the ^{7}Li(p, $n_{0,1}$)^7 Be($Q \approx 1.8$ MeV) reaction. This reaction goes either to the ground state of ^{7}Be (n_0) or its first exited state at 0.429 MeV (n_1). The peak fraction varies with incoming proton energy but comprises typically 50% of the produced neutrons. Below the peak, a break-up continuum from ^{7}Li(p, nx) reactions stretches down to zero neutron energy. The remaining proton beam is bent away towards a beam dump and the neutron beam is formed by various arrangements of collimators. Suppression of the low-energy neutron tail can be achieved using time-of-flight techniques since the incoming proton beams are pulsed.

2.1   The Svedberg Laboratory

A description of the neutron fields provided at The Svedberg Laboratory (TSL) is given in Ref. [5]. The proton beam from the cyclotron is pulsed with 45 ns between pulses at 180 MeV which increases as the energy decreases. Below 100 MeV the proton current can reach 10 µA, while above 100 MeV the maximum current is about 300 nA. The average energies of the peaks of the fluence distributions for the neutron beams which can be provided range from 11 MeV to 175 MeV, with a full width half maximum of about 2-5 MeV, depending on beam energy and the chosen Li target thickness. The neutron beam is monitored by means of a thin film breakdown counter (TFBC) and an ionization chamber monitor (ICM). A special feature of the facility is the availability of collimator openings ranging from 1 to 30 cm. Thus beam diameters of more than 1 m can be reached at the largest distance. The peak neutron fluence reaches 10^6 cm^{-2} s^{-1} for proton energies below 100 MeV and 10^5 cm^{-2} s^{-1} for higher energies.

2.2   iThemba Labs

Detailed information on the QMN fields at the iThemba LABS facility in Cape Town, South Africa can, *e.g.*, be found in Ref. [6]. Neutron peak energies ranging from 60 to 200 MeV are available. A special feature of this facility is that measurements can be made in 4 degree steps from 0 to 16 degrees. Spectra at 16 degrees can be subtracted from the 0 degree spectra to increase the monoenergetic component. Four beams at neutron peak energies of about 65, 97, 147 and 197 MeV at 0° and 16° have been studied carefully by time-of-flight spectrometry, using a proton recoil telescope for the measurement of the fluence in the peak for the lower energy and a ^{238}U fission chamber for the higher. For the 97 MeV beam, the full width half maximum of the peak is about 4 MeV. The beam is pulsed with 33 ns between pulses at 200 MeV which increases as the energy decreases. The time between pulses can be optionally increased using pulse selection up to 1 in 7. Peak neutron fluences at 8 m distance from a 6 mm ^{nat}Li target can reach a few times 10^4 cm^{-2} s^{-1} for proton currents of 3 µA.

2.3 TIARA

The Takasaki Ion Accelerators for Advanced Radiation Application (TIARA) in the Takasaki Advanced Radiation Research Institute (TARRI) of JAEA, offers QMN fields ranging from 40 to 90 MeV [7]. Protons from the AVF cyclotron are transported to a ^{7}Li disk. Protons passing through the target are bent into a Faraday cup shielded by an iron beam dump. The Faraday cup and ^{238}U and ^{232}Th fission chambers around the target are used as beam monitors. Also, a thin plastic scintillation monitor is used to monitor the neutron beams at the collimator exit. The produced neutrons are guided to an experimental room through about 3 m thick collimator. The flight path can be up to 18 m. The irradiation area has uniformity within 3.5% for the central beam area. Reference calibration fields with neutron peak energies of 45, 60 and 75 MeV are available. The peak energy and energy spectrum of the neutron beams have been measured by the time-of-flight (TOF) method using an organic liquid scintillation detector [8].

2.4 CYRIC

The CYRIC facility at Tohoku University can supply QMN beams from 20 to 90 MeV. The basic structure is similar to the field of TIARA. Intense neutron beams with the fluences exceeding 10^6 cm^{-2} s^{-1} are available [9]. The proton beam dump is embedded in the shielding wall. The magnet itself acts as shield to reduce the volume of shielding materials. The beam is monitored using a Faraday cup in the proton beam dump. The beam profile was measured by neutron activation of a 2-mm thick aluminium plate through the ^{27}Al (n, α) ^{24}Na reaction. A spatial uniformity within < 10% was found. The energy spectra where determined from TOF measurements using liquid organic scintillation counters.

2.5 RCNP

The third facility in Japan offering QMN fields is located at the Research Center for Nuclear Physics (RCNP) of Osaka University. Here fields with peak energies ranging from 100 to 400 MeV have been developed. The protons are accelerated up to 65 MeV in an AVF cyclotron and can be boosted up to 400 MeV in the Ring Cyclotron. The protons are transported to a 10 mm thick ^{7}Li target placed in a vacuum chamber within the swinger magnet. Protons passing through the target are directed towards a Faraday cup inside the beam dump to monitor the current. Since the Li target can be moved to different positions inside the swinger magnet, the neutron beam can, similar to the situation at iThemba Labs, be extracted at angles up to 30 degrees relative to the proton beam. For this purpose, the collimator wall can be shifted to different positions. A clearing magnet in the collimator reduces secondary charged particles in the beam. The experimental room behind the collimator has a length of about 100 m, which allows precise measurements of energy spectra with the TOF method [10]. The highest neutron fluence reaches 1.0×10^{10} sr^{-1} µC^{-1}.

Currently, the National Metrology Institute of Japan (NMIJ) in the National Institute of Advanced Industrial Science and Technology (AIST) is performing intensive studies for standardization of the quasi-monoenergetic neutron fields of TIARA in comparison with CYRIC. NMIJ has also started characterization studies for the high energy neutron fields of RCNP for energies of 140, 250, 350 and 392 MeV [11].

3. NFS

The future SPIRAL-2 facility, currently under construction at GANIL, Caen (France) will produce radioactive ion beams in the mass range from A = 60-140 from neutron-induced fission of ^{238}U [12]. A high-power superconducting driver LINAG (LINear Accelerator of Ganil) will deliver high-intensity proton, deuteron and heavy ions beams. The LINAG beams will also be used for other purposes: Neutrons For Science (NFS) [13], and the Super Separator Spectrometer for SPIRAL-2 stable beams (S3). Starting 2013, NFS will offer intense pulsed white neutron and QMN beams in the 100 keV-40 MeV energy range, as well as an irradiation facility with neutrons and light charged particles. The neutron beams will be delivered to a 30 m time-of-flight area, allowing for good neutron energy resolution.

For the white neutron beam, the LINAG will deliver deuterons with energies up to 40 MeV which will interact with a 1 cm thick rotating converter target (made of Be or C). The resulting neutrons will have a continuous energy spectrum with an average energy close to 14 MeV at 0 degrees. The QMN beam will be produced from a thin Li target (1-2 mm thick).

The LINAG beam frequency of 88 MHz is not well adapted for time-of-flight measurements. To avoid the overlap of neutrons generated during several bursts the neutron beam frequency will be reduced below 1 MHZ using a fast beam chopper. Therefore, the maximum ion beam intensity delivered to NFS will be 50 µA which corresponds to a maximum power deposition in the converter of 2 kW.

This facility will be a very powerful tool for physics from fundamental research to applications like the transmutation of nuclear waste, design of future fission and fusion reactors, nuclear medicine or the test and development of new detectors.

4. NIGISOL

The IGISOL (Ion Guide Isotope Separator On-Line) facility is a high-resolution mass separator that has been operating at the University of Jyväskylä for many years [14]. The facility is currently moving to a new location next to the recently installed

MCC30/15 high intensity cyclotron. This cyclotron produces a 30 MeV proton beam with a current up to 200 μA. It is planned to construct a neutron source that will make use of this high proton intensity and allow for measurements of independent fission yields. First experiments are expected for 2012.

In an initial stage thermal and fast fission of ^{235}U would be measured for calibration purposes, followed by measurements of ^{238}U, ^{239}Pu and ^{240}Pu. Furthermore, the facility can be used for systematic measurements on potential actinides of relevance in various Generation IV scenarios, such as ^{232}Th, ^{233}U, ^{237}Np, and ^{243}Am.

The initial target design studies have been performed mainly with the MCNPX code and independent double checks with the FLUKA code. The approach so far was to utilize the experience with the ANITA white neutron source which is being used at TSL [15, 16]. At ANITA, neutrons are produced through the W(p,nx) reaction. Tungsten is a neutron rich nuclide, has a high melting point, high thermal conductivity, and produce only limited residual radioactivity. The neutron source will produce a fast neutron spectrum, optionally moderated to thermal energies using a water sphere with radius 5 to 10 cm [17].

Ongoing tasks for the target design are:

- Optimization of the neutron spectra with respect to energy and intensity in order to make them as similar as possible to LWR and FR spectra, by modifying the target geometry and moderator thickness. Besides water, other moderator materials are considered.
- Due to the intense proton beam, up to 200 μA, different cooling options need to be considered.
- The target area needs to be accessed on a regular basis, so activation of material needs to be investigated. The possibility of moving the target assembly into a shielded area will be considered.
- Other kinds of neutron sources are considered, using, *e.g.*, the available deuteron beam. Mono-energetic neutron sources such as portable D-T or D-D generators are relatively easy to handle, but they give lower neutron yields than the proton beam. On the other hand they have the advantage of giving neutrons at a well defined energy.

5. CONCLUSION

There are currently only five QMN facilities in the world which often have limited availability. To meet the needs of the nuclear data community and, eventually, the end-users of nuclear data, new dedicated QMN facilities, offering peak energies from below 20 MeV to several hundred MeV, with a well-characterized neutron energy spectrum, and low, well characterized background are necessary. From experience with the facilities listed above, some conclusions leading to a wish list for an ideal QMN facility can be drawn:

- stable, short-pulsed proton beam of sufficient intensity
- adjustable time between subsequent proton pulses (pulse selection/beam kicker)
- efficient shielding between the neutron production area and the measurement area
- several neutron beam lines at various angles (from 0 up to about 25 degrees)
- large measurement area/experimental hall to allow for usage of TOF methods
- sunken floor and raised roof in the experimental area to minimize ambient background

Two new neutron facilities are currently under construction in Europe, NFS and nIGISOL. NFS will offer both QMN and white neutron beams of high quality, but limited to energies below 40 MeV. At nIGISOL both thermal neutrons and neutrons with a fast reactor-like energy spectrum will be produced and used for fission studies and basic nuclear physics. Both new facilities will become operational around 2012-2013 and offer good research opportunities for the nuclear data community.

Acknowledgements

Several people have contributed to the work reported in this paper. I would like to thank especially R. Smit and H. Yasuda (from EURADOS WG11), X. Ledoux (NFS), and H. Penttilä, M. Lantz and D. Rados (IGISOL). Financial support from the Swedish Nuclear Fuel and Waste Management Co and the Swedish Radiation Safety Authority within the AlFONS project, the Swedish Research Council, and the International Symposium on Accelerator and Radiation Physics Research (ISARP-2011), is gratefully acknowledged.

References

1. http://www.world-nuclear.org
2. http://ec.europa.eu/energy/technology/set_plan/set_plan_en.htm
3. http://www.snetp.eu/
4. http://www.snetp.eu/www/snetp/images/stories/Docs-AboutSNETP/sra2009.pdf, page 71-72.
5. A V Prokofiev, J Blomgren, O Byström, C Ekström, S Pomp, U Tippawan, V Ziemann, M Österlund, Radiat. Prot. Dosim. 126 **18** (2007)
6. M Mosconi, E Musonza, A Buffler, R Nolte, S Röttger, F.D. Smit., Rad. Meas. **45** 1342 (2010)
7. M Baba, Y Nauchi, T Iwasaki, T Kiyosumi, M Yoshioka, S Matsuyama, N Hirakawa, T Nakamura, Su Tanaka, S Meigo, H Nakashima, Sh Tanaka, N Nakao, N., Nucl. Instrum. Methods. Phys. Res. A **428** 454 (1999)

8. Y Shikaze, Y Tanimura, J Saegusa, M Tsutsumi, Y Yamaguchi, Y Uchita, Radiat. Prot. Dosim. **126** 163 (2007)

9. M Baba, H Okamura, M Hagiwara, T Itoga, S Kamada, Y Yahagi, E Ibe, Radiat. Prot. Dosim. **126** 13 (2007)

10. S Taniguchi, N Nakao, T Nakamura, H Yashima, Y Iwamoto, D Satoh, Y Nakane, H Nakashima, T Itoga, A Tamii, K Hatanaka, Radiat. Prot. Dosim. **126** 23 (2007)

11. H Harano, T Matsumoto, Y Tanimura, Y Shikaze, M Baba, T Nakamura T., Rad. Meas. **45** 1076 (2010)

12. "Report of the SPIRAL 2 Detail Design Study", available on http://www.ganil.fr

13. http://pro.ganil-spiral2.eu/spiral2/instrumentation/nfs

14. H Penttilä, K Karvonen, T Eronen, V-V Elomaa, U Hager, J Hakala, A Jokinen, A Kankainen, I D Moore, K Peräjärvi, S Rahaman, S Rinta-Antila, V Rubchenya, A Saastamoinen, T Sonoda, J Äystö, Eur. Phys. J. **44** 147 (2010)

15. A V Prokofiev, J Blomgren, M Majerle, R Nolte, S Röttger, S P Platt, C X Xiao, A N Smirnov, Nuclear and Space Radiation Effects Conference (NSREC'2009), 2009 IEEE Radiation Effects Data Workshop, Quebec, Canada, July 20-24, 2009, pp. 166-173

16. Y Naitou, Y Watanabe, S Hirayama, A Prokofiev, A Hjalmarsson, P Andersson, R Bevilacqua, C Gustavsson, S Pomp, V Simutkin, H Sjöstrand, M Österlund, M Tesinsky, U Tippawan., Proceedings of the 2009 Annual Symposium on Nuclear Data, Ricotti, Tokai, Japan, Nov. 26-27, 2009

17. M Lantz, D Gorelov, B Lourdel, H Penttilä, S Pomp, D Rados, I Ryzhov, V Simutkin, E Tengborn, to appear in the proceedings of "NEMEA-6: Exploring the frontiers of nuclear data and measurements, their uncertainties and covariances", Krakow, Poland, 25-28 October, 2010, and reference therein.

Ludwig Eduard Boltzmann, Transport Equation, Microscopic Reversibility and Macroscopic Irreversibility

K.P.N. Murthy* and S. Siva Nasarayya Chari
School of Physics, University of Hyderabad, P.O. Central University,
Gachibowli, Hyderabad 500046
E-mail: *kpnmsp@uohyd.ernet.in

ABSTRACT

The Boltzmann Transport Equation(BTE) has played an important role in basic and applied sciences. It is an integro-differential equation for the phase space density of molecules of a dilute gas. It was derived by Boltzmann in 1872 and even today it remains as an important theoretical method for investigating non-equilibrium systems. Linear version of BTE provides an exact description of neutron transport in reactor cores and shields and it forms the backbone of the nuclear industry. In this paper, we discuss on non-linear transport equation and Boltzmann's quest to model the macroscopic irreversibility from microscopically reversible dynamics of the system by invoking *stosszahlansatz*. We talk on his famous *H* function, which connects the time-asymmetric macroscopic behavior of the system with the time-symmetric Newtonian dynamics of its microscopic constituents. Then a brief discourse is given on a few recent developments in non-equilibrium statistical mechanics, consisting dynamical measure, non-equilibrium work fluctuations followed by the fluctuation theorems.

Keywords: Boltzmann transport equation, Boltzmann's *H* function, Fluctuation theorems.

Pacs No.: 05.20.Dd, 51.10.+y, 05.45.-a, 05.60.-k, 05.70.-a

1. INTRODUCTION

In the year 1872, Ludwig Eduard Boltzmann derived [1] an integro-differential equation for the phase space density of molecules of a dilute gas, which we call the Boltzmann Transport Equation (BTE). Even today, it remains as an important theoretical method to investigate non-equilibrium systems. BTE plays an important role in basic and applied sciences. Linear version of BTE forms the backbone of the nuclear industry.

In this paper we consider the non-linear transport equation, derived by Boltzmann purely for comprehending the emergence of time asymmetric behavior in a macroscopic object from the time-symmetric behavior of its microscopic constituents. A macroscopic object is made of microscopic constituents like atoms and molecules. Each molecule obeys time-symmetric Newtonian dynamics. But collectively they violate time-symmetry. In the synthesis of a macro from its micro, when and why the time reversal invariance is broken?. This remains an open question, even today. To deal with a macroscopic object which consists large number of molecules ($\approx 10^{23}$), we need to track every individual particle. This is clearly impossible. Hence one would go for the statistical description. Perhaps we can say the second law emerges as a consequence of our inability to track a large number of molecules. Hence it must be of statistical origin.

But Boltzmann's hunch was that the second law is of dynamical origin. He derived the second law from Newtonian dynamics by invoking *stosszahlansatz*, see below.

2. THE BOLTZMANN TRANSPORT EQUATION AND HIS H FUNCTION

The phase space density of molecules, each of mass m, at the 6-dimensional phase space point $(\vec{r}, \vec{p})$ and at time t, is denoted by $f(\vec{r}, \vec{p}, t)$. Aim is to find an equation of motion for this function. When there are no collisions, a molecule present at $(\vec{r}, \vec{p})$, at time t, will stream to a point $(\vec{r} + \vec{p}\,\Delta t/m, \vec{p} + \vec{F}\Delta t)$, during an interval Δt, where $\vec{F}$ denotes the external force acting on the molecules. Hence, in the absence of collisions, representative points simply flow from one phase space volume element to another, like an incompressible fluid maintaining constant phase space density.

$$f\left(\vec{r} + \vec{p}\,\frac{\Delta t}{m}, \vec{p} + \vec{F}\Delta t, t + \Delta t\right) = f(\vec{r}, \vec{p}, t). \tag{1}$$

In presence of collisions, change in $f(\vec{r}, \vec{p}, \vec{t})$ due to collisions has to be taken into account; this leads to,

$$f\left(\vec{r} + \vec{p}\,\frac{\Delta t}{m}, \vec{p} + \vec{F}\Delta t, t + \Delta t\right) = f(\vec{r}, \vec{p}, t) + \left(\frac{\partial f}{\partial t}\right)_{\text{col}}\Delta t. \tag{2}$$

By Taylor expanding the left hand side of the above, up to first order in Δt and in the limit $\Delta t \to 0$ we obtain,

$$\frac{\partial f}{\partial t} = -\frac{1}{m}\,\vec{p}\cdot\vec{\nabla}_r f - \vec{F}\cdot\vec{\nabla}_p f + \left(\frac{\partial f}{\partial t}\right)_{\text{col}}, \tag{3}$$

where $\vec{\nabla}_r$ and $\vec{\nabla}_p$ are the gradient operators with respect to position and momentum respectively. Boltzmann modeled the collision-term considering only binary collisions, which is realistic for a dilute gas. Two molecules collide together at spatial location $\vec{r}$, with momenta $\vec{p}_1$ and $\vec{p}_2$, and bounce-off with momenta $\vec{p}_1'$ and $\vec{p}_2'$ after collision. He further assumed *stosszahlansatz* of Maxwell; *"the momenta of any two molecules entering a collision are uncorrelated"*.

$$f(\vec{r}, \vec{p}_1, \vec{p}_2, t) = f(\vec{r}, \vec{p}_1, t) \times f(\vec{r}, \vec{p}_2, t) \tag{4}$$

This is also referred as assumption of molecular chaos. Thus, the full nonlinear integro-differential, Boltzmann transport equation reads as,

$$\frac{\partial f}{\partial t} = -\frac{1}{m}\,\vec{p}\cdot\vec{\nabla}_r f - \vec{F}\cdot\vec{\nabla}_p f + \int d^3 p_2 \int d^3 p_1' \int d^3 p_2' \,\Sigma\,(\vec{p}_1, \vec{p}_2, \vec{p}_1', \vec{p}_2') \times [f(\vec{p}_1')\,f(\vec{p}_2') - f(\vec{p}_1)\,f(\vec{p}_2)],$$
$$\tag{5}$$

where $\Sigma\,(\vec{p}_1, \vec{p}_2, \vec{p}_1', \vec{p}_2')$ is the rate of transition from $(\vec{p}_1, \vec{p}_2)$ to $(\vec{p}_1', \vec{p}_2')$. For the detailed derivation of the Boltzmann collision term, see [2]. Boltzmann defines his famous H function,

$$H(t) = \int d^3 p\, f(\vec{p}, t)\,[\log f(\vec{p}, t)] \tag{6}$$

A density $f(\vec{p}, t)$, that solves the transport equation obeys,

$$\frac{dH}{dt} \le 0, \tag{7}$$

which is Boltzmann's H-theorem. The above is clearly time asymmetric. In contrast to the Newtonian dynamics, which does not distinguish future from the past, the H function has a well defined direction of time. Thus, Boltzmann has derived time asymmetric macro from time-symmetric micro. The objections that were raised against H-theorem has led to paradoxes. They are *"Loschmidt's reversibility paradox"* [3] and *"Zermelo's recurrence paradox"*, [4] see [Appendix. I].

However, the crucial point that had been overlooked in the formulation of Boltzmann collision term is the use of *stosszahlansatz* after collision. The momenta of the molecules will get highly correlated after the collision, as enforced by the law of momentum conservation. Boltzmann conceded that, perhaps the *stosszahlansatz* has smuggled in an element of stochasticity into his otherwise purely dynamical derivation of transport equation. He then proposed a stochastic approach to the problem of macroscopic irreversibility [5].

3. BOLTZMANN ENTROPY

Consider an isolated macroscopic system of N particles, represented by a point in $6N$ dimensional phase space moving incessantly along a trajectory. Boltzmann interpreted the probability of finding a system in a region of phase space as the fraction of observation time the dynamical trajectory spends in that region. Let $\vec{x}$ be the $6N$ dimensional vector denoting the phase space point of the system. $\vec{r}(\vec{x}, t)\, d^{6N}x$ refers probability of finding the system in an infinitesimal volume $d^{6N}x$ centered at $\vec{x}$, at time t.

If we coarse grain the phase space in terms of hyper cubes each of volume h^{3N}, the discrete representation of phase space density is given by $\{\rho_i\}$. Boltzmann's H function can be written as,

$$H(t) = \sum_{i=1}^{\vec{\Omega}} \rho_i \, \log \, (\rho_i) \tag{8}$$

where Ω is the total number of microstates accessible to the system under the constraints of energy U, volume V and number of particles N. Then Boltzmann defines entropy as,

$$S \, (U, \, V, \, N) = -k_B \sum_{i=1}^{\vec{\Omega}} \rho_i \, \log \, (\rho_i), \tag{9}$$

where k_B is now called the Boltzmann's constant. With the assumption that all the microstates are equally probable ($\rho_i = 1/\Omega, \, \forall i$), we get famous formula for Boltzmann's entropy,

$$S = k_B \, \log \, (\Omega), \tag{10}$$

engraved on his tomb in Zentralfriedhof, Vienna. This formulation of entropy depends only on Ω. Thus Boltzmann has liberated entropy from its thermal confines and his expression for entropy laid the foundation for statistical mechanics.

4. DYNAMICAL MEASURE: RECENT TRENDS IN NON-EQUILIBRIUM STATISTICAL MECHANICS

Edward Norton Lorenz (a meteorologist and mathematician) made a chance observation [6] in the year 1963. He was solving three coupled first order nonlinear differential equations on a computer, which were intended to provide an approximate description of atmospheric behavior. He discovered that he had two different numerical solutions for the same problem with almost identical initial conditions. This is due to chaotic dynamics: two phase space trajectories of a chaotic system starting off from arbitrarily close phase space points diverge exponentially and become completely uncorrelated asymptotically. All systems that obey the laws of thermodynamics are chaotic. Thus chaos provides a link between dynamics and stochastics.

The phase space density ρ remains constant along a trajectory; this is called Liouville theorem. A single trajectory visits all the phase space regions of an energy surface. This is called ergodicity. Boltzmann's ergodicity has been generalized by Sinai, Ruelle and Bowen, in their SRB measure [7, 8]. The SRB measure on the attractor expressed in terms of phase space volume is analogous to the Liouville measure on the energy surface of an equilibrium isolated system. In the phase space of the statistical mechanics, a system in equilibrium or otherwise can be represented by a point; any process quasi-static or otherwise can be represented by a trajectory. Such a generalization permits assignment of dynamical weight (a measure of instability) to a trajectory based positive Lyapunov exponents. Thus the dynamical measures of the recent times has liberated entropy from its equilibrium and quasi-static confines.

5. FLUCTUATION THEOREMS

We have fluctuation theorems of Entropy, Heat and Work. Here we briefly discuss about Work and entropy fluctuation theorems, see [9, 10].

5.1 Work Fluctuation Theorem

If a closed system in equilibrium at a temperature T, draws Q amount of heat from the reservoir by a reversible process, the entropy of the system increases by,

$$dS = \frac{Q}{T} \tag{11}$$

The work done(W) during this process is calculable from first law of thermodynamics as,

$$W = dU - Q, \tag{12}$$

where U is the internal energy of the system. For an isothermal reversible process, Q can be replaced by TdS [from Eq. (11)] and right hand side of the above becomes dF, where F denotes the Helmholtz free energy. Reversible work thus equals the free energy change. Whereas, for an irreversible isothermal process, the increase in entropy of the system is given by the second law inequality,

$$dS > \frac{Q}{T} \tag{13}$$

From the above, we obtain the relation between W and dF as,

$$W > dF \tag{14}$$

In statistical mechanics, we take average of a quantity over a suitable ensemble to make correspondence with thermodynamics. Let W be a random variable associated with work done during a switching process[1]. Different switching experiments all carried out with the same protocol, can in principle yield different work values. Thus we have to deal with an ensemble of work values, denoted by Ω. It is quite possible that some of the values of Ω, shall be less than ΔF. We term these as corresponding to the second law violating switching. The fraction of the work values in Ω, that are less than ΔF, is usually called as the probability of violation of second law, denoted by the symbol $p(\tau)$, where τ refers to switching time. Formally we define,

$$p(\tau) = \int_{-\infty}^{\Delta F} \rho(W, \tau)\,dW, \tag{15}$$

where $\rho(W, \tau)$ is the probability distribution of W describing the ensemble Ω, for a given switching time (τ). From Callen-Welton theorem [11], we say that for nearly quasi-static processes, $\rho(W, \tau)$ is Gaussian. In the quasi-static limit ($\tau \rightarrow \infty$), $\rho(W, \tau)$ tends to a Dirac

---

[1]A process carried out according to a predefined protocol, which switches the system from initial equilibrium state to a final non-equilibrium state, by changing the value of a macroscopic variable.

delta function in W, peaked at ΔF. This implies that $p(\tau)$ should increase with τ, and reach one-half asymptotically (as $\tau \to \infty$). Though this seems to be counter intuitive, it can be understood from Callen-Welton theorem,

$$\langle W \rangle = \Delta F + \frac{1}{2}\, \beta \sigma_w^2 \tag{16}$$

The angular brackets in the above stands for the average over the ensemble Ω. Let us define $W_d = \langle W \rangle - \Delta F$, which is referred as dissipation during the process and it goes to zero in the reversible limit. Hence from Eq. (16) we have $\sigma_w^2 \propto W_d$. This means $\sigma_W \to 0$, slower than W_d. Hence we expect $p(\tau)$ to increase with τ. For more details, see [12].

5.2   Entropy Fluctuation Theorem

Let S_τ corresponds to a dynamical entropy obtained from observing the phase space expansion or contraction. $\Pi(S_\tau)$ is the probability of S_τ. Then entropy fluctuation theorem [13, 14] states as,

$$\frac{\Pi(S_\tau)}{\Pi(-S_\tau)} = \exp[\tau\, S_\tau] \tag{17}$$

Entropy fluctuation theorem helps us to calculate the probability for the entropy to change in a way opposite to that dictated by the Second law. This probability of violation of second law is very small for large systems and for long observation times.

6.   CONCLUSION

Fluctuation theorems predict and more importantly quantify, second law violation in small systems, on small time scales of observation and the predictions of the fluctuation theorems have since been verified experimentally. The fluctuation theorems, especially the ones on entropy, has shown that Boltzmann's original hunch about the dynamical origin of entropy was indeed right.

Acknowledgements

We dedicate this paper to P.K. Sarkar on the occasion his becoming sixty years young.

SSNC acknowledges CSIR for providing financial support.

References

1. L. Boltzmann, in S.G. Brush [13] Vol. 2, pp. 88-175 (Pergamon: Oxford) (1966)
2. Huang, K. *Statistical Mechanics,* 2nd ed. (New York: Wiley) (1987)
3. J. Loschmidt, Weiner Berichte **73** 128 (1876); **75** 287 (1877); **76** 209 (1877)

4. E. Zermelo, in S.G. Brush [13] Vol. 2, pp. 208-217 (Pergamon: Oxford) (1966)

5. L. Boltzmann, Weiner Berichte **76** 373 (1877)

6. E. N. Lorenz, *J. Atm. Sci.* **20** 130 (1963)

7. D. Ruelle, *Ann. N.Y. Acad. Sci.* **357** 1 (1980)

8. J.P. Eckmann and D. Ruelle, *Rev. Mod. Phys.* **57** 617 (1985)

9. E.N. Lorenz, *J. Atm. Sci.* **20** 130 (1963)

10. C. Jarzynski, *Phys. Rev. Lett.* **78** 2690 (1997)

11. Callen and Welton, *Phys. Rev.* **83** 34 (1951)

12. M. Suman Kalyan, G. Anjan Prasad, V.S.S. Sastry and K.P.N. Murthy, *Physica A* **390** 1240 (2011)

13. D.J. Evans, E.G.D. Cohen and G.P. Morris, *Phys. Rev. Lett.* **71** 3616 (1993)

14. E.G.D. Cohen and G. Gallavotti, *J. Stat. Phys.* **95** 367 (1999)

15. S.G. Brush, *Kinetic Theory*, Vol. 1, (Pergamon: Oxford) (1965); Vol. 2, (Pergamon: Oxford) (1966)

APPENDIX. I

(a) Loschmidt's reversibility paradox: Let an isolated system evolve from time $t = 0$ to τ and H function decreases during this evolution. At $t = \tau$, reverse the momenta of all the molecules and let the system evolve from $t = \tau$ to 2τ. At $t = 2\tau$, reverse once again the momenta of all the molecules. Since the system obeys the time-reversal Newtonian dynamics, it will end up at the same phase space point it started from. Hence there would be an increase of H function during the evolution from $t = \tau$ to 2τ, contrary to the Boltzmann's claim.

(b) Zermelo's Recurrence Paradox: An isolated system, will return arbitrarily close to its initial point in phase space after Poincaré recurrence time. If there is a spontaneous increase in the entropy during an interval of time, there would be a spontaneous decrease of entropy during the interval of Poincaré recurrence. Hence this contradicts Boltzmann's claim. Poincaré recurrence is easy to observe for systems with few degrees of freedom. The recurrence time increases exponentially with the system size. Hence Poincaré recurrence is seldom observed in thermodynamic systems.

Radiological Safety of J-PARC

H. Nakashima[1*], T. Shibata[1], T. Miura[2], and Radiation Safety Section of J-PARC center

[1]J-PARC Center, Japan Atomic Energy Agency, Tokai, Naka, Ibaraki 319 1195, Japan
[2]J-PARC Center, High Energy Accelerator Research Organization, Oho, Tsukuba, Ibaraki 305 0801, Japan
E-mail: *nakashima.hiroshi@jaea.go.jp

ABSTRACT

The first phase of J-PARC (Japan Proton Accelerator Research Complex) is completed and J-PARC is operated now. This paper reviews radiological safety aspects of high-energy and high-intensity accelerator facilities at J-PARC during construction and operation.

Keywords: J-PARC, Accelerator, High energy, High intensity, Radiation, Safety, Shielding, Activation.

1. INTRODUCTION

The first phase of J-PARC (Japan Proton Accelerator Research Complex) is completed in 2009 which is the joint program of Japan Atomic Energy Agency (JAEA) and High Energy Accelerator Research Organization (KEK). J-PARC consists of three accelerators: Linac, 3 GeV rapid cycle synchrotron and 50 GeV synchrotron, and three major experimental facilities: Material & Life Science Facility (MLF), Hadron Experimental Facility and Neutrino Facility, as shown in Fig. 1. Transmutation Experimental Facility is still planning as the second phase. Various secondary particles produced by high energy proton beam are used for a variety of sciences such as material and life science by muons and neutrons, and nuclear and particle physics including neutrino as well as accelerator driven nuclear transmutation technology. [1], [2]

From the viewpoint of the radiological safety, the characteristics of J-PARC are summarized as high beam power of 1 MW, high beam energy up to 50 GeV and large-scale accelerator complex. These characteristics create many radiological problems during construction and operation which must be overcome. [3-6] This paper reviews radiological safety aspects of J-PARC.

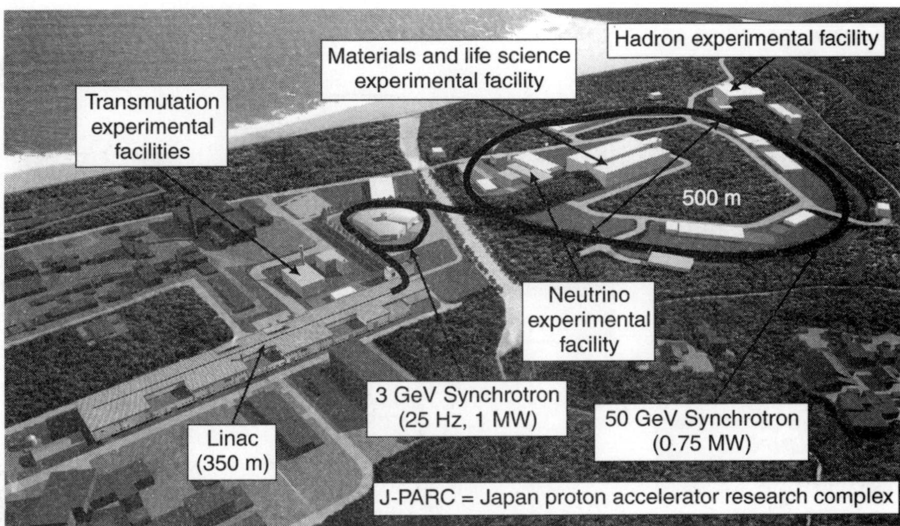

Fig. 1 *Bird's eye view of J-PARC facilities. [1]*

2. SHIELDING DESIGN AND BENCHMARKING

In order to overcome a difficulty, that is, the radiation sources are widely distributed in a large-scale accelerator complex a calculation system employing simplified and detailed methods was used for the shielding design and safety analyses needed for licensing of J-PARC. In the detailed calculation flow-diagram, Monte-Carlo codes such as PHITS [7], MCNPX [8], MARS [9] are used for high-energy particle transport calculation above 20 MeV for neutrons and above 1 MeV for charged particles and mesons. The MCNP-4 code [10] together with a nuclear data set, JENDL-3.3 [11], is applied to calculate low-energy neutrons up to 20 MeV and photons. The DCHAIN-SP 2001 code [12] together with the FENDL-Dosimetry file [13] is used to estimate induced radioactivity and doses due to residual radioactive nuclei in the machine components and in the walls of the accelerator room.

To study the accuracy of the design methods, some benchmark problems on thick target neutron yield, beam dump, bulk shielding and streaming, which were based on JAEA/TIARA, LANL/WNR, BNL/AGS, KEK/PS and NIMROD shielding experimental data, were analyzed by the design methods. Resultantly, a safety factor of two was adopted for the shielding calculation. [14]

As an example of bulk shielding problems, we show analysis on a series of experiments with a mercury spallation target using high-peak-power GeV proton-beam from the Alternating Gradient Synchrotron (AGS) of Brookhaven National Laboratory (BNL) is shown. In the experiment, a mercury target was bombarded with 1.6-, 12- and 24-GeV-protons. [15] A shielding experiment carried out using steel and concrete shields set around the mercury target. Fig. 3 shows the arrangement of lateral and forward

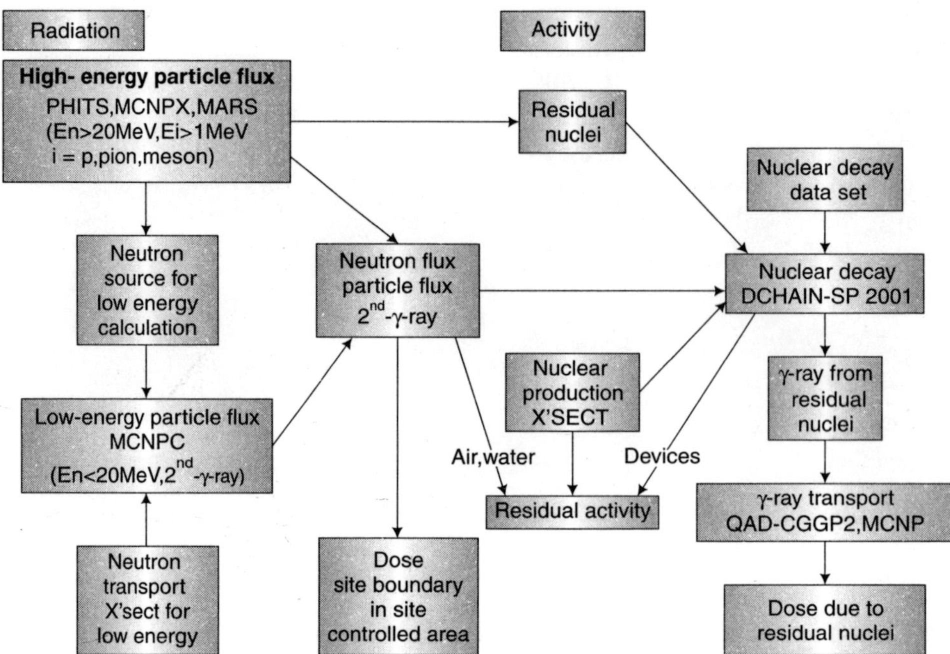

Fig. 2 *Calculation flow-diagram of radiation and activity used in the J-PARC shielding design. [3]*

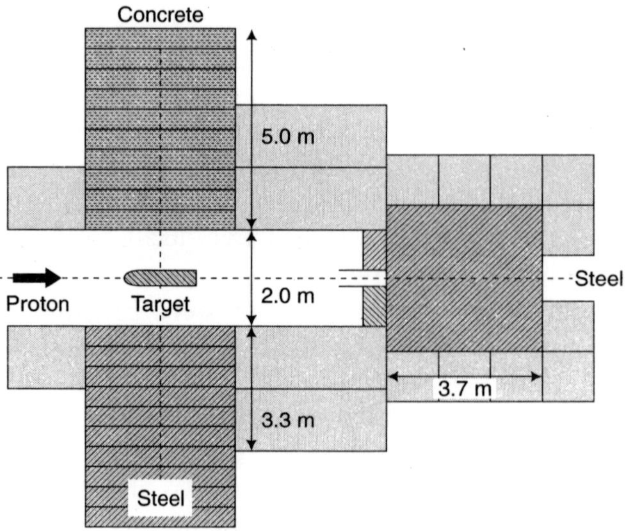

Fig. 3 *Horizontal cross sectional view of shield arrangement for shielding experiment at BNL/AGS.[15], [16]*

shields made of steel of 3.3 m thick and ordinary concrete of 5.0 m. Measured spatial distribution of neutron reaction rates of the ^{209}Bi(n, 6n)^{204}Bi reaction (E$_{th}$: 38.1 MeV) inside the steel shield is compared with the PHITS calculations and the MCNPX calculations at 24 GeV protons in Fig. 4 as an example. The PHITS calculation agrees very well with the measurement almost at all positions. The MCNPX calculations show the same tendency of neutron attenuation and agree with the measurement within a factor of 2, although the calculations yield slightly lower values than the measurement.[16]

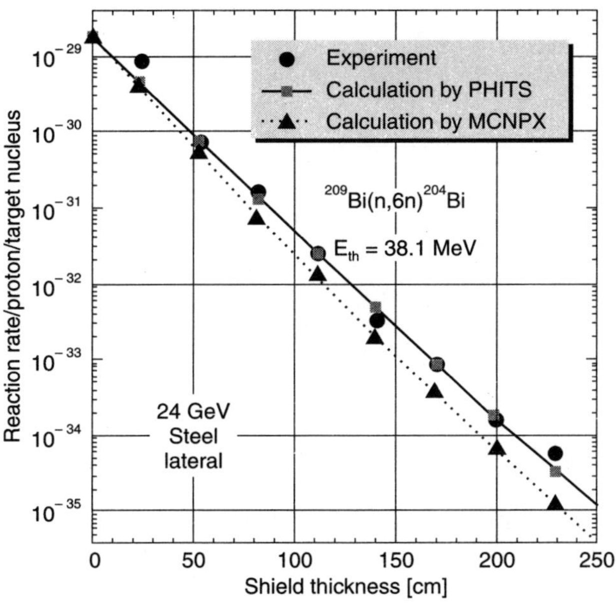

Fig. 4 *Comparisons on distributions of 209Bi(n, 6n)204Bi reaction rates in steel shield among calculations and measurement at BNL/AGS.[15], [16]*

3. ACTIVATION ESTIMATION AND HANDLING

In high intensity particle accelerators such as J-PARC, activation is one of serious problems. However, it is difficult to estimate exactly the radioactivity induced in accelerator devices, wall, air and cooling water in advance. The primary and secondary particles at the beam loss point activate materials with several modalities.

The main radio nuclides produced in air are ^{3}H, ^{7}Be, ^{11}C, ^{13}N, ^{15}O, and ^{41}Ar. The half-lives of nuclides except for ^{3}H are shorter than 2 hours and ^{7}Be is easily trapped by an air filter such as the HEPA filter. The best way to dispose of these nuclei is to confine the air during operation, and release it after operation when the concentrations of these nuclei decrease below the regulated maximum level. Thus, a buffer region is installed between the accelerator room and the external area, as shown in Fig. 5.

Although the cables and pipes are carefully sealed by putty, the activated air in the accelerator room will leak into the buffer region. The buffer region is kept in negative pressure and the radioactivity concentration in this region is monitored during operation of the accelerator.

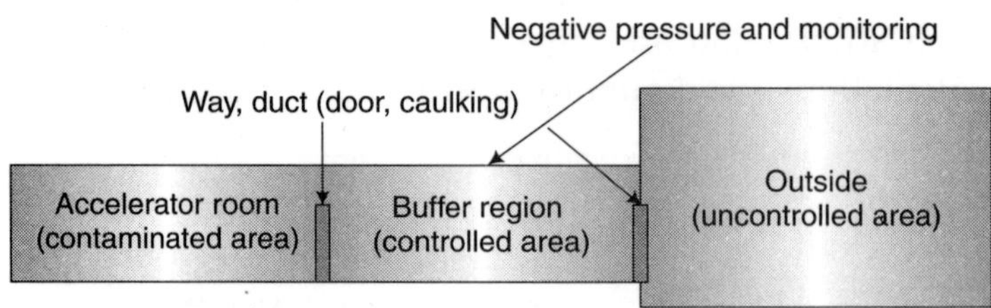

Fig. 5 *Buffer region between accelerator room and outside to confine the activated air.[5]*

The cooling water is in a closed loop with ion exchange system to maintain purity. The radioactive nuclei produced in the water are expected to be captured in the system except for Tritium. The cooling water is stored and monitored in a disposal tank. The water is discharged before the tritium concentration reaches the regulated maximum value.

4. ENVIRONMENTAL RADIATION PROTECTION

Most of high energy accelerator facilities are constructed under the ground and use soil and rock as shields. Neutrons and so on may activate the soil and rock, and the radioactive nuclides may flow to the residential area. Thus, the environmental effect of activated soil and rock should be estimated. In this estimation, the groundwater should be regarded as uncontrolled water. Based on the documents "Principles for the Exemption of Radiation Sources and Practices from Regulatory Control" (IAEA safety series No.89) [17] and "International Basic Safety Standards for Protection against Ionizing Radiation and for the Safety of Radiation Sources" (IAEA safety series No.115)[18], for J-PARC, a criteria was set for the dose due to intake of the groundwater at a site boundary.

In order to estimate the activity contributing for the dose, transition of the nuclides to the site boundary was calculated. First of all, the production of radioactivity was calculated with the detailed calculation system shown in Fig. 2. Then, the dissolution was estimated with the evaluated distribution coefficients from soil to water for each nuclide. Finally, the movement to the outside of the site was estimated dependently on the distance from the accelerator facility to the site boundary, the velocity of groundwater, retardation factors, and so on. In this estimation, although the contributions of ^{3}H and ^{22}Na to the total production were about 50 and 20% at the surface of the concrete shield under the ground, the contributions of ^{3}H and ^{14}C to the dose at the site boundary were about 80 and 20%. Design constraints on the radioactivity production rate and the neutron dose rate on the external surface of the shield under the ground was set for J-PARC, which is equivalent to 10 μSv y^{-1} as the dose due to intake of the groundwater at the site boundary. [5]

5. HIGH POWER TARGET HANDLING AT MLF

A high power target system of MLF is mainly composed of radiation and thermal hydraulic components as well as maintenance components. The cross sectional view of the target system is shown in Fig. 6. [19]

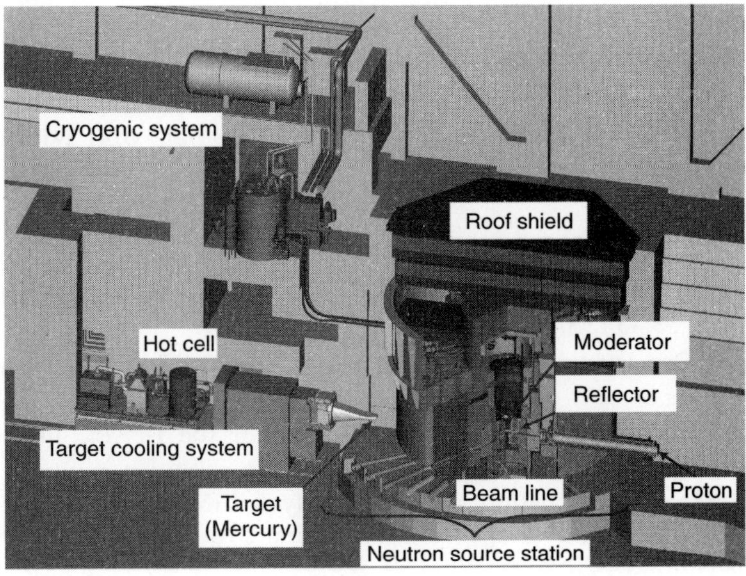

Fig. 6 *Cross sectional view of the target system at MLF. [9]*

The shielding design was carried out by using the calculation flow-diagram shown in Fig. 2. The shields in proximity of the mercury target are in iron about 4 m in thickness and in magnetite concrete about 1.6 m in thickness, and the dose rate distribution was estimated outside shields for both the primary proton and the secondary neutron beam lines.

The nuclear heat and the peak heat density of the target are about 60% and 690 W cm^{-3} at the beam power of 1 MW, and the remainder is absorbed in the shields, reflector and moderators surrounding the target. Thus, the target must be cooled and a high heat capacity is required. Finally, mercury was selected as the target material. [19]

The target structural integrity against radiation decides the life of the target, and rules the maintenance frequency. For its estimation, the displacement per Atom

(DPA) is used as an index of the radiation damage. The maximum DPA of the target vessel was estimated to be about 10 under the condition of 1 MW beam power and 5000 hours of operation. The target vessel (made of SUS316L) will have to be substituted every 6 months. On the other hand, the target system is highly activated by the primary beam and the secondary neutrons. Thus, the maintenance will be completely carried out by a remote handling system inside the hot cell connected to the target system. [19], [20]

6. SUMMARY

The radiological safety aspects of high-intensity and high-energy accelerators are complicated compared with those relating to conventional accelerator facilities. In order to ensure safety, the safety estimation and management must be performed with well-established estimation methods based on reliable data and with stable systems using reliable devices.

References

1. Tanaka, S., High Intensity Proton Accelerator Project in JAPAN (J-PARC), Radiat Prot Dosimetry, **115**, 33-43 (2005).

2. Nagamiya, S., Construction status of the J-PARC project, J. of Nucl. Materials **343**, 1-6 (2005).

3. Nakashima, H., *et al.*, Radiation Safety Design for the J-PARC Project, Radiat Prot Dosimetry, **115**, 564-568 (2005).

4. Nakashima, H., *et al.*, Radiation shielding study for the J-PARC project, Proc. of 14th Biennial Topical Meeting of the ANS Radiation Protection and Shielding Division, 267-282 (2006).

5. Nakashima, H., *et al.*, Benchmark Calculation for the J-PARC Shielding Design, Proc. on 8th International Topical Meeting on Nuclear Applications and Utilization of Accelerators (AccApp'07), Pocatello, USA, 267-282 (2007).

6. Nakashima, H., OPERATIONAL RADIATION PROTECTION ISSUES SPECIFIC TO HIGH-INTENSITY BEAMS, Radiation Protection Dosimetry, ORPARM SPECIAL ISSUE, 1-16 (2009).

7. Niita, K., *et al.*, PHITS: Particle and Heavy Ion Transport code System, Version 2.23, JAEA-Data/Code 2010-022, JAEA (2010).

8. Waters, L. S., (Ed.), MCNPX Users Manual – Version 2.1.5, TPO-E83-G-UG-X-00001, LANL (1999).

9. Mokhov, N.V. and Striganov, S.I., Hadronic Shower Code Inter-Comparison and Verification, Fermilab-Conf-07-009-AD (2007), Hadronic Shower Simulation Workshop AIP Proceedings 896.

10. Briesmeister, J.F., (Ed.), MCNP – A General Monte Carlo N-Particle Transport Code, Version 4A, LA-12625, LANL (1993).

11. Shibata, K., *et al.*, Japanese Evaluated Nuclear Data Library Version 3 Revision-3: JENDL-3.3, J. Nucl. Sci. Technol. **39**, 1125 (2002).

12. Kai, T., *et al.*, DCHAIN-SP 2001: High Energy Particle Induced Radioactivity Calculation Code, JAERI-Data/Code 2001-016, JAERI (2001) (in Japanese).

13. Pashchenko, A. B., IAEA Consultants' Meeting on Selection of Evaluations for the FENDL/A-2 Activation Cross Section Library, Summary Report, INDC(NDS)-341, IAEA (1996).

14. Matsuda, N., *et al.*, Analyses of benchmark problems for the shielding design of high intensity proton accelerator facilities, JAEA Technology 2008-030, JAEA (2008).

15. Nakashima, H., *et al.*, Research Activities on Neutronics under ASTE Collaboration at AGS/BNL, J. Nucl. Sci. Technol. Suppl.2, 1155-1160 (2002).

16. Nakashima, H., *et al.*, Current Status of the AGS Spallation Target Experiment, Proc. of OECD/NEA Workshop on Shielding Aspects on Accelerator, Target and Irradiation Facilities, SATIF 6, NEA No. 3828, 27-36 (2004).

17. Principles for the Exemption of Radiation Sources and Practices from Regulatory Control, Safety series No. 89, IAEA, Vienna (1988).

18. International Basic Safety Standards for Protection against Ionizing Radiation and for the Safety of Radiation Sources, Safety series No. 115, IAEA, Vienna (1996).

19. High Intensity Proton Accelerator Project (J-PARC) Technical Design Report Material & Life Science Experimental Facility, JAERI-Tech 2004-001, JAERI (2004) (in Japanese).

20. http://j-parc.jp/MatLife/en/source/neutronsource.html

Real-time Neutron Dosimetric Techniques in Particle Accelerators

Sunil C.

Health Physics Division, Bhabha Atomic Research Centre, Mumbai 400085, India

E-mail: *sunilc@barc.gov.in*

ABSTRACT

TNeutron dose measurement in particle accelerators is a complex problem due to the high energy, anisotrpic emission, pulse nature and high fluence rate encountered. Active techniques such as time of flight and proton recoil telescopes can be used to measure fluence distribution with high precision. However, the sophistication of the setup and the expertise required to conduct such experiments rule out their usage on a regular basis. For regular monitoring of workplace environments, conventional dose rate meters are still preferred due to their portability and ability to give instantaneous results. They however have poor dose response and limited useful energy range, both imposing restrictions on the dose measurements particularly in an accelerator environment. LET based TEPC monitors are better suited for such measurements as the LET spectra also constitutes a basic quantity for radiation protection. These can be configured in coincidence-anticoincidence modes to separate out charged and neutral particles. Dose estimation can also be carried out by flux measurements using flat response instruments such as a 4π counter based on moderated BF_3 proportional counters. Some of these techniques are discussed with experimental and simulations results.

Keywords: Neutron dosimetry, Particle accelerator, LET, Neutron spectrometry, FLUKA.

Pacs No.: 87.55. N-Radiation monitoring, Control, and safety

87.53. Bn Dosimetry/exposure assessment

87.15. ak Monte Carlo simulations

29.30. Hs Neutron spectroscopy

1. INTRODUCTION

Neutron dose estimation in particle accelerators are particularly difficult due to the wide dynamic energy and fluence range, their pulsed nature and highly anisotropic nature of emission. For example, in a high energy hadron machine the neutron energy inside the machine room can range from thermal to GeV. In lower energy accelerators, low and intermediate energy neutrons will dominate outside the shield. The neutron fluence rate too can vary since the emission rate in an accelerator is directly proportional to the beam current. Since beam current can vary from few nA to hundreds of mA depending upon the application of the accelerator, the neutron fluence also can vary proportionality by several orders of magnitude. Pulsed nature of neutron radiation environment can also pose problems for the radiation protection specialist as the pulsed repetition rate can be as low as few Hz or as high as MHz. Conventional dose rate meters widely used in routine monitoring of work places will misread the actual dose rate under all these circumstances. Often combination of detectors and techniques are required to correctly interpret the situation. Passive techniques rely on accumulation of signals that are conveniently read later. The flip side is no information is available on a real time basis for instantaneous decision making. Active techniques on the other hand fulfill this aspect but can have problems related to gamma interference, dead time effects and energy dependent efficiencies. The inherent quest for dose estimation with increasing precision and the fact that the definition of dose equivalent can be changed by ICRP necessitates techniques for the measurement of invariant quantities such as energy deposition and energy-fluence distribution. This is further accentuated by the anisotropy in the neutron emission. While such anisotropy is well known in high energy hadron accelerators due to intra nuclear cascades and direct reactions, it is a lesser known fact that fair amount of anisotropy exits even in low energy heavy ion interactions [1] where compound nucleus emission is predominant. Active techniques such as time of flight and proton recoil telescopes can be used to measure fluence distribution with high precision. However, the sophistication of the setup and the expertise required to conduct such experiments practically rule out their usage on a regular basis. For regular monitoring of workplace environments, conventional dose rate meters are still preferred due to their portability and instantaneous results obtained in terms of dose equivalent. They however have poor dose response and limited useful energy range, both imposing restrictions particularly in an accelerator environment. LET (linear energy transfer) based tissue equivalent proportional counter (TEPC) are better for such measurements as the LET spectrum is also a basic aspect of the radiation protection quantity. These can be configured in coincidence-anticoincidence modes to separate out charged and neutral particles.

2. CONVENTIONAL DOSE RATE METERS

The so called rem meters have been around for several years and have proved to be handy in all types of installations where neutron monitoring is necessary. Due to the ad-hoc nature of the design they suffer from severe under and over responses. In Fig. 1, the fluence (left ordinate) and dose response (dotted line, right ordinate) of such an instrument is shown [2]. The instrument over-responds from 30 keV to 80 keV and from 1 MeV to 5 MeV while under-responding at all other energies. The integral response of the instrument for an incident spectrum can be estimated using the fractional weight in every energy bin folded by the corresponding dose response. Thus, for a neutron spectrum emitted from a ^{252}Cf fission source, the instrument has been found to overestimate the actual dose by about 10% while for the EMPIRE [3] calculated neutron spectrum from 145 MeV ^{19}F incident on a thick Al target this is about 20% under response. While the neutron spectrum from the Cf source extends from thermal to about 10 MeV, the EMPIRE spectrum extends from 1 MeV to 20 MeV. The large under and over responses thus can somewhat compensate each other for an incident neutron spectrum that extends all across its measurement range. The limited energy range of a conventional rem meters have been overcome by the recent developments in the field of extended range rem meters such as WENDI [4] and LINUS [5]. Since they use heavy metals as converters, the weight rules out their use as portable instruments.

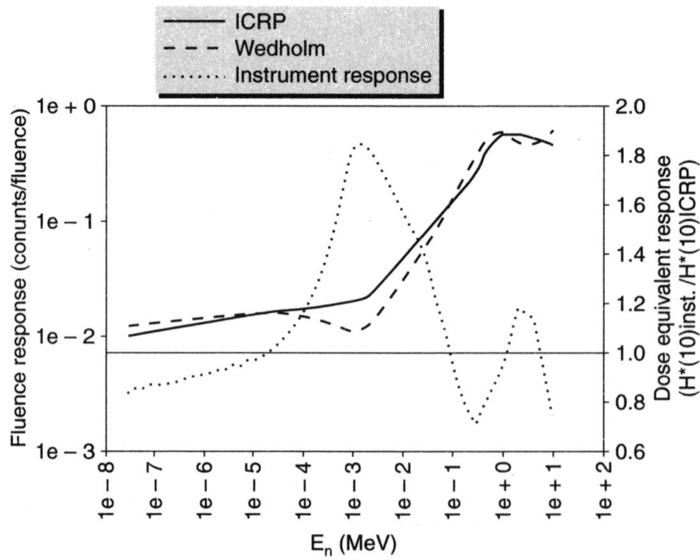

Fig. 1 *The fluence and dose responses of a neutron rem meter along with the ICRP fluence to dose response coefficients.*

3. PORTABLE NEUTRON SPECTROMETER

Portable neutron spectrometers are now commercially available [6]. In one such instrument, two probes are used to cover the energy region extending from thermal to 20 MeV The region from thermal to 0.8 MeV is covered by a spherical ^{3}He proportional counter while above 0.8 MeV the neutron probe is a cylindrical organic liquid scintillator of 51 mm × 51 mm size which uses the pulse-shape discrimination technique to reject gamma signals. The pulse height data is unfolded using an internal algorithm and the final spectrum from thermal to 20 MeV is given in predetermined 18 energy bins. The resulting neutron spectrum is then converted to dose equivalent using the ICRP fluence to dose conversion coefficients. In Fig. 2, spectra obtained from a 14 MeV neutron generator and a research reactor measured using one such spectrometer are shown.

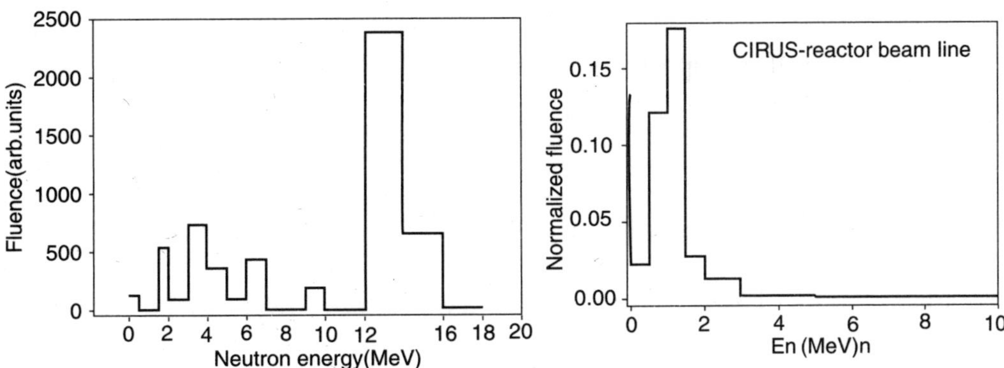

Fig. 2 *The neutron spectra obtained from a 14 MeV neutron generator (left) and from a research reactor (right).*

A good test for any spectrometer, especially one that utilizes few channel methodology and unfolding technique, is to reproduce a mono energetic source. The instrument fairly reproduces the 14 MeV peak (left Figure) as well as the Maxwellian shape in the research reactor (right) and is therefore a good-albeit an expensive-radiation monitor.

4. LET BASED MONITORS

Rossi type tissue equivalent proportional counters have also been used by the radiation protection community for quite some time. The advantage is the readout in terms of LET and easy n-γ discrimination since photons induced events will be always of low LET (less than ~ 10 keV/µm) [7] while neutrons give rise to higher LET events. In Fig. 3 the event spectra for neutrons emitted from a ^{252}Cf source (left Figure) obtained by a LET based TEPC neutron monitor is shown. The events are above the cutoff of about 10 keV/µm which is the set threshold for n-γ discrimination. A sharp drop in the event spectra, known as the proton edge, can also be seen that about 150 keV/mm. The maximum energy deposition by a secondary proton happens when it is just able to traverse the maximum chord length. For tissue equivalent material it is 320 keV/µm.

Since the mean chord length of a sphere is two third of its diameter, the proton edge channel number in the event spectra will depend upon the diameter of the detector expressed in terms of tissue equivalent thickness. However, the event spectra alone

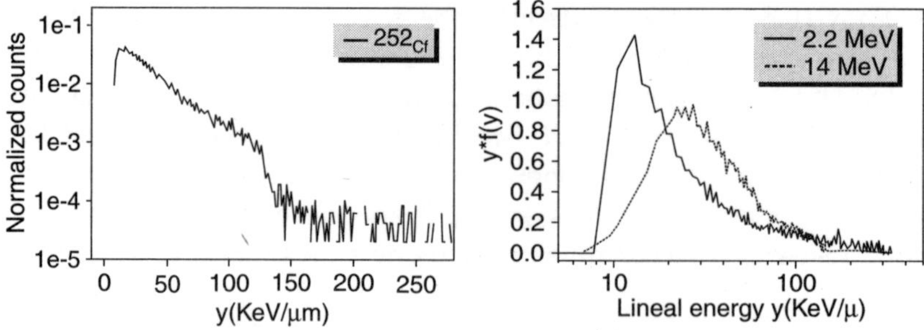

Fig. 3 *The LET event spectra from photons (left) and neutrons emitted from a 252Cf source (right).*

does not convey enough information that can be used to distinguish various neutrons sources. When the ordinate is plotted as a product of lineal energy transfer (keV/um) and the normalized event spectra, the sources can be clearly distinguished. In Fig. 3 (right), the 14 MeV source is clearly seen as low LET events while the 2.2 MeV neutron source is harder in the LET scale.

A sophisticated LET based in-flight dosimetry developed by PTB [8] uses coincidence anticoincidence (CACS) techniques to distinguish charged and neutral particle events and can be handy in high energy particle accelerators where the secondary particle yield can be a mixture of several types of charged particles, neutrons and photons.

In Fig. 4, the schematic of the instrument is shown (left picture) while the dose equivalent measured using this instrument at a heavy ion accelerator along the results obtained from FLUKA Monte Carlo code [9] simulations are shown (right plot). The FLUKA simulations and the experimental results agree quite well for the neutral particle dose equivalent measured with the (πDOS) except at extreme forward angles. This could possibly be due to the increased fluence of the charged particles at forward angles (mainly protons and He isotopes), which produce secondary neutrons in the surroundings of πDOS [10].

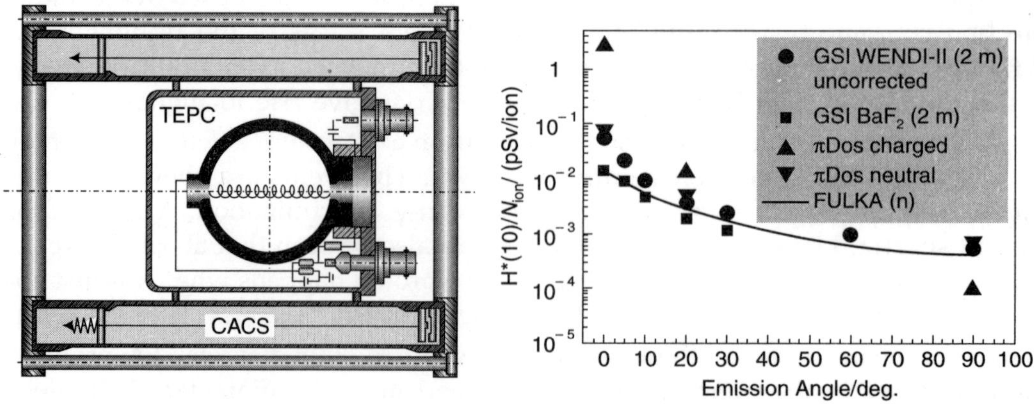

Fig. 4 *The PTB inflight dosimeter (pDOS) (left) with the coincidence-anticoincidence (CACS) and a comparison of the results obtained by measurements and FLUKA simulations.*

5.   TIME OF FLIGHT SPECTROMETRY

The neutron energy spectra is considered to be an absolute quantity for dose estimation since it remains invariant. While the ICRP coefficients are subjected to change, the knowledge of the spectra can be used to recalculate the dose equivalent with the new set of coefficients simply by folding the spectra with ICRP fluence to dose conversion coefficients. This is not possible with the direct reading dosimeter with their ad-hoc mechanism of measurement and hence not based on any physical quantity. Time of flight experiments are based on some event generating signal such as a prompt gamma ray emission or a signal from the accelerator that indicates the arrival of the reaction initiating projectile on target. Such experiments are usually carried out using pulsed beam and a signal from the buncher is usually available for the experimentalist. Neutron detection here is usually carried [11] out using tailor made scintillators such as NE213 or BC501 that have excellent pulse shape discrimination properties. However, it is also possible to carry out such experiments with E-ΔE telescopes [12] that can discriminate particles on the basis of energy deposition in the ΔE detector. Excellent energy resolution is possible using large distances between the detector and the target but with the disadvantage of a lower solid angle and therefore larger counting time.

In Fig. 5 (left), a typical block diagram of a multi parameter data acquisition system is shown. The pulse shape discrimination (PSD) circuit discriminates the neutrons from gamma photons while the CFD-TAC combination measures the flight time (TOF). With PSD and TOF, superior n-γ separation can be achieved as can be seen in the 2d plot in Fig. 6 (right). Since the photons arrive first and at a fixed known time, the photon peak can be assigned to a channel number. A time calibration procedure will give the conversion from channel number to time. With appropriate software gates, the list mode data collected can be analyzed to obtain the flight time of the neutrons. With the known distance, it is possible to get the energy information. The counts can then be re-grouped into the required energy bin size. This energy information can now be used to calculate the dose equivalent using the fluence to dose conversion coefficients given by ICRP. In Fig. 6 the neutron spectra obtained by such an experiment is given along with the calculated results using the EMPIRE nuclear reaction model code.

6.   CONCLUSION

Neutron dosimetry is a challenging problem in particle accelerators due to the wide range of energy and fluence present, pulsed nature, the anisotropy of emission and the insufficient instrumentation. Since dose equivalent is a defined quantity, the task is even more difficult as no physical quantity that can be measured physically is involved. No single method can solve this problem and a combination of detection techniques will have to deployed to accurately estimate the neutron dose. Conventional dose response

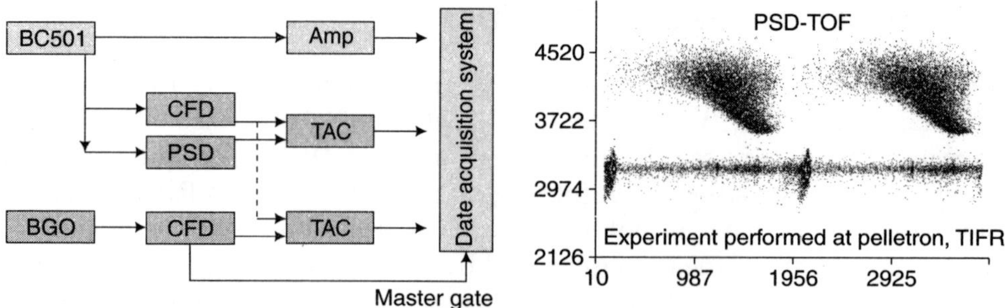

Fig. 5 *A typical block diagram used to discriminate neutrons from gamma using the pulse rise technique and neutron energy measurement using time of flight technique (left). On the right is a 2d plot that shows the separation of neutrons from gamma in pulse shape (ordinate) and time of flight (abscissa) axes.*

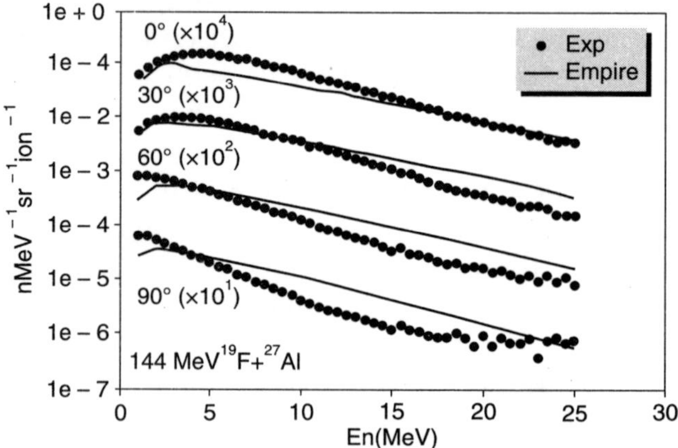

Fig. 6 *The neutron energy spectra from 144 MeV ^{19}F projectile incident on a tick Al target obtained by the time of flight method.*

instruments can somewhat satisfy the routine operational monitoring requirements. LET based neutron monitors have an advantage over the conventional dose meter since they measure a physically meaningful quantity for radiation protection. For calculations involving radiation shielding, induced activity etc, or for accidental dosimetry, accurate information of the neutron dose is essential. This can be achieved by measuring the neutron energy spectrum which is an invariant physically measurable quantity. Further, the dose equivalent also can be derived from it using the ICRP fluence to dose conversion coefficients.

References

1. Sunil C, Nandy M, Sarkar PK, *Phys. Rev.* **C 78**, 064607 (2008)
2. Sunil C., A.A. Shanbhag, M. Nandy, T. Bandyopadhyay, S.P. Tripathy, C. Lahiri, D.S. Joshi and P.K. Sarkar *Radiat. Prot. Dosim.* **143**, 4 (2011)

3. Herman, M., Capote, R., Carlson, B. V., Oblozinsky, P., Sin, M., Trkov, A., Wienke, H. and Zerkin, V. *Nucl. Data Sheets* **108**, 2655 (2007)

4. Olsher, R.H., Hsu, H.H., Beverding, A., Kleck, J.H., Casson, W.H., Vasilik, D.G. and Devine, R.T. *Health Phys.* **70**, 170 (2000)

5. Birattari, C., Ferrari, A., Nuccetelli, C., Pelliccioni, M. and Silari, M. *Nucl. Instrum. Methods* **297**, 250 (1990)

6. H. Ing, S. Djeffal, T. Clifford, R. Machrafi, and R. Noulty, *Radiat. Prot. Dosim.* **126**, 238 (2007)

7. James C. Liu, Workshop on Neutron Field Spectrometry in Science, Technology and Radiation Protection, Pisa, Italy, 2000 SLAC-PUB-8258.

8. Wissmann F, Langner F, Roth J, Schrewe U *Radiat. Prot. Dosim.* **110** 347 (2004)

9. Fasso, A., Ferrari, A., Ranft, J. and Sala, P.R. FLUKA: a multi-particle transport code. CERN-2005–10 (2005), INFN/TC_05/11, SLAC-R-773.

10. F. Wissmann, U. Giesen, D. Schardt, G. Martino, Sunil C, *Radiat Environ Biophys* **49**, 331 (2010)

11. Sunil C, Saxena A, Choudhury R K, Pant L.M., *Nucl. Instrum. Methods A* **534**, 518 (2004)

12. Gunzert-Marx K, Iwase H, Schardt D, Simon RS. *New J Phys* **10**:075003 (2008)

Neutron Measurement in Accelerator Environment using Track Detectors

S.P. Tripathy* and P.K. Sarkar

Accelerator Radiation Safety Section, Health Physics Division,
Bhabha Atomic Research Centre, Mumbai 400085, India
E-mail: **sam.tripathy@gmail.com*

ABSTRACT

Neutron measurements in a pulsed and mixed radiation field such as particle accelerators require specific detectors. In this aspect, the response of solid polymeric track detectors (SPTD), due to their insensitivity to low LET radiations and integrating nature of signal registration, are found to be effective and convenient. Permanent record of the neutron-induced latent tracks via recoil (H, C, O), as well as nuclear reactions ((n, p), (n, α), (n, f), etc.) facilitates offline analysis conveniently. Physical characteristics like dimension, weight, mechanical strength, etc., and resistance to environmental parameters add to the increasing application of these detectors. These detectors are also being used in combination with selective fissile foils to cover a wide neutron energy range (response is from ~0.1 to ~10^3 MeV). One of the major disadvantages with these detectors is prolonged processing time, which can now be overcome with the recently developed microwave-induced chemical etching (MICE). In this work, neutron response, MICE technique and generation of neutron spectrum with CR-39 detector, will be presented.

Keywords: Track detector, Neutron dosimetry, Bismuth fission, Fast etching.

Pacs No.: 87.55.N, 87.53, 25.85.Ec, 81.40.Wx

1. INTRODUCTION

Increase in number of high-energy particle accelerators and their applications in various fields leads to involvement of more users and operators. The directional, dynamic, pulsed and mixed nature of the complex (secondary) radiation field, generated by the primary beam hitting a target, poses a unique challenge for the radiological safety aspects, specially the neutron component contributing a significant dose even beyond the shields [1, 2]. Measurement of wide energy neutron component mixed with photons require specific detectors covering the entire energy that can either be insensitive to photons or should be able to discriminate both components during analysis.

A large number of detectors, both active and passive, are in use for this purpose. Active detectors for dosimetry mainly consist of GM counters, ionisation chambers, scintillation counters, rem counters, silicon detectors and tissue-equivalent proportional counters. These can provide real time results immediately in the measurement procedure, very useful particularly for monitoring abrupt strong radiation such as solar events. However, these require complicated electronics and power supplies which in some cases will be inconvenient and even interfere with the original radiation environment if the detector system is large and heavy. Whereas, the passive detectors include the solid polymeric track detectors (SPTD), TLD, OSL, radiochromic films, etc. which are almost 2 dimensional and light-weight detectors. Out of these, the SPTDs are free from the effects due to low LET components, RF-wave interference and pulsed field, small in size and light-weight (do not interfere with the radiation field to be measured), suitable for long-time integral mode exposure, electronics free, cheaper, easy to process, even though it requires longer time for offline analysis including the long time track development procedure by chemical processing (etching).

Extensive studies on the formation and development of latent etchable tracks in SPTDs are reported by Fleischer et al. and Durani and Bull [3,4]. It has been established that, a polymer with electrical resistivity > 2000 Ωm and thermal diffusivity < 0.06 cm^2 s^{-1} can register nuclear tracks. However, if the stopping power (dE/dX) of the ion in the detector is less than the critical stopping power $(dE/dX)_c$ for that detector, then no track is formed, even if the ion possesses considerable kinetic energy [5].

2. DEVELOPMENT OF THE LATENT TRACKS

The latent tracks (~10 nm) registered in these detectors can be viewed under microscope only if these are properly developed up to micrometer level by using etching procedure(s), so that the track diameters exceed the wavelength of the light used for its microscopic observation. The faster etch rate along the nuclear track (V_t) compared to the uniform and slower etch rate of the detector material (V_b) lead to formation of a conical track, cone axis representing the particle trajectory. However, the etching procedure is time consuming and tedious, especially when a large number of detectors are to be analyzed routinely. Extensive investigations are still being carried out to reduce the etching time [6] and the references therein, [7] with the aim to attain better track profiles thereby improving the detector efficiency. The efficiency obviously depends on both the steps, viz. etching procedure and counting technique. The track profile depends on the competition between V_t and V_b, which are controlled by the concentration and temperature of the etchant, the etching period and the strength of stirring during etching. Since the critical angle of particle registration is not only a function of the particle energy but also of the etching time, therefore, an optimization of the etching conditions is necessary and the track registration properties of the detectors are to be determined for the optimum conditions.

Recently we have developed a new technique called microwave-induced chemical etching (MICE) [6] based on dielectric heating, where the detectors immersed in the etchant gets continuously exposed to microwave. In the process, a large amount of heat is developed instantly due to severe vibration of dipoles and the induced friction in it. This friction due to continuous rotation and vibrations may potentially induce an effect similar to the stirring effect but at the atomic level, which can facilitate easy and faster opening of the damaged trails. The simultaneous action of heat localized along the damaged trail and continuous stirring effect resulted in faster etching preferably along the polarized ion path which significantly reduced the etching time of the detectors with improved V_t/V_b ratio. The detector surface and the track profiles were also observed to be better due to preferred rapid etching along the polarized ion path leading to an overall improvement in the track revelation process. This improves the accuracy in subsequent counting or analyzing procedures. Other advantages of this method are:

- Less surface damage and homogeneous etching.
- and get stabilized, as in case of chemical etching.
- Does not involve complicated appliances like constant temperature unit, high-voltage, etc.

Fig. 1 shows the recoil tracks in CR-39 detector developed using MICE technique, which required only 450 s at 600 Watt of microwave power. As evident from Fig. 1, the detector surface is smooth and the tracks seem to have better profiles leading to more accuracy in the measurement of different track parameters by automatic image analyzing systems. To understand the complete mechanism, the information on common properties such as the dipolar moment, heat capacity, dielectric constant, polarizability and dielectric loss constant with the rate of heating and temperature profile is required to be known precisely.

3. NEUTRON MEASUREMENT USING SOLID POLYMERIC TRACK DETECTOR (SPTD)

The primary mechanism through which neutron-induced events get registered in SPTDs are recoil (mainly, H, C, O), as well as nuclear reactions ((n, p), (n, α), (n, f), etc.). In combination with various fissile radiators, moderators, and converters, these detectors can be used over wide neutron spectrum. Commonly used SPTDs for neutron measurement are PADC (polyallyl diglycol carbonate) or CR-39, polycarbonates (Lexan, makrofol), cellulose nitrate (LR-115), etc. The elastic scattering cross-sections for H, C, O

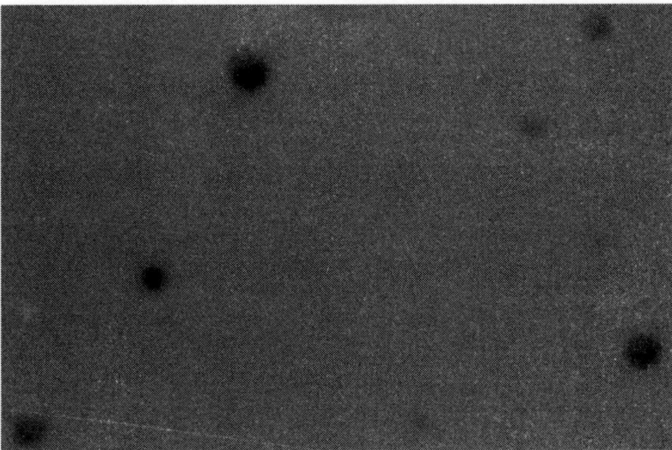

Fig. 1 *Recoil tracks in CR-39 (Intercast) developed using MICE (600W, 450 s)*

(σ_{recoil}) and for CR-39 detector ($N\sigma_{recoil}$) are shown in Fig. 2, where N is the number of atoms present in CR-39.

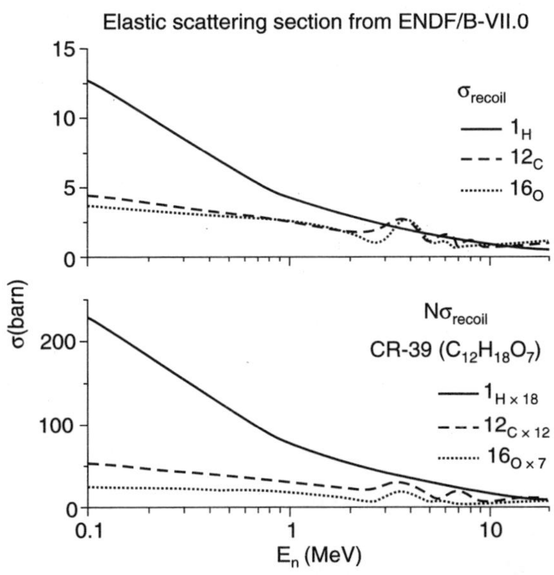

Fig. 2 *σ_{recoil} for H, C, O and $N\sigma_{recoil}$ for CR-39 detector, taken from ENDF/B-VII.0*

As can be seen from the Fig., the cross-section for proton recoil falls off about 4 to 5 times at higher energies compared to the lower side, where as the carbon and oxygen recoil cross-sections remain reasonably flat and becomes comparable to that of proton at the higher energy side contributing to the tracks with larger diameters.

The efficiency (tracks per neutron) of CR-39 detector per unit neutron fluence were obtained from the measurements performed with five different monoenergetic sources at PTB, Germany (0.565, 1.2, 5.0, 8.0, 14.8 MeV) as well as in IPR (14.2 MeV), India. The values are plotted in Fig. 3 along with the reported values from the measurements done by Benton et al [8]. The reduction in efficiency with increase in energy is essentially due to the fall in recoil cross-section. The response values extrapolated up to 20 MeV can be used for unfolding procedures to generate fast neutron spectra from the track density distribution based on suitable track parameters such as track area, diameter, etc. [9-11]. The bin size would essentially depend on the requirement of the unfolding code.

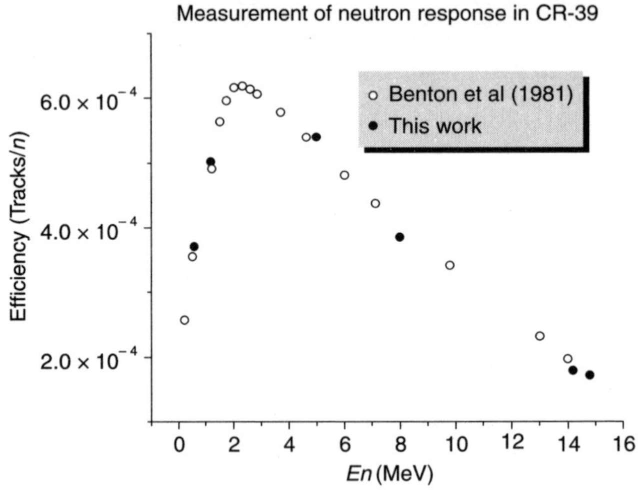

Fig. 3 *Neutron fluence response (Tracks per unit fluence) measured with CR-39 detectors.*

4. MEASUREMENT OF HIGH ENERGY NEUTRONS BY SPTD

In combination with fissile radiators (as shown in Fig. 4), the response of these solid polymeric track detectors can be extended to higher energy neutrons, the threshold energy is usually obtained from the reaction cross-section values (Fig. 4, lower half). The response (counts per unit neutron fluence) of bismuth fission detector stack up to 250 MeV is plotted in Fig. 5, as obtained from the reported values [12 and references therein].

A similar arrangement, but with ^{10}B on CR-39, is also in use for thermal neutron measurement via (n, α) reaction. The ultimate information derived from these detectors are: charged particle fluence (track density), direction (track angle), energy loss (track length or track opening), particle charge and energy (measurements in several consecutive detector sheets).

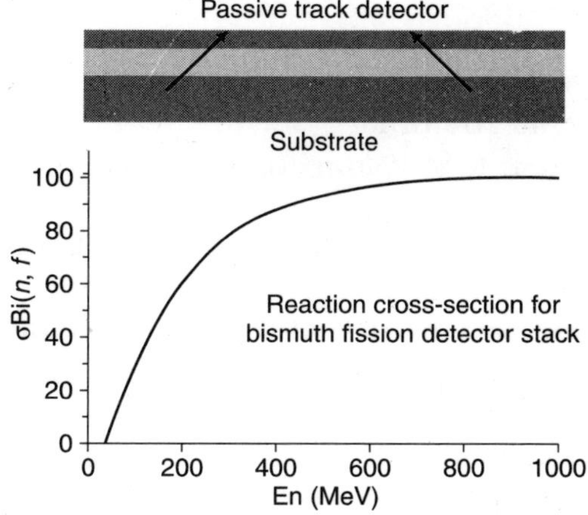

Fig. 4   *Bismuth fission detector stack along with its cross-section. Sensitive to neutron component only above 50 MeV.*

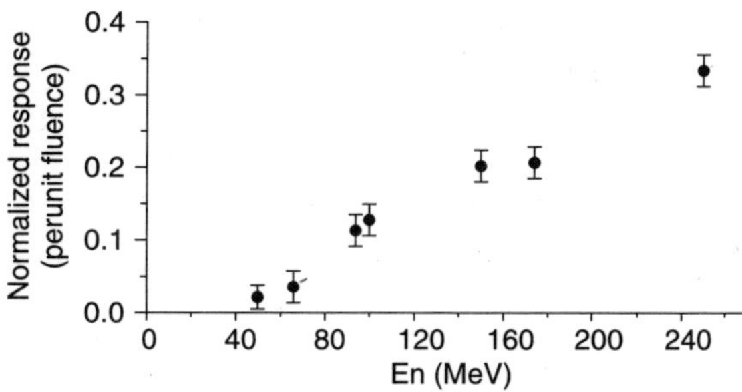

Fig. 5   *Normalized neutron response per unit fluence for bismuth fission detector stack.*

5. SPECTROMETRY

Various track parameters such as diameter, area, etc. can be used to generate the neutron spectrum, where a predetermined, either measured or calculated, energy response as a function of any of the parameters serve as the response matrix and the distributed track parameters serve as input file in the unfolding procedure. Very few investigators have reported the neutron spectrum generated using CR-39 track detectors [9, 10]. Alternatively, the track length method can also be followed for fast neutron spectrometry, where the length is directly proportional to energy loss in the detector medium [13].

Progressive development of etching techniques and image analyzing softwares improves the accuracy in track profile and automatic counting of the track parameters, which helps in overcoming the tedious counting and measurement task as well as the prolonged etching procedures. This may introduce new scopes and applications of these detectors.

References

1. PK Sarkar *Radiat. Meas.* 45 1476 (2010)

2. S Agosteo *Radiat. Meas.* 45 1171 (2010)

3. RL Fleischer, PB Price and R M Walker *Nuclear Tracks in Solids* (Berkely, USA: Univ. of California Press) (1975)

4. SA Durrani and R K Bull Solid State Nuclear Track Detection, Pergamon Press, Oxford, 1987

5. KK Dwivedi, S Mukherji *Nucl. Instrum. Meth. in Phys. Res.* **161** 317 (1979)

6. SP Tripathy, RV Kolekar, C Sunil, PK Sarkar, KK Dwivedi, DN Sharma, *Nucl. Instrum. Meth. in Phys. Res. A* **612** 421 (2010)

7. V Chavan, PC Kalsi, VK Manchanda. *Nucl. Instrum. Meth. in Phys. Res. A* 629 145 (2011)

8. EV Benton, RA Oswald, AL Frank and RV Wheeler, *Health Phys.* **40** 801 (1981)

9. F d'Errico, M Weiss, M Luszik-bhadra, M Matzke, L Bernardi and A Cecchi, *Radiat. Meas.* **28** 823 (1997)

10. M. Luszik-bhadra, E Dietz, F d'Errico, S Guldbakke and M Matzke *Radiat. Meas.* **28** 473 (1997)

11. SP Tripathy, Sunil C, GS Sahoo, DS Joshi, F Wissman and PK Sarkar, *Presented in this conference 'ISARP-2011, SINP, Kolkata, 16-18 February 2011'*

12. L Tommasino, RK Jain, D O'Sullivan, AV Prokofiev, NL Singh, AN Smirnov, SP Tripathy and P Viola, *Radiat. Meas.* **36** 307 (2003)

13. A R El-Sersy, SA Eman and NE Khaled *Nucl. Instrum. Meth. B* **226** 345 (2004)

Radiological Safety Issues at the Synchrotron Radiation Sources Indus-1 and Indus-2

Haridas G.[1*] and P.K. Sarkar[2]

[1]*Health Physics Unit, Raja Ramanna Centre for Advanced Technology,*
[2]*Health Physics Division, Bhabha Atomic Research Centre, Mumbai 400085, India*
E-mail: *haridas@rrcat.gov.in*

ABSTRACT

Indus-1 (450 MeV) & Indus-2 (2.5 GeV) are two electron synchrotron radiation sources (SRS) at Indore, India with critical energies 61A° and 2 A° respectively. Indus-1 is operational with 5 beam lines and Indus-2 is in an advanced stage of commissioning. Provision for 27 synchrotron radiation (SR) beam lines exists in Indus-2, of which 3 are in regular operation, 3 beam lines are under trial operation and several are under installation. Radiation environment of these sources comprises of mainly of bremsstrahlung photons and photo-neutrons. However at the SR beam lines of Indus-2 in addition to bremsstrahlung photons synchrotron radiation in the X-ray region down to few KeV contributes significantly to the radiation environment.

As the bremsstrahlung photons are of high energy nature in these sources, accurate measurement of dose rate with conventional instrument underestimates the dose rates by several factors. Similarly at synchrotron radiation hutches, since SR is of low energy nature (from few keV to tens of KeV), underestimation of dose rate results with conventional monitors due to self absorption in the detector wall and directionality of the radiation. The present talk describes the radiation environment and problems in dose measurement in the high and low energy radiation environment at the sources, Indus-1 & Indus-2.

Keywords: Accelerator, Bremsstrahlung, Synchrotron, Radiation monitoring.

Pacs No.: <87.56.bd, 03.50.-Z, 41.60.Ap, 87.55.N->

1. INTRODUCTION

The radiation environment of electron synchrotron radiation sources like Indus-1 & Indus-2 [1] consists mainly of bremsstrahlung radiation (BR) and to lesser extent photo-neutrons. They are produced as a result of the interaction of relativistic electrons either with the vacuum chamber or with the residual molecules with in the vacuum chamber. When relativistic electrons (450 MeV in case of Induss-1 & 2.5 GeV in case of Indus-2) interact with matter the phenomena called electromagnetic cascade occurs [2]. During the process electron loses energy by radiation, where high energy bremstrahlung photons, having a wide spectrum up to the electron energy is produced. This high energy photons interact through (e^-, e^+) pair production, which in turn interact through radiative loss and the process repeats resulting in the development of an electromagnetic cascade as shown in Fig. 1. The storage ring is usually housed in a well shielded structure whereas the SR front end (interface between storage ring and SR beam line) of SR beam lines protrude out through the shielding to the experimental hall, where the beam line extends to several tens of meters and a variety of experiments are performed.

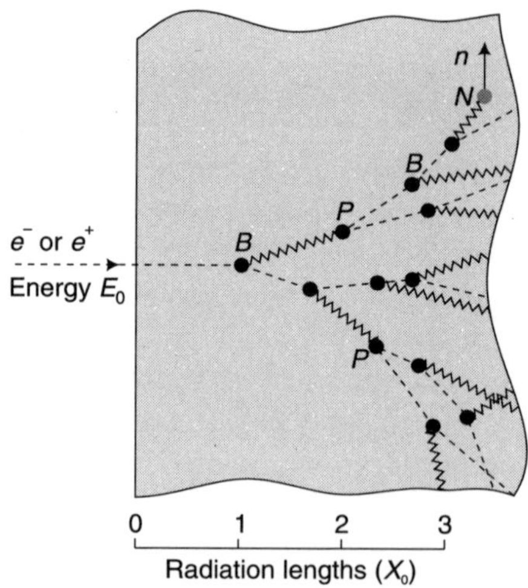

Fig. 1 *Schematic diagram of the generation of electromagnetic cascade generation*

In such facilities Bremsstrahlung radiation generated as a result of the generation of electro-magnetic cascade dominates the radiation field in comparison with other components which is depicted in the Fig. 2

Since the energy of the BR is high, extending up to hundreds of MeV, commercially available radiation monitors do not give the correct dose, rather they may underestimate. Some measurements are carried out at the Indus-1 storage ring is presented in the following section to find out the underestimation factor which can be used as the radiation monitor response correction factor.

As far as the Synchrotron radiation (SR) safety is concerned, at Indus-1 the SR is produced in the vacuum ultraviolet region which is well contained in the vacuum chamber itself. However at Indus-2, since the critical wavelength is in the x-ray region

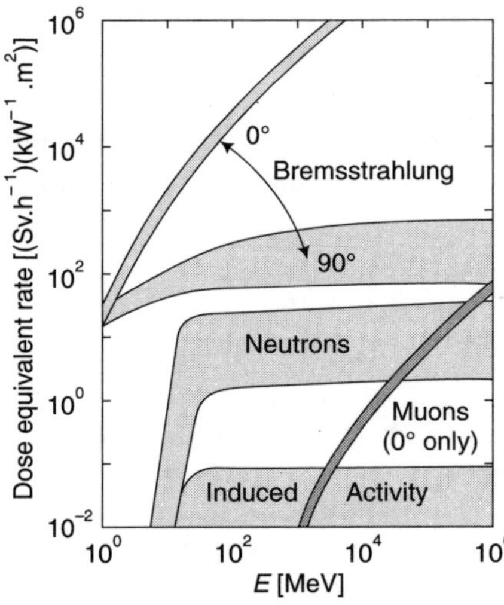

Fig. 2   *Comparison of various radiation hazards in high energy electron accelerators [3]*

(~2 A°), the radiation is brought out in air for experiments which poses radiation hazard to users and hence specially designed beam line hutches are required to contain the radiation hazards. Radiation safety issues of synchrotron radiation facilities are summarized by J.C.Liu [4]. The present paper also describes the measurements performed to quantify the radiation hazards in SR beam lines of Indus-2.

2.   DEDUCTION OF HIGH ENERGY DOSE CORRECTION FACTOR FOR BR

Initial random dose build up factor was observed in different media like copper, mild steel, lead, water etc. around Indus-1 storage ring. This was performed by inserting the media in front of the radiation detector/survey meter. The random buildup factors obtained from around 30 observations varied from 2 to 4.5 [5] is shown in Fig. 3. Later on systematic measurements of depth dose in water was performed at the front end exit and near a bending magnet, close to the ring which also confirmed the dose build up. The depth where the maximum dose occurs was found to be in tens of cm. Fig. 4 shows the depth dose measured with a victoreen survey meter and water phantom (water slabs) at front end and near the bending magnet location. The reason for the variation in build up factor and the depth at which maximum dose occurs even in the same medium, is attributed to the bremsstrahlung spectral variation at the measurement location. Though the measured build up factor is about 4, to find out a conservative correction factor for safety purpose, monte-carlo simulation of the depth dose was performed using EGS nrc [6]. For this bremstrahlung spectra was generated from different thickness of copper target and the depth dose curve was generated by allowing the spectrum to fall on semi-infinite slabs of water in broad beam geometry. The build up factor obtained is plotted as a function of the depth at which maximum dose occurs (shower maximum) is shown in Fig. 5. The average energy of the incident bremsstrahlung photon spectra is also shown in the Figure.

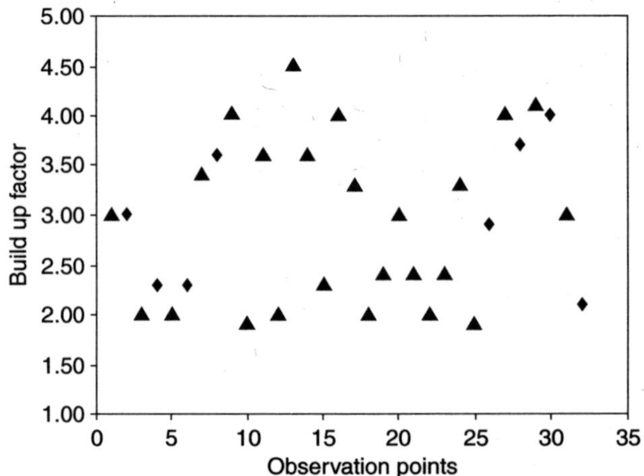

Fig. 3   *Random build up factors measured around Indus-1 in different media*

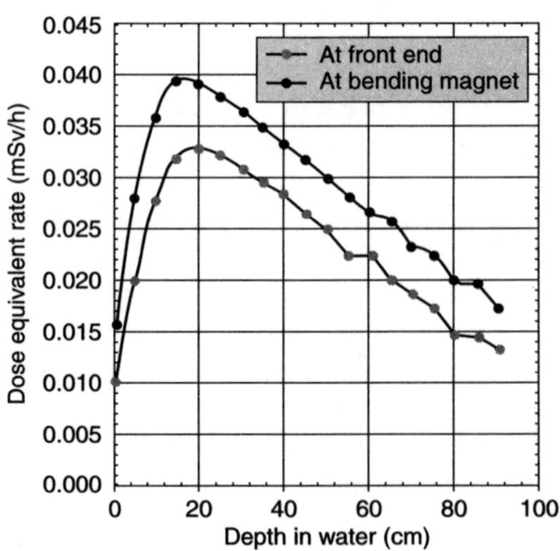

Fig. 4   *Depth dose measured in water at front at different locations*

The calculated build up factor was matching well with the measured results. From the fit for the build up factor, a theoretical build up factor is deduced to be 6.4 for 30 cm depth, which is close to the human body size. This can be used as a

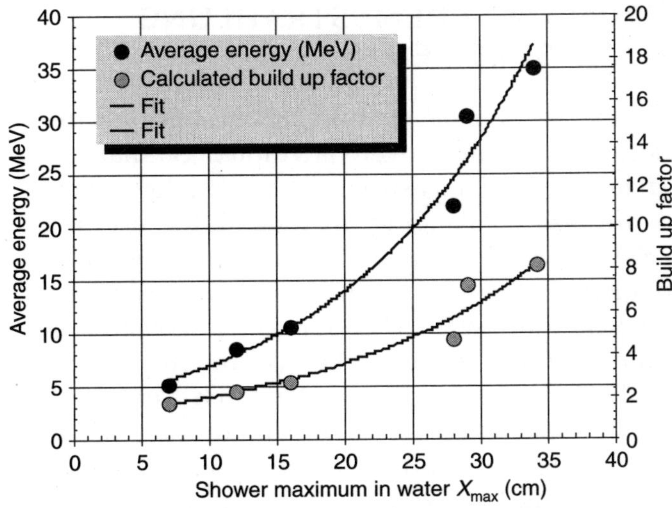

Fig. 5 *Simulated does build up facter in water*

conservative dose buildup correction factor for radiation dosemeters, exposed in the radiation environment of Indus-1 [7]. Based on the measured dose build up factor, the annual dose limits for Indus-1 is reduced by a factor of 4 by the National Regulatory Authority, Atomic Energy Regulatory Authority, India.

3. RADIATION SAFETY ISSUES AT SYNCHROTRON RADIATION BEAM LINES OF INDUS-1 AND INDUS-2

The critical energy of synchrotron radiation at Indus-1 being in the vacuum ultraviolet region, does not pose significant radiation protection problems during the beam users. Moreover all the beam ports (five in number) are connected with beam lines though their front ends and experiments are carried out in ultra high vacuum environment.

However in Indus-2, since the critical energy is in the x-ray region and experiments are carried out in air, radiation safety issues arise. Out of 26 beam lines planned, 3 beam lines are in regular operation, 3 are under trial operation and several are under installation stage.

The radiation environment of Indus-2 beam line consists of direct and scattered synchrotron & gas bremsstrahlung radiation and photo-neutrons (negligibly small). Due to the interaction of synchrotron radiation with air, ozone gas production also takes place. For radiation safety reasons, all the beam lines are housed in specially designed shielded hutches. Door interlocks, radiation interlocks, ventilation, radiation monitors (for synchrotron within the hutch and Bremsstrahlung outside the hutch) is provided to ensure safety of users, beam line scientists and other workers associated with beam lines.

4. SYNCHROTRON AND BREMSSTRAHLUNG RADIATION MEASUREMENTS

In order to quantify the SR and BR channeling through the front-ends to the beam line, preliminary dose measurements were attempted. Similar to the problems of high energy radiation detection and measurement issues, low energy photon detection and measurements are also challenging. In Indus-2 the useful SR photon energy is in the range 5-50 KeV. Normal detectors like ion chamber used for the monitoring of *x*-or gamma radiation always give a safe reading in low energy SR photons due to self absorption in the chamber wall. Therefore for low energy synchrotron radiation measurements a thin window ion chamber (1 mg/cm^2 Mylar) was developed in house (Electronics Division, Bhabha Atomic Research Centre, Mumbai). SR intensity measurements were initially carried out by mounting the detector in front of a Be-window at the front-end of an SR beam line (BL-07). Maximum transmission of SR was achieved by creating localized bump in the beam orbit by beam physicists. The bump was created by varying the current in steps of 0.2 A on a suitable vertical steering coil supply in the storage ring. On each current setting ion current from the thin window ion chamber is measured. The experimental set-up is shown in Fig. 6. The SR intensity in terms of ion current as a function of the bump is shown in Fig. 7. It has been observed that maximum x-ray transmission occurs at a steering coil current setting of 1.8 A, beyond which the stored current started dropping. The average pressure in the ring during the experiment was 5.8 E-10 mbar.

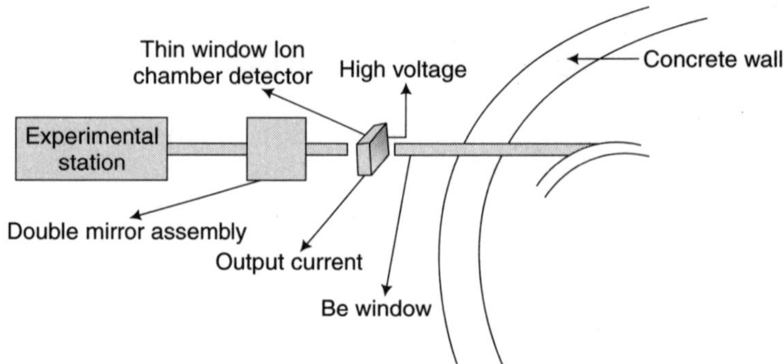

Fig. 6 *Schematic diagram of the experimental set-up used for SR & BR measurements*

The ion current at the maximum SR transmission is converted into exposure rate using a sensitivity figure (269 pA/Sv/h) obtained for Co-60. The exposure rate normalized to beam current obtained is 3.83 × 10^3 Sv/h-mA @ 2.0 GeV. The bremsstrahlung radiation dose rate was attempted to measure by interposing copper absorbers of area 70 mm × 70 mm and thickness of 3 mm. Cu absorbers are inserted one by one and the corresponding ion current is observed.

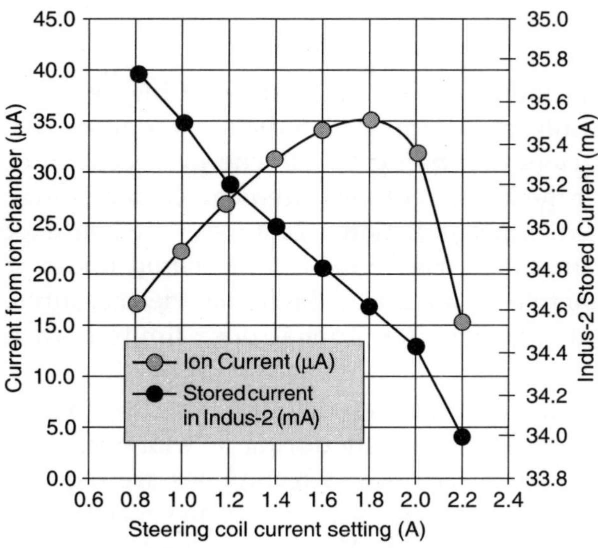

Fig. 7 *SR intensity (in terms of ion current) for various bump settings*

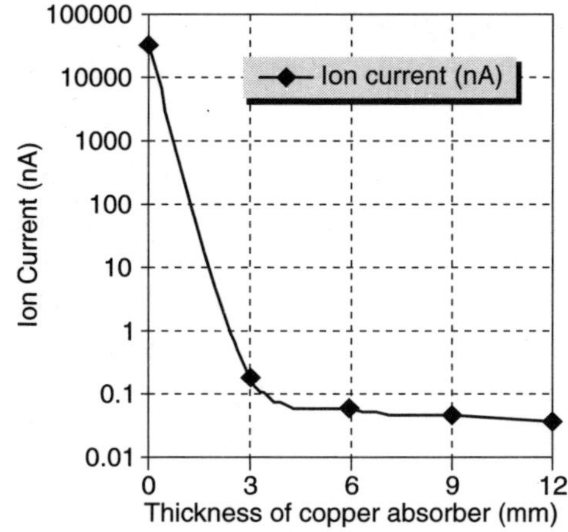

Fig. 8 *SR intensity (in terms of ion current) for various bump settings*

It is observed that there is a drastic drop in the ion current when an absorber of 3 mm was inserted and after 6 mm, there is no substantial variation in the measured ion current. It infers that within 3-6 mm Cu all the SR is contribution is absorbed and the transmitted is only the BR component. The normalized exposure rate due to BR calculated from the ion current at 6 mm thick copper using a sensitivity of 4 nA/Sv/h (for Co-60 full exposure to the detector) is 0.42 mSv/h-mA.

5. CONCLUSION

Radiation dose measurements in the environment of high energy electron accelerators has to be done carefully as commercial radiation monitors may underestimate the dose rate by many factors depending on the energy spectrum of the bremsttrahlung radiation prevailing at the location of measurement. Underestimation can result in area radiation monitors, personnel radiation monitors etc. when exposed to high energy bremsstrahlung radiation field. For Indus-1, a measured underestimation factor up to about 4 in water phantom is obtained. Based on the measurements, the regulatory authority has decided to reduce the annual dose limits by a factor of 4 for Indus facilities.

Dose measurements in synchrotron radiation beam lines to quantify the synchrotron and bremsstrahlung radiation especially during its trial operation/commissioning also needs more care as special detectors and techniques are required to be developed to quantify the dose/dose rate. The exposure rate due to synchrotron radiation normalized to the stored beam current @ 2.0 GeV, at an average pressure of 5.8×10^{-10} mbar measured using a thin window chamber is 3.83×10^3 Sv/h-mA. The bremsstrahlung dose rate normalized to beam current obtained from the ion chamber using the attenuation technique is 0.42 mSv/h-mA.

Acknowledgements

We are grateful to Dr. P.D. Gupta, Director, RRCAT for encouragement and support given in the radiological physics studies at Indus Accelerator Complex. We acknowledge the encouragement given by Dr. A.K. Gosh, Director, H & S Group, BARC for the present work. We are thankful to Shri Gurnam Singh, Head, IOAPDD, Dr. S.K. Deb, Head, SU & MRD guidance and permission to carry out the present measurements at storage ring Indus-1 & 2 and at SR beamlines. We are also thankful to Shri. M.K. Nayak & Shri. Vipin Dev for help in preparing the manuscript.

References

1. http//www.cat.gov.in
2. Basic aspects of high energy particle interactions and radiation dosimetry *ICRU Report 28,* International Commission on Radaition Units and Measurements (1978)
3. Swanson. W.P, IAEA Technical Report Series No.188, Radiological safety aspects of the operation of electron linear accelerators (1979)
4. Liu. J.C and Vylet. V, Radiation protection at Synchrotron Radiation Facilities, Radiation Protection Dosimetry **96** (4) (2001) 345-357
5. Haridas. G, Thakkar. K.K, Pradhan. S.D, Nayak. A.R and Bhagwat. A.M, Response of monitoring instruments to high energy photon radiation, Nucl.Instrum. Methods A **449** (2000) 624-628

6. Rogers. D.W.O, Kawrakow. I, Seuntjens. J.P, Walters. B.R.B and Mainegra-Hing. E, NRC user codes for EGSnrc, NRCC Report PIRS-702 (Rev.B) (2003)

7. Haridas G. Nair, MK Nayak, Vipin Dev, KK Thakkar, PK Sarkar and DN Sharma Dose build up correction for radiation monitors in high energy bremsstrahlung photon radiation fields. Radiation protection dosimetry, **118**, (3), 233-237 (2006)

Experimental Observation of Z-dependence of Albedos Characterizing Multiply Backscattering of Gamma Photons

Arvind D. Sabharwal, B. Singh, and B.S. Sandhu*

Physics Department, Punjabi University, Patiala, India
E-mail: *balvir@pbi.ac.in and balvir99@indiatimes.com*

ABSTRACT

In present measurements, to extract information about backscattering probability of gamma photons, interacting with different atomic number (Z), number (AN), energy (AE) and dose (AD) albedos are experimentally evaluated. Experimentally evaluated response function in the form of a matrix converts the observed pulse-height distribution of a NaI(Tl) scintillation detector to a true photon spectrum for better quantitative analysis. For incident gamma photon energy of 1.12 MeV, the numbers of these multiply backscattered events show an increase with increase in target thickness, and then saturate for a particular thickness of the target called saturation thickness (depth). It has been observed that the number, energy and dose albedos are saturating for the same thickness where the numbers of multiply backscattered events saturate. The energy albedos are found to be decreasing with increase in atomic number of the target material. The dose albedos do not differ significantly from the energy albedos.

Keywords: Number, Energy and dose albedos, Response function, Multiple backscattering

Pacs No.: 32.80.t, 13.60.r, 78.70.b

1. INTRODUCTION

The multiply backscattering (or reflection) of gamma photons is a problem of fundamental importance in Compton profile, radiation shielding, industrial and medical applications. The parameters called the number and energy albedos [1-2] give the effect of multiply backscattered radiations on the original signal (incident photon flux) and characterize the reflection probability of a material for gamma photon flux. Our recent measurements ([3] and references therein) provide a survey of analytical and Monte Carlo simulation approaches to study the multiple backscattering, available experimental data, and number, energy and dose albedos on these processes.

The importance of multiply backscattered radiation diminishes with increase in the atomic number because the probability of photoelectric absorption increases with the atomic number, thus the number and energy albedos should approach a value characteristic for singly scattered radiations. So, there is need to investigate the effect of multiply backscattered radiations on the various incident photon energies and atomic number materials. The present measurements provide measurement of saturation thickness of multiply backscattered photons and Z-dependence of albedos for 1.12 MeV incident gamma photons.

2. EXPERIMENTAL SET-UP AND MEASUREMENTS

The experimental set-up used for the present measurements (Fig. 1), performed for 1.12 MeV incident gamma photon energy obtained from a small isotropic radioactive source of ^{65}Zn of activity 222 MBq. Experimental set-up is so designed to generate maximum numbers of backscattered photons towards the direction of NaI(Tl) detector and does not have any dead space of measurements [2]. The targets used are elements of various atomic numbers (Carbon, Aluminium, Iron, Zinc, Stanus) having 80 × 40 mm^2 surface areas, of varying thickness. The backscattered radiations are detected by a properly shielded and collimated NaI(Tl) scintillation detector, optically coupled to photomultiplier tube, is cylindrical in shape having dimensions 51 mm in diameter and 51 mm thickness. The backscattered beam is further collimated by a cylindrical collimator made of lead and lined with aluminum having opening of radius 10 mm and thickness 30 mm. The field of view of NaI(Tl) detector is confined to target only. The experimental data are accumulated on a PC based ORTEC Mastreo-32 MCA.

3. RESULTS

A typical observed pulse-height distribution (curve-a) from the tin target (thickness 15 mm), exposed to 1.12 MeV incident gamma photon is given in Fig. 2. The target related observed spectrum consists of intensity distribution of singly as well as multiply backscattered photons. Curve-b is the observed pulse-height distribution without the target. The subtraction of events under curve-b from those observed under curve-a accounts for events originating from interactions of the primary gamma beam photons with target material only. The events under the observed backscattered peak (curve-c) of such a target related spectrum are composite of singly as well as multiply backscattered events.

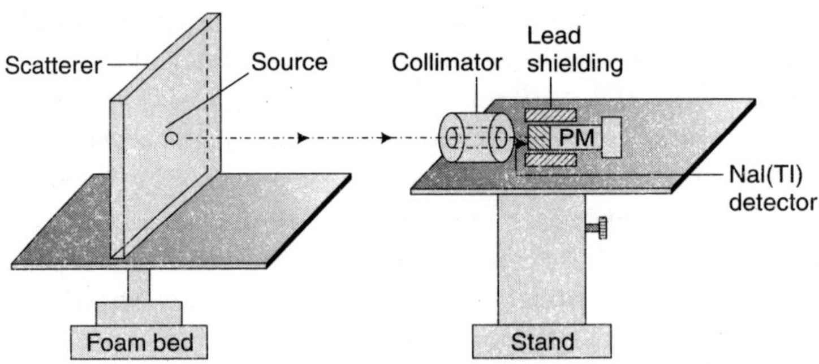

Fig. 1 *Experimental set-up*

The singly scattered events (curve-d) under the backscattered peak, details provided in measurements [4], are obtained by reconstructing analytically the singly backscattered peak using the experimental determined parameters, such as counts at the photo-peak and FWHM and the detector efficiency of the detector corresponding to the backscattered energy. The curve-c is converted to a photon energy spectrum with the help of an experimentally constructed [4] inverse response matrix of order 16 × 16. The solid curve-e is the resulting calculated histogram obtained from the curve-c. The response matrix enables the low pulse-height counts resulting from partial absorption of higher energy photons to shift to the backscattered peak energy region. The events under curve-d of Fig. 2 are divided by peak-to-total ratio of the gamma ray detector and then their subtraction from the events under the calculated histogram (curve-e) in the specified energy range results in events originating from multiple backscattering but having the same energy as in singly Compton scattering process.

The plots of observed number of multiple backscattered events, having energy equal to the signal from singly scattered gamma photons, for different atomic numbers as a function of target thickness is shown in Fig. 3. The present experiments are also simulated with the Monte Carlo package developed by Bauer and Pattison [5]. The solid and dotted curves provide the best-fit curves to the present experimental data (filled circles) and simulated data (hollow circles) respectively. The measured statistical uncertainties lie within the diameter of filled circles representing the experimental data. The present experimental results show that for each of the targets made of different elements, the numbers of multiple backscattered events increases with increase in target thickness and then saturate beyond a particular value of target thickness, called saturation thickness (depth). The saturation of multiple backscattered photons is due to the fact that an increase in target thickness results in a higher number of scattering centres for the interaction of primary gamma rays with the target material. Futhermore, the probability for absorption within the target sample is also enhanced with increasing target thickness. Thus a stage is reached when the thickness of target becomes sufficient to compensate for the above increase and results in a decrease

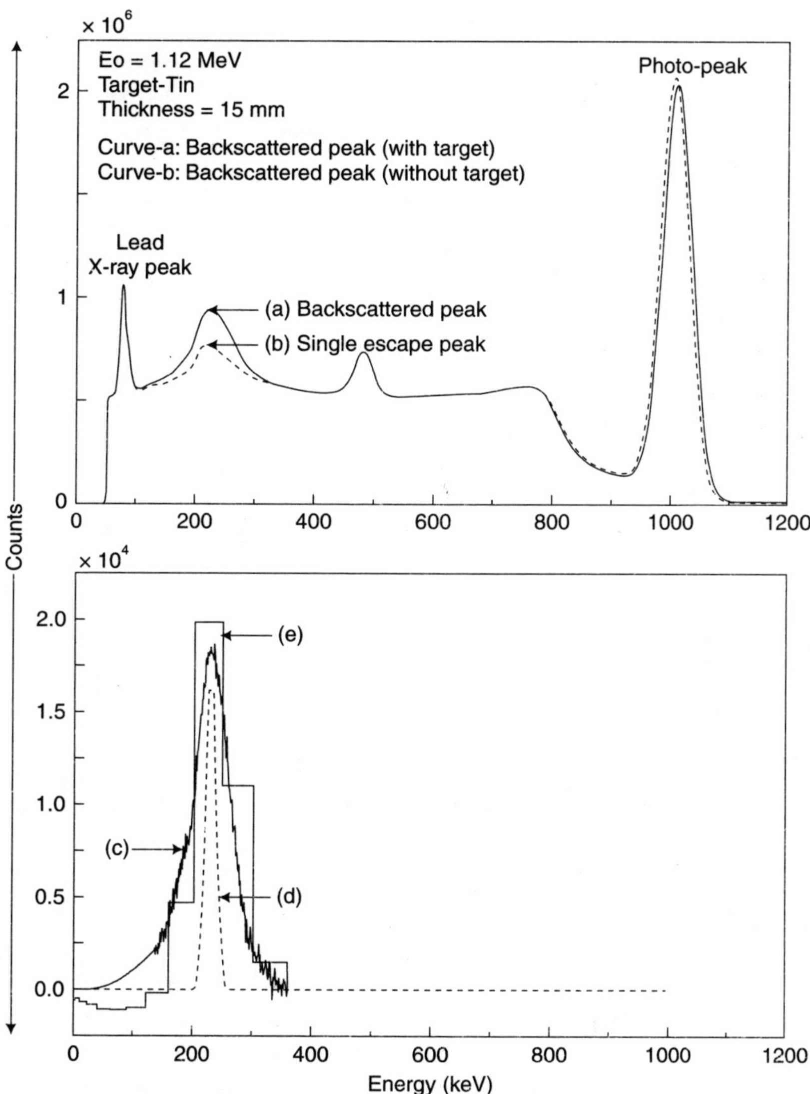

Fig. 2 *A typically observed pulse-height distribution with (curve-a) and without (curve-b) tin target in the primary gamma beam. Experimentally observed pulse-height distribution (curve-c) obtained after subtracting background events. Normalized analytically reconstructed singly backscattered full energy peak (curve-d) and resulting calculated histogram (curve-e) of converting pulse-height distribution to a true photon spectrum.*

of the number of multiple backscattered photons, and hence the number of multiple backscattered photons coming out of the target saturates.

The evaluated data for number and energy albedos from the present experimental results is provided in Table 1. It has been observed that for a particular target material,

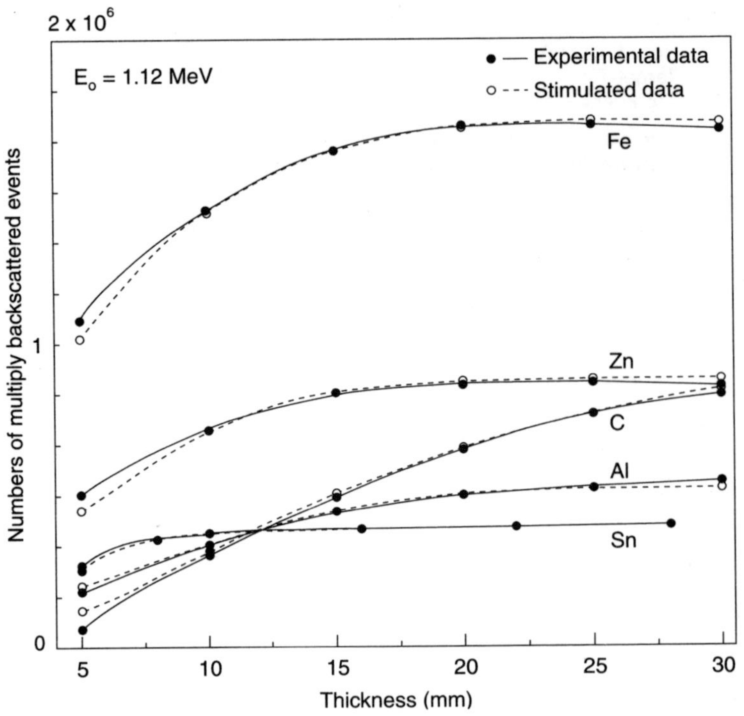

Fig. 3 *The experimentally measured (solid-symbol) and Monte Carlo simulated (hollow-symbol) numbers of multiply backscattered events from targets of C, Al, Fe, Zn and Sn for 1.12 MeV incident gamma photons.*

both the number and energy albedos show an increase with increasing target thickness and finally saturate at the same thickness for which numbers of multiply backscattered events saturates. This wok is supported by experimental measurements of Bulatov and Garusov [6], Biswas *et al.* [7] and by Monte Carlo simulations by Berger and Raso [8] at different gamma photon energies and different Z-numbers. Dose albedo, A_D, is defined as energy albedo weighted in proportion to the response of the scintillation detector to different photon energies. The dose albedos are also found to be saturating for the same thickness where the numbers of multiply backscattered events saturate.

Table 1 *Experimentally measured values of number and energy albedos for different Z-numbers at 1.12 MeV gamma photons under narrow beam geometry.*

Z	Thickness (mm)	Number albedo (An)	Energy albedo (AE)
	5	2.48E-04	5.12E-05
	10	2.58E-04	2.32E-05
6	15	2.64E-04	5.32E-05
	20	2.68E-04	5.44E-05

Contd...

Contd...

	25	2.71E-04	5.53E-05
	30	2.71E-04	5.58E-05
	5	2.61E-04	4.93E-05
	10	2.68E-04	5.05E-05
13	15	2.71E-04	5.05E-05
	20	2.74E-04	5.13E-05
	25	2.75E-04	5.18E-05
	30	2.75E-04	5.19E-05
	5	2.74E-04	5.80E-05
	10	2.93E-04	6.26E-05
26	15	3.06E-04	6.26E-05
	20	3.08E-04	6.47E-05
	25	3.09E-04	6.52E-05
	5	2.16E-04	5.50E-05
	10	2.89E-04	5.79E-05
30	15	3.00E-04	5.79E-05
	20	3.04E-04	5.98E-05
	25	3.05E-04	6.07E-05
	5	2.69E-04	5.62E-05
	10	2.75E-04	5.74E-05
50	15	2.77E-04	5.77E-05
	20	2.78E-04	5.81E-05
	25	2.78E-04	5.81E-05

4. CONCLUSION

The number and energy albedos increase with an increase in target thickness and saturates at the saturation thickness of multiply backscattered photons. The energy albedos are found to be decreasing with increase in incident gamma ray energy. This behaviour supports the experimental work of several authors [6, 7 and 9]. The energy albedos are found to be decreasing with increase in atomic number of the target material. This work supports the Monte Carlo calculations of Berger and Raso [8]. It is further required to carry out measurements using HPGe detector that provides more faithful reproduction of the shape of distribution of the photons originating from the interactions of incident photons with the target.

References

1. E Hayward and J H Hubbell *J. Appl. Phys.* **25** (1954) 506.
2. T Hyodo *Nucl. Sci. & Engg.* **12** 178 (1962)

3. A D Sabharwal, B S Sandhu, B Singh *Physica Scripta.* **83** 025303 (2011)

4. A D Sabharwal, M Singh, B Singh, B S Sandhu *App. Rad. Isot.* **66** 1467 (2008)

5. G.E.W. Bauer, P. Pattison, Compton scattering experiments at the HMI (HahnMeitner institute fur kernforschung, Berlin). HMI-B 364 (1981) 1.

6. B P Bulatov and E A Garusov *J. Nucl. Energy A* **11** 159 (1960).

7. M Biswas, A K Sinha and S C Roy *Nucl. Instr. & Meth.* **159** 157 (1979).

8. M J Berger and D J Raso *Rad. Res.* **12** 20 (1960).

9. J F Perkins *J. Appl. Phys.* **26** 1372 (1955)

Investigation of Cross Sections for Electron Scattering with GeH$_4$ Molecules

A.K. Jain[1,*], Harsh Mohan[1], Gurpreet Kaur[1,2], Parjit S. Singh[2] and Sunita Sharma[3]

[1]*Department of Physics, M.L.N. College, Yamuna Nagar, India*
[2]*Department of Physics, Punjabi University, Patiala, India*
[3]*Department of Chemistry, M.L.N. College, Yamuna Nagar, India*
E-mail: *arvindjain61@gmail.com*

ABSTRACT

Spherical complex optical potential (SCOP) approach in the fixed nuclei approximation has been used to compute the differential, total (elastic + inelastic) and momentum transfer cross sections for electron scattering with GeH$_4$ molecules at incident energies up to 300 eV. The real part of the optical potential consists of three potentials namely, the static, the exchange and the polarization. For the imaginary part of the SCOP, we employ semi-empirical model absorption potential. Molecular charge density has been determined from single-centre wave functions with enough terms in the expansion of each orbital. The present calculated differential cross sections (DCS) exhibit all important features (such as forward peaking, dip at middle angles and enhanced backward scattering) observed in other theoretical calculations and experimental measurements.

Keywords: Electron, Scattering, Charge density, Potentialg

Pacs No.: 34.80.-i, 34.80.Bm, 34.90.+q

1. INTRODUCTION

In matter, energy deposition is due to the interaction of radiation with atoms and molecules. So investigation of electron scattering cross section is an interesting area of radiation physics. The cross sections for electron/positron collisions with a large number of atoms and polyatomic molecules have been measured from low (~1 eV) to keV energy region [1-6]. These cross sections have imperative applications in space, plasma, laser, biomedical science and the environment. Many approaches ranging from as simple as IAM [7] to *ab initio* methods [8] have been projected and developed. In case of polyatomic molecules, *ab initio* calculations are quite intricate to perform. The system studied in the present paper, the germane (GeH_4) molecule, is a typical example since it exhibits structural similarities with methane (CH_4) and silane (SiH_4) and, at the same time, because of its high symmetry and large number of electrons (36) suggests a complicated interaction which can be treated analogously to heavier atomic systems. Therefore, our goal here is to apply a spherical complex optical potential (SCOP) approach for calculating the differential cross section (DCS), total (elastic + inelastic) cross section (σ_t) and momentum transfer cross section (σ_m) for electron scattering with a spherical molecule i.e. GeH_4 molecule up to 300 eV.

2.   THEORETICAL METHODOLOGY

The total interaction of electron molecule system can be represented by a local complex optical potential

$$V_{opt}(r) = V_R(r) + i\, V_{Abs}(r) \tag{1}$$

where the real part $V_R I$ of the optical potential, consists of the static potential V_{st}, the exchange potential V_{ex} and the parameter free correlation polarization potential V_{cop}. The complex optical potential is obtained by adding a model absorption potential $V_{Abs}I$ as imaginary part to the real potential. All the four potentials can be generated easily once the target charge density $\rho_0 I$ is known. We determine $\rho_0 I$ with enough terms in the expansion of each orbital. $P_0 I$ is expanded in terms of symmetry-adapted functions belonging to the totally symmetric $^1A_1$ irreducible representation of the molecular point group

$$\rho_0(r) = \Sigma\, \overline{\rho}_{LH}(r)\, X_{LH}^{A_1}(\rho) \tag{2}$$

In the spherical approximation, we only need the first term ($L = 0$, $H = 1$) of the expansion of equation (2), in order to evaluate all the potential terms, i.e. V_{st}, V_{ex}, V_{cop} and V_{Abs}. Explicit expression for V_{st} and V_{ex} (HFEGE) is given in the literature [9]. The attractive term i.e. polarization effect arises due to distortion in the target charge cloud because of incoming charge particle. This effect is invoked by a parameter free correlation polarization potential, which is based on the correlation energy of the target molecule. The calculation of an actual polarization potential which provides the second and higher order effects in the electron-molecule interaction is rather a difficult task. The correlation polarization potential (cop) for the e-GeH_4 system is calculated from the expression given in the literature by Padial and Norcross [10] and Gianturco *et al.* [11].

The imaginary part of the optical potential is referred as absorption potential, which takes into account the flux going through all possible inelastic channels. An exact determination of V_{Abs} is very difficult. Thus a number of empirical and semi-empirical models have been proposed. In the present work, it is determined *ab initio* as the function of charge density of the target molecule following the procedure of Staszewska *et al.* [12].

Finally, a partial wave analysis approach is adopted to compute the ℓth partial wave phase shifts for the solution of radial Schrodinger equation. We have employed a variable-phase approach (VPA) [13] to find its solution. The corresponding quantities (i.e. DCS, st and sm) are then easily obtained from S-matrix at each energy. All our cross sections are converged with respect to the number of partial waves up to a value of 0.000 01 radians only.

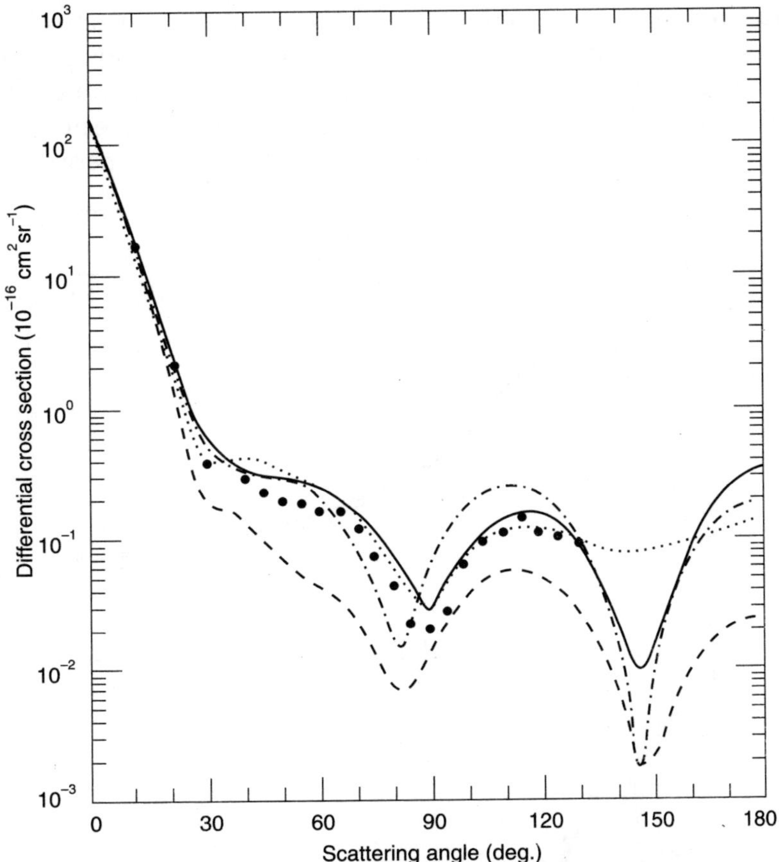

Fig. 1 *Differential cross section for scattering of electrons from GeH4 molecule at 100 eV. Calculations: ____, present results; _ _ _, SEP results from Jain et al [15]; _ . _ . _, relativistic results from Kumar et al [14]; _ _ _ _, CMS results from Dillon et al [16]. Experiment: •, Dillon et al [16].*

3. RESULTS AND DISCUSSION

Figure 1 display our calculated differential cross sections (DCS) at 100 eV using SCOP approach together with relativistic calculations of Kumar *et al.* [14], the non-relativistic

SEP results of Jain *et al.* [15], the theoretical results from the CMS approach of Dillon *et al.* [16] and also with their experimentally measured (rotationally summed) values. It is clearly seen that the present results show good agreement with the measurements of Dillon *et al.* [16] and also with their CMS results in the entire angular region for which data points exist. On the other hand, calculated non-relativistic SEP result from Jain *et al.* [15] agrees with experimentally measured DCS only in the forward direction for a scattering angle $\theta < 30°$ and thereafter it shows a quantitative agreement both in shape and magnitude.

Finally we show our computed results for the total (elastic + inelastic) (σ_t) and momentum transfer (σ_m) cross sections in Fig. 2(a) and (b) respectively. As observed from the figures the present calculated cross sections are in good agreement with the experimental measurements of Dillon *et al.* [16] due to the inclusion of imaginary part of the interaction in the calculations. For an electron energy greater than 10 eV, the present σ_t and σ_m both fall monotonically with an increase in energy.

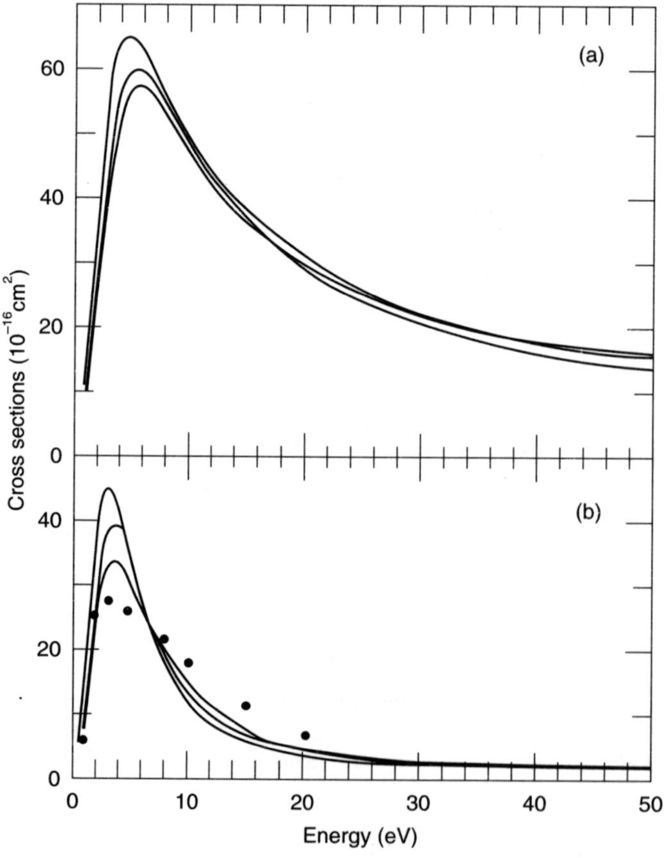

Fig. 2 *Cross section for scattering of electrons from GeH$_4$ molecule. (a) Total (σ_t) cross section, (b) Momentum transfer (σ_m) cross section. Calculations: ___, present results; _ _ _, SEP results from Jain et al [15]; _ . _ . _, relativistic results from Kumar et al [14]. Experiment: •, Dillon et al [16].*

4. CONCLUSION

The calculated DCS, σ_t and σ_m cross sections reproduce the angular shape and size of the experimental results. The neglect of the angular asymmetry in the present interaction potential appears to be well justified as the target molecule remains spherical for the incoming electron.

References

1. E McEachran and A D Stauffer *J. Phys. B: At. Mol. Opt. Phys.* **42** 075202 (2009)

2. S J Buckman, T Maddern, J Francis-Staite, L Hargreaves, M J Brunger, G Garcia, J C Lower, S Mondal, J P Sullivan, A Jones, P Caradonna, D Slaughter, C Mackochekanwa and R P McEachran R P *J. Phys.: Conference Series* **133** 012001 (2008)

3. J R Francis-Staite, T Maddern, M J Brunger, S J Buckman, C Winstead, V McKoy, M A Bolorizadeh and H Cho *Phys. Rev.* A **79** 052705 (2009)

4. C M Surko, G F Gribakin and S J Buckman *J. Phys. B: At. Mol. Opt. Phys.* **38** R57 (2005)

5. C Szmytkowski *J. Phys. B: At. Mol. Opt. Phys.* **43** 055201 (2010)

6. A K Jain, I Dubey and H Mohan *J. Phys.: Conference Series* **199** 012017 (2010)

7. Y Jiang, J Sun and L Wan *Phys. Rev.* A **52** 398 (1995)

8. F A Gianturco, R R Lucchese and N Sanna *J. Chem. Phys.* **100** 6464 (1994)

9. F A Gianturco and A Jain *Phys. Rep.* **143** 347 (1986)

10. N T Padial and D W Norcross *Phys. Rev.* A **29** 1590 (1984)

11. F A Gianturco, A Jain and L C Pantano *J. Phys. B: At. Mol. Phys.* **20** 571 (1987)

12. G Staszewska, D W Schwenke, D Thirumalai and D G Truhlar *J. Phys.* B **16** L281 (1983)

13. F Calagero *Variable Phase Approach to Potential Scattering* (New York : Academic) (1974)

14. P Kumar, A K Jain and A N Tripathi *J. Phys. B: At. Mol. Opt. Phys.* **28** L387 (1995)

15. A Jain, K L Baluja, V Dimartino and F A Gianturco *Chem. Phys. Lett.* **183** 34 (1991)

16. M A Dillon, L Boesten, H Tanaka, M Kimura and H Sato *J. Phys. B: At. Mol. Opt. Phys.* **26** 3147 (1993)

Study of 1 A GeV ^{56}Fe Ions Interactions in Polyethylene, Nextel and Aluminium using Geant4

Ashavani Kumar and Summit Jalota

National Institute of Technology, Kurukshetra, India

E-mail: *ashavani@yahoo.com*

ABSTRACT

The response of 968A MeV 56Fe ions in high density polyethylene, Nextel and Aluminium shielding materials is studied using a simulation tool GEANT4. The Bragg's peaks are observed at 24.1, 30.5 and 25.0 g/cm^2 in Polyethylene, Nextel and Aluminium respectively. Polyethylene provides higher attenuation in contrast to Nextel and Aluminium. The shapes of the Bragg curves are in well agreement with the experimental results and the initial energy loss is not constant and followed by the Bragg peak in the end. The minimum dose before Bragg peak, Bragg peak position, 50% entrance dose thickness beyond the peak entrance dose are presented. In case of polyethylene, the entrance dose and peak height is more than half as compared to Nextel and Aluminium.

Keywords: Aluminium, Depth-dose deposition, Geant4, Nextel, Polyethylene, Radiation protection shielding.

Pacs No.: 61.80

1. INTRODUCTION

Galactic Cosmic Rays (GCR) are one of the main health risks for space exploration. Interaction of heavy ions at high energies is of great interest in the study of GCR as these ions are used in many experiments from few decades back. Even though the presence of heavy ions in GCR is about 1% of the total flux, HZE-particles contribute significantly up to 25% due to their high relative biological effectiveness (RBE); their effect can't be ignorable. Among all HZE, the contribution of effective dose equivalent for ^{56}Fe ions is highest [1]. The radiation due to cosmic rays on Earth to the human beings can be ignore to the acceptable limits, but the radiations and shielding for high energy heavy ions in space becomes highly problematic because of nuclear fragmentation of shielding materials and other exotic particles. It is well known that for shielding, low Z materials are good. So, hydrogenated materials like polyethylene, PMMA, water, CR39 and Nextel etc. are preferred. However, several materials are in use for the construction of spacecrafts, exploration vehicles and other space missions [2], there are no clear properties available regarding fragmentation of heavy ions and target materials. Only few studies about fragmentation of shielding materials as targets and heavy ions fragmentations have been done so far for such materials which incorporated the effective dose equivalent, Bragg's peak positions and nuclear fragmentation etc. for various ions and energies. It is therefore important to study the theoretical as well as experimental response of the materials to energetic heavy ions at various particle acceleration facilities and simulation using Monte Carlo (MC) codes and compare the results to have better understanding of radiation on human beings.

2. METHODOLOGY

Present simulation study is concern with the passage of energetic charged Fe particles in various medium by considering electromagnetic interactions, elastic and inelastic scattering processes using GEANT4 a simulation toolkit [3]. 1A GeV ^{56}Fe ion beam may be a useful simulation of the actual GCR heavy ion spectrum. The models included in present work are standard electromagnetic model, nuclear elastic model and binary cascade model for heavier ions. Low-energy parameterized model is included for elastic scattering. The major categories of processes provided by Geant4 are electromagnetic, hadronic, and transportation.

To study the recoiling reactions and reaction products, we incorporated elastic scattering interaction and inelastic process respectively and the Binary Cascade model [4-5] for one-on one collision.

Bragg curves plotted for Fe^{26+} at 968 MeV/n in Aluminium, Nextel and polyethylene shielding materials. The broad beam is used here for estimating the dose deposition. The detector chosen is of simple geometry of 3D box, whose two sides are kept same and one dimension, parallel to the direction of beam kept longer, in order to get range and deposition along the track of ion. The materials chosen are of uniform geometry and moderate beam number is selected for better statistics.

3. RESULTS AND DISCUSSION

We have calculated the response of 968A MeV ^{56}Fe ions in high density polyethylene, Nextel and Aluminium shielding materials using GEANT4 and compare the results with C. Lobascio, *et al.*, 2008 [6]. To compare the simulated results with the experimental observations, the primary ion beam energies on the target is consider as 968 A MeV instead of 1A GeV beam. These energies are decreasing on the target due to the air gap between source and target. The linear energy depositions are normalized to entrance dose and dose profile between normalized dose and areal density is shown in Fig. 1. The Bragg's peaks are observed at 24.1, 30.5 and 25.0 g/cm^2 in Polyethylene, Nextel and Aluminium respectively.

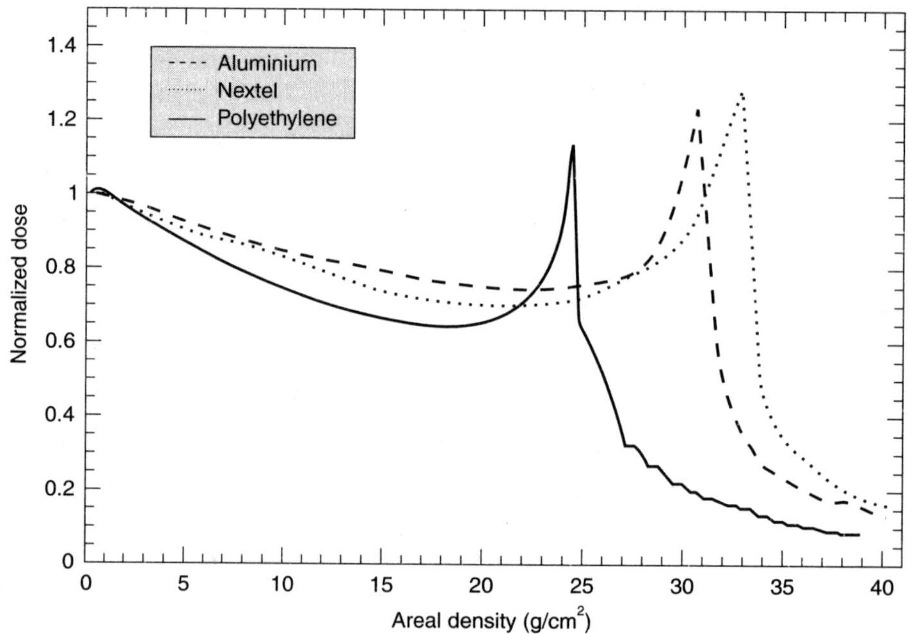

Fig. 1 *Bragg curves of 968A MeV ^{56}Fe ions measured in polyethylene, nextel and aluminium.*

Density used for Polyethylene, Nextel and Al are 0.97 g/cm^3, 2.70 g/cm^3 and 2.70 g/cm^3 respectively. The minimum dose before Bragg peak, Bragg peak position, 50% entrance dose thickness beyond the peak entrance ratio are presented in Table 1. The shapes of the Bragg curves are in well agreement with others [7-9] and the initial energy loss is not constant and followed by the Bragg peak in the end. The Bragg peak positions are also comparable to the experimental results [6]. The difference in the results for Nextel and Aluminium may be due to slight variations in the material compositions, air gaps in the stacks (not included in the present work) and incorporated constants in the program and also due to experimental errors in beam energy on the target.

Table 1

	Technique	Polyethylene	Nextel	Aluminium
Minimum (normalized) dose before	Experimental	0.54	0.65	0.76
Bragg peak	Geant4	0.64	0.68	0.75
Bragg peak position	Experimental	24.2	28.5	30.0

Contd...

Contd...

(g cm^{-2})	Geant4	24.1	30.5	32.6
50% entrance dose thickness beyond the peak (g cm^{-2})	Experimental	24.9	29.0	31.8
	Geant4	25.0	31.5	33.0

4. CONCLUSION

The range of ^{56}Fe ions of energy 968A MeV is minimum in Nextel and maximum in polyethylene. By mass comparison, we conclude that the polyethylene required less mass for reduction of same dose radiation due to cosmic rays in comparison to Nextel and Aluminium. Polyethylene provides higher attenuation in contrast to Nextel and Aluminium. In case of polyethylene, the entrance dose and peak height is more than half as compared to Nextel and Aluminium.

References

1. J.W. Wilson, *et al., Health Phys.,* **68**, 50 (1995)
2. S. Guetersloh, *et al., Nucl. Instrum. Meth B,* **252**, 319 (2006)
3. C. Zeitlin, S. Guetersloh, L. Heilbromn, J. Miller, *Nucl. Instrum. Meth. B,* **252**, 308 (2006)
4. S. Agostinelli, *et al., Nucl. Instrum. Meth. B,* **506**, 250 (2003).
5. S Sharma, H Khan and S K Sharma *Indian J. Pure Appl. Phys.* **41** 301 (2003)
6. L. Sihver, D. Matthia, T. Koi and D. Mancusi, *New J. Phys.,* **10** , 105019 (2008)
7. C. Lobascio, *et al., Health Phys.* **94,** 242 (2008).
8. C. Zeitlin, L. Heilbronn, J. Miller, *Radiat. Res.,* **149**, 560 (1998).
9. D.I. Lowerstein and A. Rusek, *Radiat.Environ. Bio. Phys.,* **149**, 91 (2007).
10. C. Lovascio et al., *Health Phys.,* **94**, 242 (2008).

Proton Induced Degradation of Aqueous Gentian Violet Dye Solution

A.A. Shanbhag[1*], S.P. Ramnani[2], Sunil[1] C., S.C. Sharma[3],
L. Varshney[2], P.V. Bhagwat[3], P.K. Sarkar[1],
R.K. Choudhury[3] and D.N. Sharma[4]

[1]*Health Physics Division, BARC, Mumbai, India*
[2]*Radiation Technology Development Division, BARC, Mumbai, India*
[3]*Nuclear Physics Division, BARC, Mumbai, India*
[4]*Radiation Safety Systems Division, BARC, Mumbai, India*
E-mail: *shanbhag@tifr.res.in

ABSTRACT

Radiation induced degradation of aqueous dye solution results in its discoloration and in most cases the degree of discoloration is proportional to the incident radiation dose. Hence aqueous dye based solutions can find application as radiation dosimeters. Here results of the discoloration of aqueous Gentian Violet dye solution by a 20 MeV proton beam are discussed. This experiment was basically planned as a dosimetry experiment but its results also confirmed the successful installation of the liquid target assembly for the 6M Irradiation setup of BARC-TIFR Pelletron Accelerator Facility.

Keywords: Proton, Dosimetry, Dye solution.

1. INTRODUCTION

Ionizing radiation may lead to modification in the physical, chemical and biological properties of materials. This characteristic of radiation was recognized very soon after the discovery of X-rays and natural radioactivity. Marie Curie-Sklodowska, a founder of radiochemistry, very soon, jointly with her husband recognized action of radiation on living organism (radium therapy) and materials (e.g. water, paper) [1]. In 1910, her student Kernbaum presented the paper on water radiolysis and radiation chemistry was founded [1]. Even 100 years after this discovery, the field of water radiolysis continues to evolve further and is now being utilized in diverse fields ranging from treatment of industrial waste waters to radiation dosimetry.

Since radiation induced degradation of aqueous dye solutions also results in discoloration, it can also find potential application as radiation dosimeter. In most cases, the color discoloration of dye based solutions is directly proportional to the irradiation dose [2]. The use of dye based solutions for radiation dosimetry is also advantageous because dyes are easily available, relatively cheap, the solution can be prepared easily and the discoloration can be measured using a UV-VIS spectrophotometer which is routinely available in most chemistry labs [3].

In this study we have used a proton beam to study the discoloration of aqueous gentian violet solution at the BARC-TIFR Pelletron Linac Accelerator Facility. Here the preliminary results of these experiments are reported.

2. EXPERIMENT

All the solutions were prepared in nano pure water. The gentian violet dye (HIMEDIA make) was of highest purity and used as received. The aqueous solution of 1000 ppm Gentian Violet (GV) was taken in a cylindrical teflon holder (10 mm diameter, 14 mm height). These target holders are a part of the liquid target assembly which is described elsewhere [4]. The range of protons in water was calculated using the SRIM program and for 20 MeV protons it is 4.17 mm. Hence 20 MeV proton beam is completely stopped within our target solution.

The liquid target assembly was mounted at the 6 m irradiation facility of the BARC-TIFR Pelletron Linac Accelerator Facility. The aliquots were then irradiated by using a 20 MeV proton beam and the current on target was maintained at 18 nA. The time of irradiation was varied from 30 seconds to 10 minutes. After irradiation, the UV-VIS absorbance spectra of both the un-irradiated and irradiated solutions were recorded using a Shimadzu make Model 4600 UV-VIS recording spectrophotometer in the range of 200-700 nm. The recorded spectra are shown in Fig. 1. The wavelength of maximum absorbance (λ_{max}) was found to be 590 nm. The extent of discoloration was calculated by measuring the absorbance in terms of optical density (OD) of unirradiated and irradiated dye at 590 nm using the following equation.

$$\% \text{ Discolouration} = [(OD)_{unirradiated} - (OD)_{irradiated}]/(OD)_{unirradiated} * 100$$

3. RESULTS AND DISCUSSION

When the irradiated aqueous dye solution was observed, it was obvious even to the naked eye that the samples were significantly discolored. It is known that ionizing radiation causes the generation of free radicals and reactive species such as hydrated electron, H atom and hydroxyl radical in water. These radical species react with GV dye molecules resulting in the formation of secondary products that are colorless, which

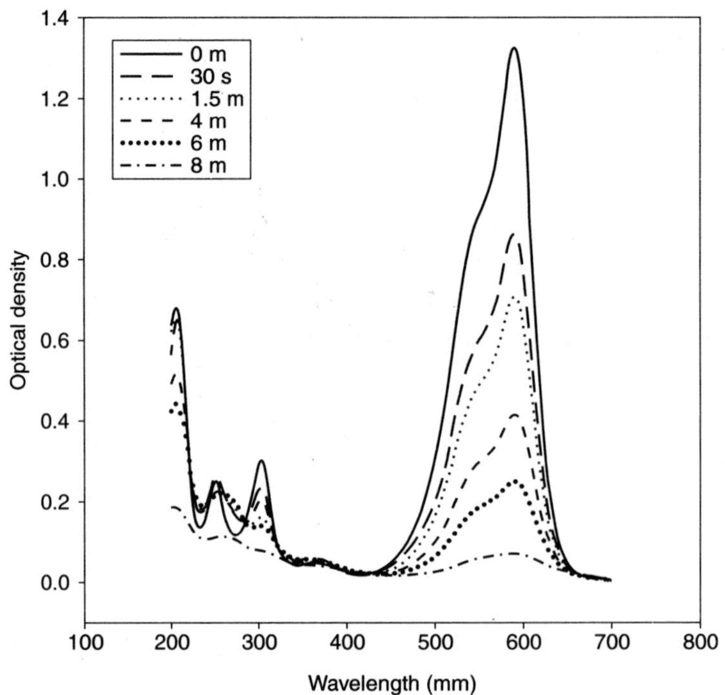

Fig. 1 *UV-VIS Absorption spectra of un-irradiated and irradiated solutions in the range of 200-700 nm*

leads to the discoloration of dye solution upon irradiation. As seen from Table 1, even 30 seconds of irradiation caused nearly 35% dye degradation. The degradation process slows down with the passage of time since the degraded products compete with the original dye molecules for further degradation.

Table 1 Variation of optical density (at 590 nm) and %discoloration with time of irradiation

Irradiation time (minutes)	Optical density (OD) (at 590 nm)	% Discoloration (x)	% Dye remaining (c_t) = (100 − x)	c_t/c_0
0	1.324	0	100	1
0.5	0.865	34.7	65.3	0.653
1.5	0.71	46.4	53.6	0.536
4	0.4156	68.6	31.4	0.314
6	0.2498	81.1	18.9	0.189
8	0.0726	94.5	5.5	0.055
10	0.0388	97.1	2.9	0.029

The plot of -ln (C_t/C_0) against time (in minutes) is a straight line passing through the origin. Hence the rate of discoloration is directly proportional to the amount of dye

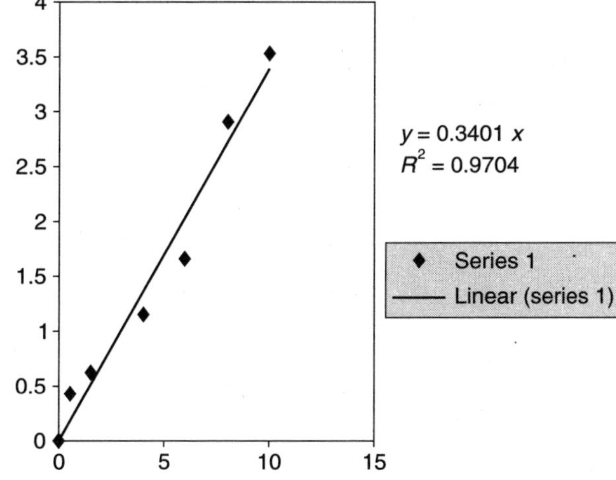

Fig. 2   *Plot of –ln(C_t/C_o) against Time (in minutes – x axis)*

present at a given time. This implies that the dye degradation reaction caused by proton irradiation follows first order kinetics. For the present experimental conditions the rate constant (k) for dye removal and R^2 values are (0.3401) min^{-1} and 0.9704 respectively.

4. CONCLUSION

The proton induced discoloration of aqueous Gentian Violet solution is reported here. The dye solution is significantly discolored after irradiation and the discoloration reaction follows first order kinetics. The results indicate that discoloration of dye solution by proton beam may find application in the development of a radiation dosimeter for high energy radiations. More experiments are planned in the near future to standardize the use of aqueous gentian violet solution as a radiation dosimeter for gamma irradiators, electron accelerators and proton beams and also to study the efficacy of each system in pollutant degradation of dye based industrial waste water. This experiment was basically planned as a dosimetry experiment but its results also confirmed the successful installation of the liquid target assembly for the 6M Irradiation setup of BARC-TIFR Pelletron Accelerator Facility.

References

1. A.G. Chmielewski, *et al.*, Radiation Physics and Chemistry, 71 16-20, (2004)

2. M. El-Banna, *et al.*, Journal of Radianalytical and Nuclear Chemistry, 264 (3) 657-664, (2005)

3. H.M. Khan, *et al.*, Radiation Physics and Chemistry, 63 713-717, (2002)

4. S.C. Sharma, *et al.*, Proceedings of DAE Symposium on Nuclear Physics, 55 766-767, (2010)

Study of L X-ray Intensity Ratios in Bi with Low Energy Protons

Harsh Mohan[1*], A.K. Jain[1], Gurpreet Kaur[1,2], Parjit S. Singh[2], and Sunita Sharma[3]

[1]Department of Physics, M.L.N. College, Yamuna Nagar, India
[2]Department of Physics, Punjabi University, Patiala, India
[3]Department of Chemistry, M.L.N. College, Yamuna Nagar, India
E-mail: *mohan_harsh@yahoo.com

ABSTRACT

The Coulomb ionization of an inner shell electron by an energetic proton is a fundamental problem in ion-atom collision physics. L X-ray intensity ratios for Bi with protons in the energy range 260 – 400 keV at the interval of 20 keV are reported. These are compared with calculations obtained on the basis of recent prevailing ECPSSR, UAECPSSR theories. Their importance in understanding this phenomenon and existing arguments in this regard will be highlighted.

Keywords: Proton induced X-ray emission cross-section; ECPSSR theory; UAECPSSR theory.

Pacs No.: 29.30 Kv, 32.70 Fw, 39.30 +w

1. INTRODUCTION

In the study of inner-shell ionization by charged particles, accelerators play an important function. A literature survey indicates that for L-shell ionization, most of the experiments have been conducted with protons having energies greater than 500 keV [1-3]. Thus, there is a lack of experimental data in this regime. At the same time large inconsistency between the experimental measurements and the theoretical calculations based on different models makes this work more interesting.

In this scenario we report L X-ray intensity ratios for Bi, namely, L_e/L_α, L_β/L_α and L_γ/L_α with protons over the energy range 260 – 400 keV. Their energy dependence and comparison with theoretical calculations will also be discussed. These analyses provide data in the low energy region which assist in better clarity of proton induced X-ray emission phenomenon.

2. THEORETICAL EVALUATION

In high energy region (> 1 MeV), First-order perturbation assumptions such as binary encounter approximation (BEA) [4], the straight-line trajectory semi classical approximation (SCA) [5] and equivalent to it, the plane wave Born approximation (PWBA) [6] are in logical agreement with the measured cross sections for L-sub shell ionization. But in low energy regime ECPSSR theory for the ionization of bound atomic electrons by incident light ions is extensively engaged, which is fundamentally a PWBA computation with a number of modifications proposed by Brandt and Lapicki [7, 8]. After that an auxiliary alteration, termed as united atom correction has been suggested [9]. It seems that, the thrust for their initiation lies in the reported differences between experimental data and the prophecy of the basic ECPSSR theory. This describes the interaction of projectile and target such that they are momentarily "united", thereby influencing the electron binding energies. According to ECPSSR theory, the binding-energy correction term is given by

$$\zeta_s\,(\xi_s) = \frac{U_{2s}\,(Z_{2s}) + \Delta U_{2s}\,(Z_{2s})}{U_{2s}\,(Z_{2s})} = 1 + \frac{\Delta U_{2s}}{U_{2s}} = 1 + \frac{2Z_1}{Z_{2s}\theta_s}\,(g_s\,(\xi_s) - h_s\,(\xi_s)) \qquad (1)$$

where U_{2s} is the subshell binding energy (s = subshell); Z_1, Z_{2s} are the projectile and effective nuclear charges, respectively; $\xi_s = (2E_1/M_1)^{1/2}\,(Z_{2s}/U_{2s}) = v_1 Z_{2s}/U_{2s}$ with E_1, v_1, M_1 being the energy, velocity and mass of the projectile, respectively. ξ_s is the reduced projectile velocity; the reduced subshell binding energy is $\theta_s = 2n^2 U_{2s}/Z_{2s}^2$, with $n = 1$(K), 2(L); and $g_s\,(\xi_s)$, $h_s\,(\xi_s)$ are analytical functions that account for the velocity dependence of the correction; $h_s\,(\xi_s)$ is related to polarization effects on the binding energy due to the presence of projectile ion in atom. Using this correction, θ_s is replaced by $\zeta_s\theta_s$. Vigilante *et al* [10] realizing that the original ECPSSR theory overvalued the binding effect for decreasing projectile velocities, proposed to "saturate" the binding correction at a value which corresponds to the binding energy of the united atom (UA); i.e. projectile plus nucleus. Eq. (1) would then be replaced by

$$\zeta_s^{UA} = 1 + \frac{\Delta U_{2s}}{U_{2s}} = 1 + \frac{U_{2s}\,(Z_1 + Z_{2s}) - U_{2s}\,(Z_{2s})}{U_{2s}\,(Z_{2s})} = \frac{U_{2s}\,(Z_1 + Z_2)}{U_{2s}\,(Z_2)} \qquad (2)$$

which is the united-atom limit for low velocities.

Yu *et al* [11] propose merging the *UA* treatment with the separated-atom approach of ECPSSR at intermediate and high projectile velocities by switching from Eqs. (1) to (2) whenever $\zeta_s\,(\xi_s) > \zeta_s^{UA}$, as ξ_s decreases. L-subshell ionization cross sections σ_{L_i}, (where $i = 1, 2, 3$) will be calculated on the basis of ECPSSR theory with its modification, termed as UAECPSSR. The UA effect is calculated in the present paper using modified equation [12]. The UA effect increases rapidly as proton energy decreases. The X-ray production cross sections can be calculated using relationship given by Close *et al* [13]. The atomic parameters such as fluorescence yields and Coster–Kronig transition probabilities, involved in theoretical calculation of X-ray production cross-sections from ionization cross-sections are used. Tabulations for emission rates are provided by Scofield [14] and that of fluorescence yields and Coster-Kronig transition probabilities are given by Campbell [15].

3. EXPERIMENTAL METHODOLOGY

The H+ beam from AN-400 Van de Graaff accelerator has been used. The beam current on the target is measured by current integrator. It has high sensitivity, accuracy, low drift and an internal calibrating source. The high purity Germanium (HPGe) detector placed at right angle to the beam is used to detect L X-rays. To ensure the precision of relative intensities, the experiment is repeated at each beam energy and spectra with good statistics for Bi-L_α are obtained. The spectrum consists of four peaks of L X-ray groups corresponding to L_e, L_α, L_β and L_γ which are well separated from each other. The details of experimental method are discussed in our earlier publications [16, 17] as the present endeavour is a part of a series directed towards these studies by low energy proton.

4. RESULTS AND DISCUSSION

To analyze, the energy variation of line ratios L_e/L_α, L_γ/L_α and L_β/L_α as function of proton energy are plotted in the Figs. 1-3.

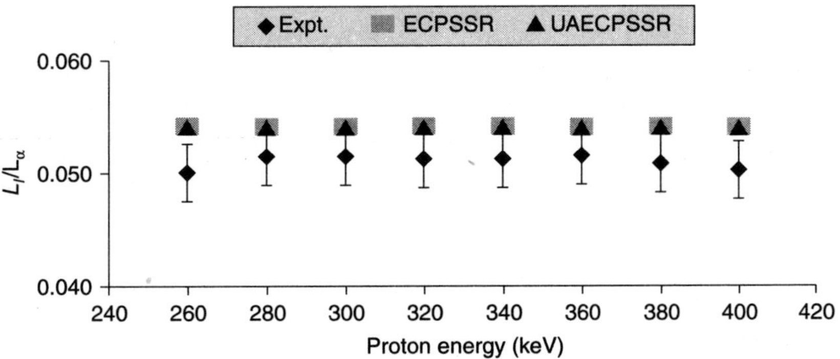

Fig. 1 L_e/L_α *as function of proton energy.*

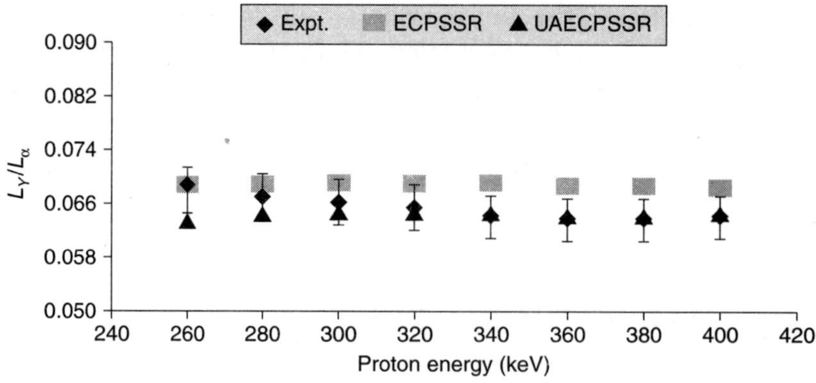

Fig. 2   *L_γ/L_α as function of proton energy.*

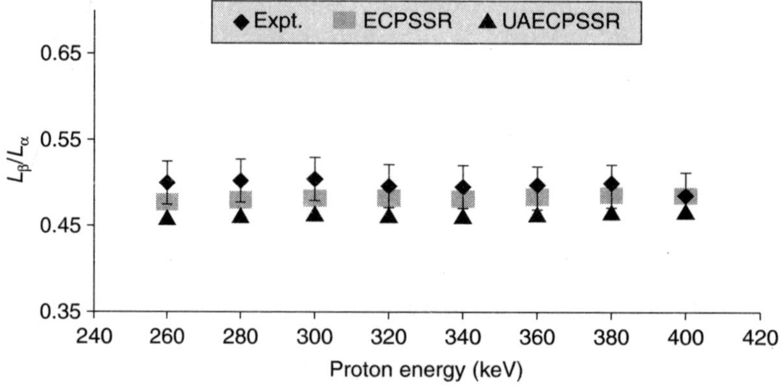

Fig. 3   *L_β/L_α as function of proton energy.*

The present experimental results are in good agreement with the theoretical values within experimental uncertainties excluding at low energy sides. Thus fine-tuning is required in the theoretical calculations especially at lower end. In recent years, Miranda *et al* [18] pointed out that multi-ionization correction is noteworthy even for proton impact in low energy region specifically, where the theoretical models have failed to predict the cross sections. This implication may prove to be vital concealed aspect to account for the discrepancies between theory and experiment in this energy range.

5.   CONCLUSION

The low energy regime (i.e. < 500 keV) provide a testing ground for the current theories. The volume of experimental data in this is energy region is very scarce. We observe a possibility that the already existing theories may require further refinements, especially at lower side. These investigations have yielded data in the low energy region, which materializes better understanding of proton induced X-ray emission phenomenon.

References

1. Sam J Cipolla *Nucl. Instr. and Meth.* B **261** 153 (2007)
2. B N Jones and J L Campbell *Nucl. Instr. and Meth.* B **258** 299 (2007)
3. G Lapicki *J. Phys.* **B**: *At. Mol. Opt. Phys.* **42** 145204 (2009)
4. J H McGuire and P Richard *Phys. Rev.* A **8** 1374 (1973)
5. J M Hansteen and O P Mosebekk *Z. Phys.* **234** 281 (1970)
6. B-H Choi, E Merzbacher and G S Khandelwal *At. Data Nucl. Data Tables* **5** 291 (1973)
7. W Brandt and G Lapicki *Phys. Rev.* A **20** 465 (1979)
8. G Lapicki *Nucl. Instr. and Meth.* B **189** 8 (2002)
9. Sam J Cipolla *Nucl. Instr. and Meth.* B **261** 142 (2007)
10. M Vigilante, P Cuzzocrea, N De Cesare, F Murolo, E Perillo and G Spadaccini *Nucl. Instr. and Meth.* B **51** 232 (1990)
11. Y C Yu, C W Wang, E K Lin, T Y Liu, H L Sun, J W Chiou and G Lapicki *J. Phys.* B **30** 5791 (1997)
12. Sam J Cipolla *Comput. Phys. Commun.* **176** 157 (2007)
13. D A Close, R C Bearse, J J Malanify, C J Umbarger *Phys. Rev.* A **8** 1873 (1973)
14. J H Scofield *At. Data Nucl. Data Tables* **14** 121 (1974)
15. J L Campbell *At. Data Nucl. Data Tables* **95** 115 (2009)
16. Harsh Mohan and A K Jain *Nucl. Instr. and Meth.* B **266** 1203 (2008)
17. A K Jain , Harsh Mohan and S Sharma *Nucl. Instr. and Meth.* B **268** 1790 (2010)
18. J Miranda, O G de Lucio, E B Tellez and J N Martinez *Rad. Phys. and Chem.* **69** 257 (2004)

Energy Absorption Build-up Factors in Bone

H.C. Manjunatha[1*], B. Rudraswamy[2] and R.S. Madhukeswara[3]

[1]*Government College for Women, Kolar, Karnataka, India*
[2]*Department of Physics, Bangalore University,Bangalore, Karnataka, India*
[3]*Government First Grade College, Pandavapura, Mandya district, Karnataka, India*
E-mail: *manjunathhc@rediffmail.com*

ABSTRACT

G.P. fitting method has been used to compute energy absorption of Bone femur and Bone cortical for wide energy range (0.015 MeV-15 MeV) up to the penetration depth of 40mean free path. The dependence of energy absorption build up factor on incident photon energy and penetration depth has also been assessed. Build up factors increases with increase of penetration depth. The computed energy absorption build-up factor is used to estimate specific absorbed fraction of energy. The relative dose is also evaluated using computed energy absorption build-up factor. The computed data of specific absorbed fractions of energy and relative dose using this method is more accurate than the data available in the literature because the variation of an effective atomic number with energy is also considered in this method.

Keywords: Bone cortical, Femur, Buildup factor, Relative dose

Pacs No.: 32.80.cy; 32.10; 32.80.-t

1.   INTRODUCTION

When gamma and x rays enter the medium/body, they degrade their energy and build up in the medium, giving rise to secondary radiation which can be estimated by a factor which is called the 'build-up factor'. The energy absorption buildup factor (B_{en}) is also defined as the buildup factor in which the quantity of interest is the absorbed or deposited energy in the interacting material and the detector response function is that of absorption in the interacting medium. The buildup factor data were computed by different codes [1-2]. American National Standards ANSI/ANS 6.4.3 [3] used G.P. fitting method and provided buildup factor data for 23 elements, water, air and concrete at 25 standard energies in the energy range 0.015-15 MeV with the penetration depth of 40 mean free paths. In the present work an attempt has been made to compute B_{en} of Bone cortical and Bone femur for 0.015 MeV-15 MeV up to the penetration depth of 40 mean free path using G.P. fitting method. The computed energy B_{en} is used to evaluate the specific fraction of absorbed energy and relative dose in bone.

2. PRESENT WORK

For computation of effective atomic number (Z_{eff}), firstly the values of mass attenuation coefficients were computed from WinXCom computer program [Gerward *et al.*, (4 & 5)]. Z_{eff} can be computed from the following equation

$$Z_{eff} = \frac{\left(\dfrac{\left(\dfrac{\mu}{\rho}\right) bio \sum\limits_i n_i A_i}{\sum\limits_i n_i}\right)}{\left(\sum\limits_i \left(\dfrac{f_i A_i}{Z_i}\right)\left(\dfrac{\mu}{\rho}\right)i\right)} \tag{1}$$

where n_i is the number of atoms of ith element in a given molecule, $(\mu/\rho)_{bio}$, the mass attenuation coefficient of bio molecule, A_i, the atomic weight of element i. $(\mu/\rho)_{bio}$ was estimated based on the chemical composition and f_i is the fractional abundance. The computed Z_{eff} is given in Fig. 1. We have evaluated the G-P fitting parameters of B_{en} (b, c, a, X_k and d) of Bone cortical and Bone femur corresponding to their Z_{eff} and energy using Lagrange's interpolation technique from the data available in literature [5]. The computed G-P fitting parameters were then used to compute the B_{en} in the energy range 0.015 MeV – 15 MeV up to a penetration depth of 40 mean free path with the help of GP fitting formula, as given by the equations

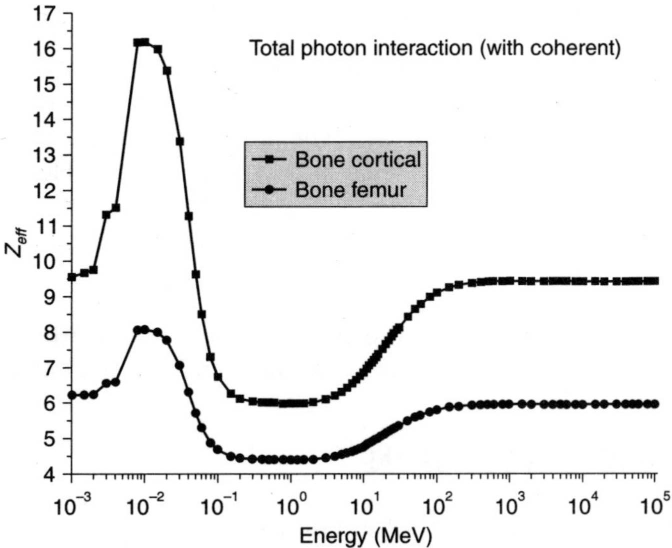

Fig. 1 *Variation of effective atomic number Z_{eff} of bone with photon energy for total photon interaction (with coherent)*

$$B_{en}(E, X) = 1 + \frac{b-1}{K-1}(K^X - 1) \text{ for } K \neq 1 \text{ and } B_{en}(E, X) = 1 + (b-1)X \text{ for } K = 1 \qquad (2)$$

$$fK(E, X) = CX^a + \frac{d \ tanh\left(\dfrac{X}{X_K} - 2\right) - tanh\,(-2)}{1 - tanh\,(-2)} \qquad (3)$$

Where X is the source-detector distance in mean free paths and b is the value of B_{en} at 1mfp. $K(E, X)$ is the dose multiplication factor.

3. APPLICATION OF THE PRESENT WORK

The specific fraction of absorbed energy (Φ) is calculated from the target to source with the point source kernel method. In this method, Φ at distance x from the point source of photon emitter is

$$\Phi(x) = \frac{\mu_{en} \exp(\mu x) \, B_{en}}{4\pi x^2 \rho} \qquad (4)$$

Here μ_{en} is linear absorption coefficient of photons of given energy, μ is linear attenuation coefficient of photons of given energy and ρ is density of the medium. The computed B_{en} values from equation (2) are used to evaluate Φ. The estimated Φ values are useful to calculate the absorbed dose at fixed distances from the point source in the infinite homogeneous medium of Bone

$$D(x) = \tau \sum \Delta_i \, \Phi(x) \qquad (5)$$

τ is the residence time of activity which is the ratio of actual activity to the administrated activity. The quantity Δ_i numerically equal to (2.13 n_iE_i) where n_i is the frequency of occurrence of emissions with energy E_i. The dose of gamma radiation depends on B_{en}, attenuation coefficient and distance from the point source. Hence the relative dose at a distance r is

$$\frac{D_r}{D_0} = e^{-\mu r}\frac{B_{en}}{r^2} \qquad (6)$$

where D_0 is initial dose delivered by the point gamma ray emitter. The multiplication of D_0 with relative dose gives actual dose of gamma D_r at a distance r.

4. RESULTS AND DISCUSSION

The variation of B_{en} with incident photon energy for Bone cortical and Bone femur is shown in Figs. 2-3. From these figures it is observed that B_{en} increases up to the E_{pe} and then decreases. Here E_{pe} is the energy value at which the photo electric interaction coefficients match with Compton interaction coefficients for a given value of Z_{eff}. E_{pe} is almost equal to 0.1 MeV. In the lower energy end photoelectric absorption is dominant photon interaction process; hence B_{en} values minimum. As the energy of incident photon increases, Compton scattering overtakes the photoelectric absorption. It results multiple Compton scattering events which increases the B_{en} up to the E_{pe} and becomes maximum at E_{pe}. Thereafter, pair production starts dominating which reduces the B_{en} to a minimum value. The variation of B_{en} with penetration depth of Bone cortical and Bone femur at 0.02 MeV, 0.1 MeV, 1 MeV, 15 MeV is shown in Figs. 4 & 5. The computed B_{en} increases with penetration depth. With increase in penetration depth, thickness of the interacting material has been increased which results in increasing the scattering events in the interacting medium. Hence it results in large B_{en} values. The variation of Φ in Bone cortical and Bone femur at 5 mfp, 10 mfp, 15 mfp, 20 mfp and 40 mfp for interacting thickness of the medium 0.01 m is given in Figs. 6-7. The variation of (D_r/D_0) with photon energy in Bone cortical and Bone femur at 5mfp for various distances (r = 0.01 m to 1 m) is given in Figs. 8-9. The variation of Φ and (D_r/D_0) with photon energy is similar to that of B_{en} which can be explained on the similar manner.

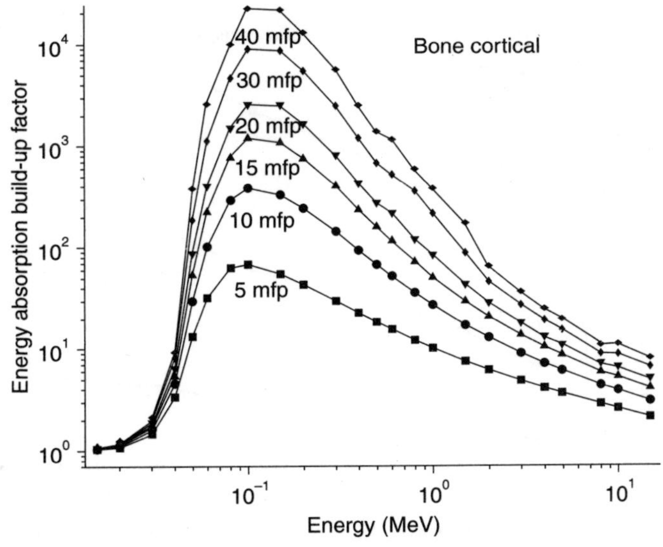

Fig. 2 *Variation of energy absorption build-up factor with photon energy for Bone cortical*

5. CONCLUSION

The maximum values of B_{en} occur in the intermediate energy region where Compton scattering is the main interaction process where as the minimum values of B_{en} occur

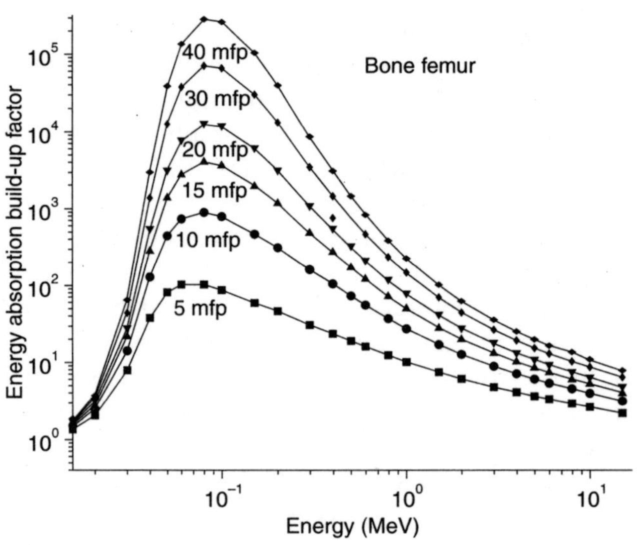

Fig. 3 *Variation of energy absorption build-up factor with photon energy for Bone femur.*

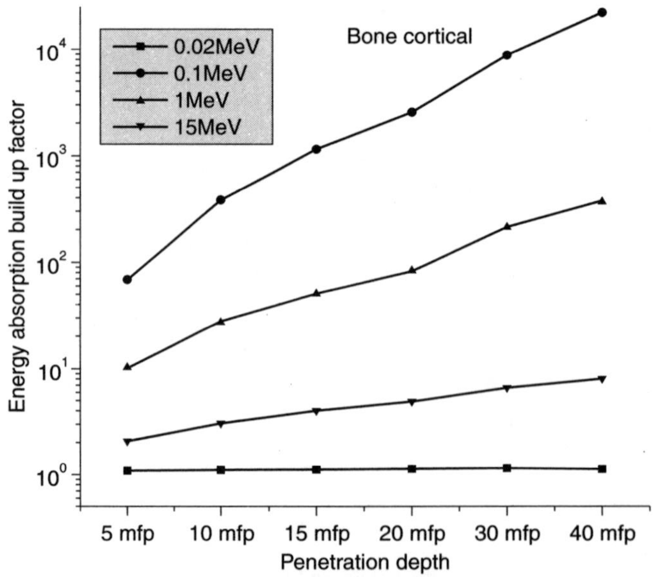

Fig. 4 *Variation of energy absorption build-up factor with penetration depth for Bone cortical.*

at lower and higher energy regions where absorption processes namely photoelectric absorption and pair production start dominating. Φ and (D_r/D_0) have been calculated using the G-P fitting formula for Bone for the first time. The computed data of Φ and (D_r/D_0) using this method is more accurate than the data available in the literature because the variation of Z_{eff} with energy is also considered in this method.

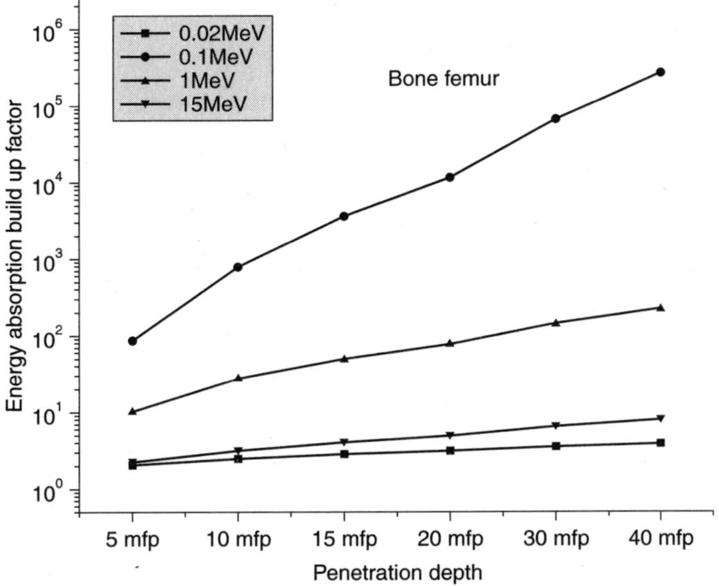

Fig. 5 *Variation of energy absorption build-up factor with penetration depth for Bone femur.*

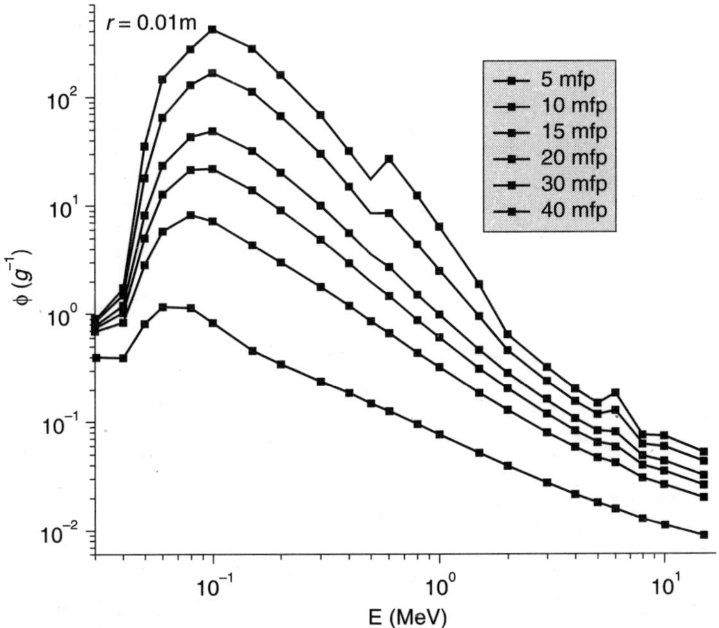

Fig. 6 *Variation F with incident photon energy in Bone cortical at r = 0.01 m and for various penetration depths up to 40 mfp*

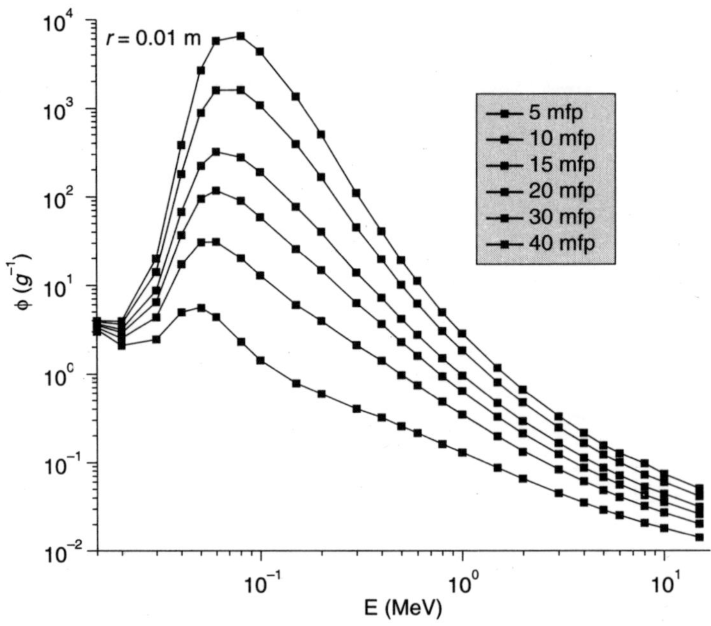

Fig. 7 *The variation Φ with incident photon energy in Bone femur at r = 0.01 m and for various penetration depths up to 40 mfp*

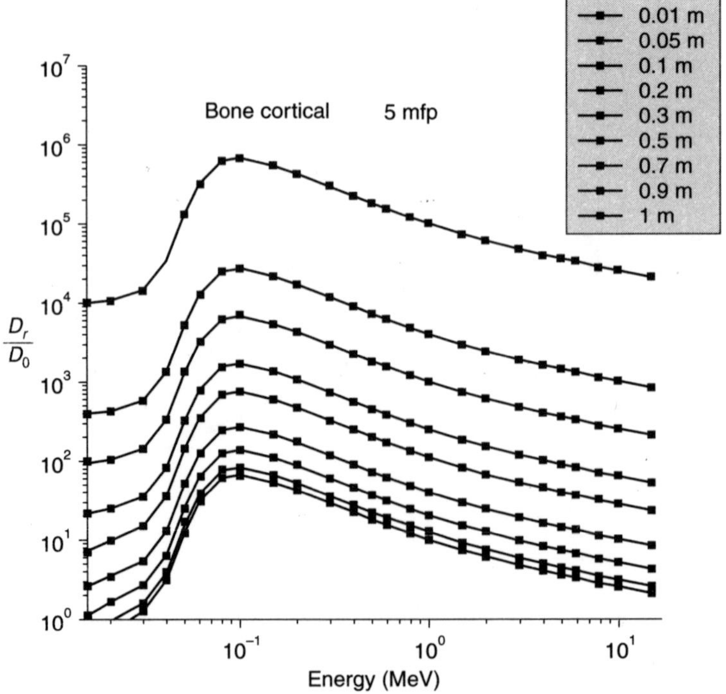

Fig. 8 *Variation of relative dose (D_r/D_o) with photon energy in Bone cortical at 5 mfp at various distances.*

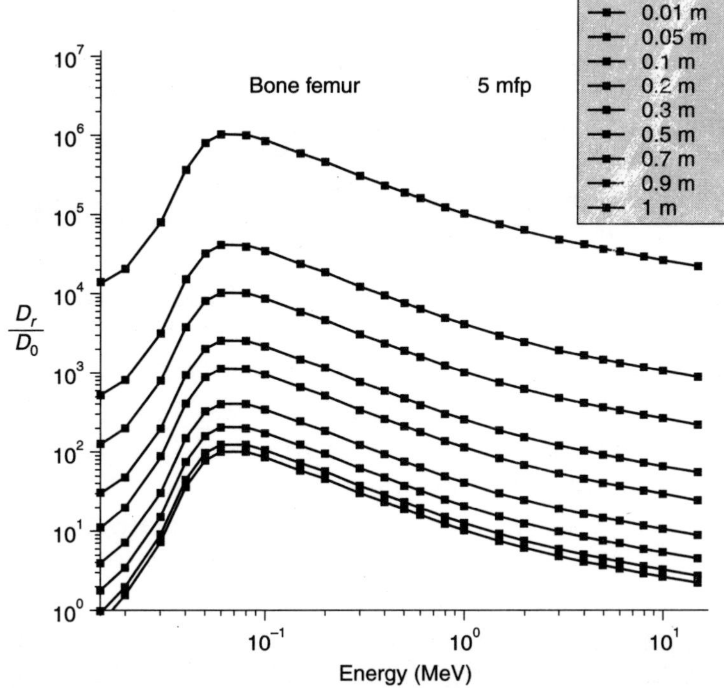

Fig. 9   *Variation of relative dose (D_r/D_o) with photon energy in Bone femur at 5mfp at various distances*

References

1. K. Takeuchi, PALLAS-PL, SP-A code dimensional transport code Paper 42 (Ship Research Institute, Japan) (1973)
2. N. Yamano, K. Minami, K. Koyama, Y. Naito, RADHEAT V-4 Japan Atomic Energy Research Institute Report 1316 (1989)
3. American National Standard (ANS), *ANSI/ANS* 6.4.3 (1991)
4. L. Gerward, N. Guilbert, K.B. Jensen, H. Levring, Radiat.Phys.Chem. **60**,23 (2001).
5. L. Gerward, N. Guilbert, K.B. Jensen, H. Levring, Radiat.Phys.Chem. **71**,653 (2004)

Coherent and Incoherent Scattering Cross Sections for Elements with $13 \leq Z \leq 50$ using ^{241}Am Gamma Rays

Latha Panakkada[1*], K.K. Abdullah[2], M.P. Unnikrishnan[3], K.M. Varier[4] and B.R.S. Babu[4]

[1]Department of Physics, Providence Women's College, Calicut, Kerala, India
[2]Department of Physics, Farook College, Feroke, Calicut, Kerala, India
[3]Department of Physics, M E S College, Valancheri, Kerala, India
[4]Department of Physics, University of Calicut, Malappuram, Kerala, India
E-mail: *lathapanak@yahoo.co.in

ABSTRACT

Coherent and incoherent scattering cross section measurements have been carried out using a HPGe detector on elements in the range of $Z = 13 - 50$ using ^{241}Am gamma rays. The cross sections have been derived by comparing the net count rate obtained from the Compton peak of aluminium with the corresponding peak of the target. The measured cross sections for the coherent and incoherent processes are compared with theoretical values and earlier reported values. Our results are in agreement with the theoretical values.

Keywords: Cross section, Coherent scattering, Incoherent scattering

Pacs No.: 32.80.–t, 32.80.Aa, 32.90.+a

1. INTRODUCTION

Measurement of differential scattering cross sections for X-rays is useful in the studies of radiation attenuation, transport and energy deposition and plays an important role in medical physics, reactor shielding, industrial radiography in addition to X-ray crystallography. Coherent (Rayleigh) scattering accounts for only a small fraction of the total cross section, contributing at the most to 10% in heavy elements, just below the K-edge energy. Incoherent (Compton) scattering accounts for the rest of the total cross section. For low Z materials this process dominates over most part of the energy range.

An extensive review of previous work on incoherent and coherent scattering has been reported by Kane [1] and Bradley et al. [2]. [241]Am is a very convenient radiation source for studies of photon interactions in the X-ray region, especially because of the relatively long half life of 450 years and its photon energy (59.54 keV) being in the vicinity of the K-edges of many elements in medium-Z region. We have embarked on a series of photon interaction studies using this versatile source. Ramachandran *et al.* [3] measured attenuation coefficients for these gamma rays in the rare earth elements with $57 < Z < 72$ and derived photoelectric cross sections therefrom. Good agreement with theoretical values based on the XCOM [4] has been reported by them. Subsequently, Abdullah et al. [5] carried out attenuation studies near the K absorption edges in rare earth elements using [241]Am gamma rays, Compton scattered at various angles by an aluminium scatterer. More recently, Abdullah *et al.* [6] made similar studies with the elements Zr, Nb, Mo and Pd. In both measurements, reasonable agreement has been observed between the experimental values and earlier results on one hand as well as with the XCOM values on the other hand.

In the present studies, we have carried out Coherent and incoherent scattering measurements on the elements in the range $13 \leq Z \leq 50$ using [241]Am gamma rays. The details of the measurements and the results obtained therefrom are given in the following sections.

2. EXPERIMENTAL DETAILS

A 1.1×10^{10} Bq [241]Am source (S) procured from Amersham England was used as the source for 59.54 keV gamma rays. The gamma rays were scattered by targets kept at an angle 45° with the incident ray. The targets used were of 99.9% purity and are in the form of thin square foils with 1.2 cm side. The mass per unit area of the targets were determined using a sensitive electronic balance. A HPGe Gamma-X detector (D) of active volume 85 cm^3 supplied by ORTEC, USA was used for detecting the scattered photon beams.

The spectra of the scattered radiations from the targets were recorded typically for about 6-9 hours such that the statistical errors in the total scattered counts were of the order of 1% or less. The detector resolution is 1.14 keV at 60 keV. The detector is arranged at an angle of 90° with the incident gamma beam. The signals from the detector were processed by the standard ORTEC modules and are then fed to a CAMAC based data acquisition system. The data analysis software, PAW, developed by CERN laboratories [7] has been used for data analysis in the present studies.

Incoherent scattering cross section is measured first with aluminium and is used as a reference value. Any cross section (coherent as well as incoherent) of interest was evaluated by comparing the net count rate obtained from the Compton peak of aluminium with respect to the corresponding peak of the target. This method was useful in eliminating the necessity to know the absolute source strength.

A target of finite dimensions can be considered to be made up of a large number of infinitesimally small elements within this target. The solid angle factor and the transmission factor vary from element to element. An average effective sum of the product $T_i \phi_D \phi_T$ over all the elements of the scatterer was used in the expression for cross section. Here T_i is the transmission factor, ϕ_T and ϕ_D are respectively the solid angles subtended by the ith element at the centre of the source and by the detector at the centre of the ith element.

3. RESULTS AND DISCUSSION

The differential coherent and incoherent scattering cross sections for the elements studied in this work are presented in figure. 1 and 2 respectively along with the theoretical values and available other experimental data from references [8, 9, 10].

Typical errors in the quoted cross sections have been estimated to be about 6%, with the major part arising from uncertainties in the target mass per unit area and minor contributions from air scattering, finite angle of acceptance of the detector etc.

The S-matrix values taken from the tabulations of Chatterjee [11] and the differential coherent scattering cross sections calculated using the relativistic modified form factor (RMFF) values [12] for all the elements under study corresponding to an angle 90° are plotted in Fig. 1. It is observed that there is better overall agreement of the present experimental values with the values calculated on the basis of RMFF. It is also observed that the S-matrix values are always higher than the present experimental values. This discrepancy with respect to the S-matrix values may be attributed to the neglect of the electron correlation effects in the theoretical S-matrix calculations, as pointed out by Chatterjee [11]. Similar discrepancies had also been noticed by [13]. We are in the midst of further investigations in this direction at lower energies with X-rays produced by the Proton Induced X-ray Emission (PIXE) technique

In Fig. 2, the differential incoherent scattering cross section is compared with the cross section calculated using non-relativistic Hartree Fock incoherent scattering function $S(x, Z)$ obtained from the tables of Hubbell [14]. Our results agree with the theoretical values.

The figures show that our experimental results are, in general, close to the theoretical values represented by the solid curves.

To the best of our knowledge sufficient experimental data is not available in literature for 59.54 keV at 90°. It is clear that the present cross section results are in good agreement with the available experimental values reported in literature. It is also worth mentioning that the results show no systematic trend in departure from

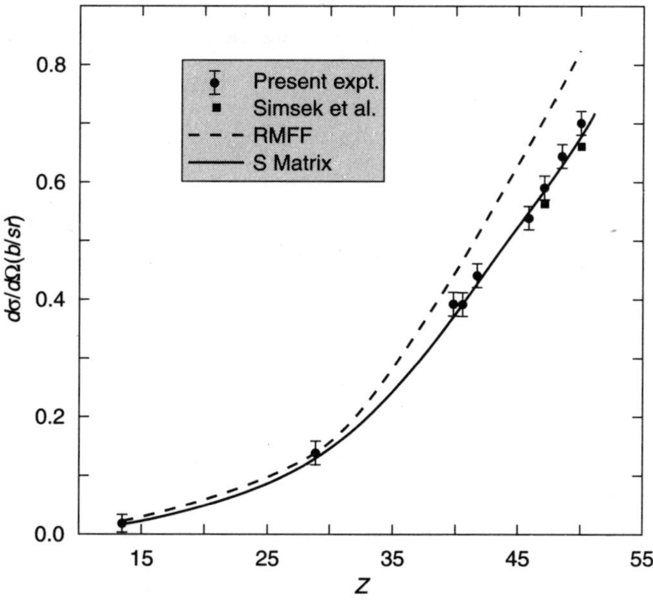

Fig. 1 *Elastic scattering cross sections at 59.54 keV for various elements*

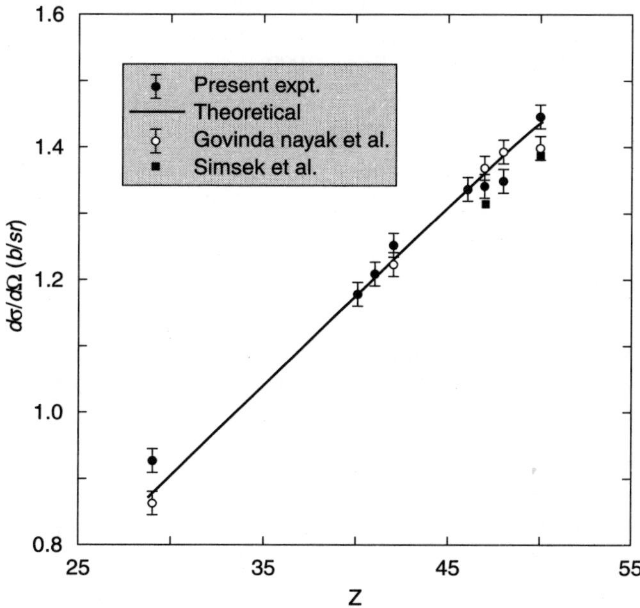

Fig. 2 *Inelastic scattering cross sections at 59.54 keV for various elements*

the theoretical values except for the S-matrix values mentioned earlier. Comparison indicates that for most cases agreement is obtained between present result and the theoretical predictions.

4. CONCLUSION

The relativistic modified form factor values of the elastic cross sections explain the present results better than the S matrix values. Calculations based on non-relativistic Hartree Fock incoherent scattering function $S(x, Z)$ explain the present results for the incoherent scattering cross sections.

References

1. P P Kane, Phys. Rep. **218**, 67 (1992)

2. D A Bradley, O D Goncalves, P P Kane, Radiat. Phys. Chem., **6**, 125 (1999)

3. N Ramachandran, K Karunakaran Nair, K K Abdullah and K M Varier, Pramana-J. Phys., **67**, 507 (2006)

4. M J Berger, J H Hubbell, S M Seltzer, J S Coursey and D S Zucker, 'XCOM: Photon Cross Section Database (version 1.2)' [Online], National Institute of Standards and Technology, Gaithersburg, MD (2003); M J Berger and J H Hubbell, XCOM : "Photon cross sections on a personal computer", Program manual, Centre for Radiation research, National Bureau of investigations Standards MD20899 (1990) and M J Berger and J H Hubbell, XCOM version 3.1 - NIST Standard Reference Data Base (1999)

5. K K Abdullah, N Ramachandran, K Karunakaran Nair, B R S Babu, Antony Joseph, Rajive Thomas and K M Varier,Pramana - J.Phys., **70**, 633 (2008)

6. K K Abdullah, K Karunakaran Nair, N Ramachandran, K M Varier, B R S Babu, Antony Joseph, Rajive Thomas, P Magudapathy and K G M Nair, Pramana-J. Phys., **75**, 459 (2010)

7. CERN laboratory in Geneva, Switzerland

8. O Simsek, J.Phys. B, **33**, 201 (2000).

9. O Simsek and M Ertugrul, Inst. SCi. and Technol., **29**, 407 (2001).

10. N Govinda Nayak, Gerald Pinto and K Siddappa, Rad. Phys. and Chem., 60, 555 (2001).B K Chatterjee and S C Roy, J. Phys. Chem. Ref. Data, **27**, 1011 (1998)

11. B K Chatterjee and S C Roy, J. Phys. Chem. Ref. Data, **27**, 1011 (1998)

12. D Schaupp, M Schumacher, F Smend, P Rullhusenan, J Phys. Chem. Ref. Data, **12** (1983) 467

13. J S Shahi, S Puri, DMehta, M L Garg, N Singh and P N Trehan, Phys. Rev., A **57**, 4327 (1998)

14. J H Hubbell, W J Veigele, E A Briggs, R T Brown, D T Cromer and R J Howerton, J Phys. Chem. Ref. Data **4**, 471 (1975).

Estimation of Streaming Radiation in High Current Medium Energy Accelerator: A Case Study

M. Sengupta Mitra*, R. Ravisankar, Tapas Bandyopadhyay and P.K. Sarkar

Health Physics Unit, HPD, BARC
Variable Energy Cyclotron Centre, 1/AF, Bidhan Nagar, Kolkata
E-mail: *mausumi@veccal.ernet.in*

ABSTRACT

The ducts are pierced through shielding wall of accelerator for engineering purposes. The streaming of radiation through duct is estimated for 30 Mev proton cyclotron. The simulation for calculation is done using multipurpose particle transport code FLUKA. The distribution of track length density of neutron and gamma are calculated at different positions.

Keywords: Proton cyclotron, Duct, Streaming.

1. INTRODUCTION

An accelerator of medium energy and high current is in demand especially to produce isotopes for radiodiagnosis and therapy. Situation demands to house this type of accelerator in public domain. It is a regulatory requirement to keep the dose within a stipulated limit following ALARA. These accelerators are housed inside reasonably thick concrete shield. Due to engineering requirement, different holes are made through this shield to provide services to the machine bunker. Streaming of radiation through these holes may cause the violation of dose limit outside the shield. It is required to estimate the streaming through these holes and plan accordingly to keep the dose in limit. A 30 Mev proton Cyclotron with 350 μA beam current is going to set up at Kolkata.

For this type of facility beam loss mechanism due to accelerated beam hitting different beam diagnostic parts of the machine results in intense neutron and gamma radiation field and there is consequent streaming of dose through the ducts required for services [1]. Beam loss occurs while beam hits beam current measuring devices which may be at the machine bunker or definitely at experimental bunkers. Therefore to assess accidental scenario as well as requirement of shielding it is necessary to have adequate knowledge about the radiation environment generated when 30 MeV proton beam is incident on machine parts. Consequent transport is done for the generated radiation at machine bunker through holes inserted in the shield wall for the services. Beam current measuring devices are made of tantalum. It is required to estimate the flux distribution of neutron as well as gamma when 30 Mev proton beam is incident on thick tantalum target. Streaming through the cylindrical duct of radius 17 cm and length 150 cm is calculated and dose estimation is done. This is to be noted that the forward direction shielding is made of 1.5 m thick concrete. Services to the experimental bunkers are provided through the wall below the beam line with cylindrical ducts of above mentioned size.

2.   DESCRIPTION OF SIMULATION GEOMETRY

We have carried out Monte Carlo simulation to estimate fluence distribution of neutron and gamma in different detectors, generated when 30 Mev proton beam is incident on 2cm thick tantalum target. In order to construct the simulation geometry we have defined different bodies and the required regions are formed with union, intersection and subtraction of different bodies according to Combinatorial Geometry package used in FLUKA [2]. The wall of the vault is constructed with 2.2 m thick concrete in all directions except the wall separating experimental cave i.e. in the direction of the beam line which is 1.5 m thick. A duct of radius 17 cm is considered in the wall separating experimental cave to calculate streaming of neutron as well as gamma through it. The cylindrical target of radius 5cm and thickness 2 cm is placed at the height of 1.3 m   from the floor of the vault. One cylindrical detector of radius 5.0 cm and thickness 2.0 cm is placed just after the target along the beam line and another detector of same dimension is placed orthogonal to the beamline at a distance of 20 cm from the target to assess radiation environment. Tracklength density of both neutron as well as photon inside the two detectors and duct are calculated.

3.   RESULTS AND DISCUSSIONS

Figures 1 and 2 show the distribution of tracklength density of neutron and photon with energy in two detectors and the duct. The detector placed just after the target is indicated as detector1 while the detector placed orthogonal to the direction of the incident proton beam at a distance of 20 cm from the Ta target is termed as detector 2. With the aid of dose conversion factors available in ICRP74 [3, 4] these flux distributions can be converted to dose.

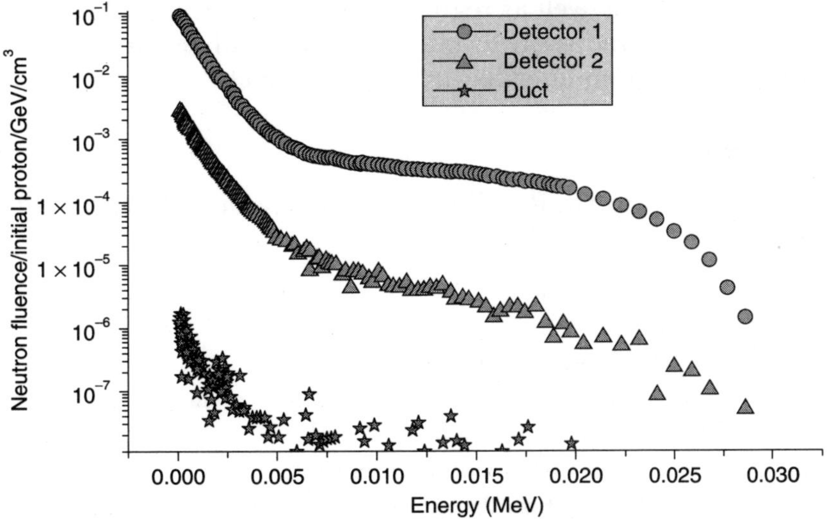

Fig. 1 *Energy distribution of neutron fluence inside detector 1, detector 2 as well as duct for 30 MeV proton beam incident on Tantalum target.*

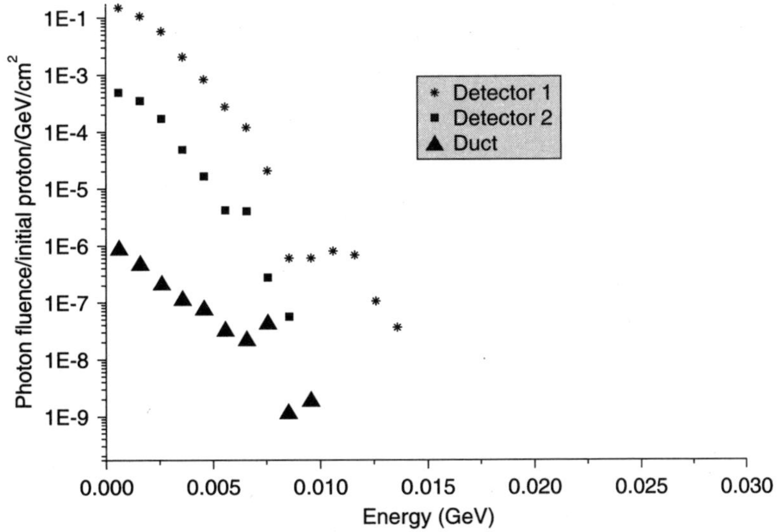

Fig. 2 *Energy distribution of photon fluence inside detector 1, detector 2 as well as duct for 30 MeV proton beam incident on Tantalum target.*

4. CONCLUSION

The knowledge regarding dose estimation at different positions inside vault as well as experimental caves is important from radiological safety point of view in order to assess accidental exposure as well as regular operating condition. Also to calculate the dose due to streaming of radiation through duct we will further continue the work placing a cylindrical phantom outside the wall near duct.

Acknowledgement

The Authors are indebted to Dr. R.K. Bhandari for continuously encouraging us to do research work.

References

1. DAE Medical Cyclotron Safety report, VECC, Kolktata
2. A. Fasso, A. Ferrari, J. Ranft, and P.R. Sala, "FLUKA: a multi-particle transport code", CERN-2005-10 (2005), INFN/TC_05/11, SLAC-R-773
3. ICRP Publication 74 – Dose Conversion Factors (1995)
4. NCRP Report No. 144 – Radiation Protection for Particle Accelerator Facilities (2003)

Neutron Dose Rate Mapping Around a 14 MeV DT Fusion Generator Source

P. Srinivasan[1,*], S. Priya[1], Vivek Bara[1], D.P. Mishra[1], Tarun Patel[2]
and R.K. Gopalakrishnan[1]

[1]*Radiation Safety Systems Division, Bhabha Atomic Research Centre, Mumbai, India*
[2]*Neutron Generator Facility, Purnima, Bhabha Atomic Research Centre, Mumbai, India*
E-mail: *praka17@yahoo.com, psri@barc.gov.in*

ABSTRACT

Compact sealed DT Generators are used as sources of 14 MeV neutrons in experimental laboratories for various applications related to nuclear physics measurements, materials research, strategic research etc. An attempt has been made to measure the neutron and gamma doserates existing around these generators and have been compared with monte carlo based calculations. The measurements of neutron dose rates were carried out using BF_3 based rem meters and neutron-gamma direct reading dosimeters. The neutron and gamma dose rates were computed by Monte Carlo simulation codes FLUKA and MCNP. It is observed that the simulation results and the actual measured values agree within 10% to 20%.

Keywords: 14 MeV neutrons, DT generators, FLUKA, MCNP, Dose rates, Shield integrity.

Pacs No.: 28.20 – v

1. INTRODUCTION

Compact sealed DT generators are used as sources of 14 MeV neutrons in experimental laboratories for various applications related to nuclear physics measurements, material research, strategic research etc. Detailed knowledge of the radiation dose rates around the DT generators are essential for ensuring adequate radiological protection of the personnel involved with the operation of DT generators and the experimenters alike. The motivation of the present work is to estimate the radiation dose rates prevalent near DT generator based 14 MeV neutron sources. This is crucial to evaluate the shielding integrity for the laboratories that will house the neutron generators. Detailed neutron dose rate mapping was carried out experimentally by using standard neutron dosemeters. Further, theoretical evaluation of the neutron dose rates near DT sources was carried out using Monte Carlo simulation in order to validate the experimental measurements.

2. EXPERIMENTAL MEASUREMENTS

The measurements of neutron dose rates were carried out using the BF_3 proportional counter based Digi-Pig neutron rem meters and MGPI make neutron Direct Reading Dosimeters (DRDs) in the Purnima facility, BARC, Trombay. The D-T neutron generator used for the experiment was a continuous neutron source with an operational flux of 10^7 n/s. Both the cumulative doses and the dose rates were registered in the monitors near 14 MeV neutron generators. Around four Digi-pig monitors and four neutron DRDs were located in the experimental positions for known time during the operation of the 14 MeV neutron generators. The accumulated dose values registered by the monitors in various locations were used in calculating the average dose rates in the respective locations. Also gamma dose rates were measured in certain experimental positions in order to verify the prompt gamma dose rates in the vicinity of the D-T generators. The ion chamber based direct reading dosimeters (DRD) were deployed for gamma dose measurements. The MGPI make neutron DRDs simultaneously register the cumulative gamma doses in addition to the neutron doses.

3. MONTE CARLO SIMULATIONS

The neutron and gamma dose rates were computed by Monte Carlo simulation code FLUKA [1] and MCNP [2]. FLUKA simulates the interaction and propagation of about 60 different particles, including photons and electrons, neutrinos, muons, hadrons (neutrons and protons) and heavy ions with energies ranging from keV – TeV. MCNP is a general purpose, continuous energy transport code which can be used to simulate the transport of neutrons, photons and electrons individually as well as in coupled mode. The neutron energy regime in MCNP normally ranges from 10^{-11} – 20 MeV whereas for photons it is 1 KeV – 100 GeV. Transport of neutrons with energies lower than certain energy is performed in FLUKA by a multigroup algorithm and in MCNP. For neutrons with energy lower than 20 MeV, FLUKA and MCNP use the reaction cross section library derived from the most recently evaluated data (ENDF). Neutron and gamma dose rates were computed by using the ambient dose equivalent factors based on ICRP-74 publication.

The DT neutron generator was assumed to be a cylindrical tube made of SS-316 with a length of 30 cm and 3 inches diameter. The SS wall thickness of 6 mm was used for the DT tube, which is assumed to be located in the center of the room. Neutrons were assumed to be emitted isotropically in all the directions from the tritium target. Source self-attenuation and attenuation and scattering in the surrounding materials were considered in the Monte Carlo computations. The room was filled with dry air of density 1.2905×10^{-3} g cm^{-3} and concrete shield of thickness 300 mm was assumed on all directions.

4. RESULTS AND DISCUSSION

The source and shield geometry, source parameters like source strength and source particle energy intensity distribution formed the input to the FLUKA and MCNP. About 2.2×10^{8} neutron histories were tracked in FLUKA to improve upon the precision of the estimated dose rates. Random statistical errors of the Monte Carlo FLUKA runs are less than 2% for the particle histories considered. In MCNP, 2×10^{7} histories were tracked with statistical errors less than 6% for the points considered. Fig. 1 shows the vertical neutron dose rate profile generated by plotting the isodose curves from the USRBIN plot of FLUKA. Doserate contour plots where generated with MCNP using the FMESH tally option. A comparison of the results obtained by the experimental measurements and that by theoretical simulation is given in Table 1. The experimental measurements were obtained using MGPI neutron DRDs and Digi-pig model neutron rem meters. Gamma dose rates registered by neutron DRDs and by pocket ion chamber gamma DRDs (IC-DRD) are given in brackets below the corresponding neutron dose rate values. It is observed that the theoretical and the experimental values agree with in 10%-20%. The deviations may be due to the uncertainties in the input data used for simulating the structural material compositions and dimensions of the DT tube, surrounding scattering media, detector holder frame, small variations in experimental distances, etc.

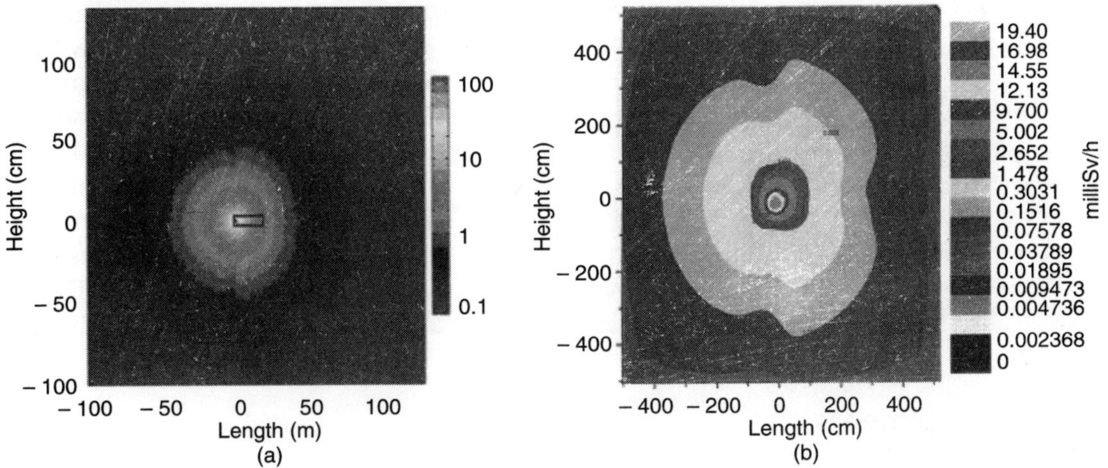

Fig. 1   *Neutron dose rate map around DT source generated by FLUKA (a), MCNP (b)*

5. CONCLUSION

Neutron and Gamma doserates around a DT neutron generator have been measured and compared with monte carlo based particle transport codes, namely, FLUKA and MCNP. A detailed knowledge of the radiation doserates around the generator will help in ensuring adequate radiological protection of the personnel involved in the operation of these generators and the experimenters alike. This exercise has also served in benchmarking the theoretical simulation methods used in shielding analysis of 14 MeV neutron sources.

Table 1 *Measured and simulated neutron and gamma dose rates*

Angular location and reference plane	Distance(cm) from source	Neutron Exp.	(Gamma) FLUKA mSv/h	Dose rate MCNP
30° with DT tube				
Length axis on HP	26	2.66	2.56	2.23
Along the length axis	25	2.82	3.11	3.41
		(0.018)	(0.023)	–
Perpendicular to the				
Tube length on HP	25	1.68	1.57	–
		(0.012)	(0.014)	
42.2° with tube on HP	40	0.68	0.64	–
10° with tube on VP				
through the DT Tube	45	0.80	0.78	–
15° with tube length				
axis on HP	69	0.226	0.250	0.250
		(0.002)	(0.0015)	
13° with tube length axis on HP	77	0.212	0.230	0.186
Perpendicular to the DT				
tube length on VP	103	0.085	0.090	–
56° with the normal				
through the tube on VP				
perpendicular to the length	266	0.0115	0.014	–
65° with the normal through				
the tube on VP perpendicular				
to the length	354	0.0013	0.0015	–

VP – Vertical Plane, HP- Horizontal Plane

Acknowledgements

The authors express their sincere thanks to Dr. D.N. Sharma, Associate Director, Health Safety and Environment Group and Dr Amar Sinha, Head, LNPS for their guidance in completing this work and Shri. R. G. Koli, RSSD for preparation of experimental setup.

References

1. A. Ferrari, P.R. Sala, A. Fasso, and J. Ranft, FLUKA: a multi-particle transport code", CERN-2005-10 (2005), INFN/TC 05/11, SLAC-R-773
2. MCNP—A General Monte-Carlo n-Particle Transport Code, version 5.Los Alamos National Laboratory Report LA-CP-03-0245, 2003

Experimental Determination of Photo Fission Cross-Section of ^{232}Th, ^{238}U and ^{237}Np using Electron Accelerator

H.G. Rajprakash[1,2]*, Ganesh Sanjeev[2], K.B. Vijay Kumar[2], H.G. Harish Kumar[3], K. Siddappa[2], B.K. Nayak[4] and A. Saxena[4]

[1]*Department of Physics, J N N College of Engineering, Shimoga*
[2]*Microtron Centre, Department of Studies in Physics, Mangalore University, Mangalagangotri*
[3]*Department of Engineering Physics, S J M Institute of Technology, Chitradurga*
[4]*NPD, Bhabha Atomic Research Centre, Trombay, Mumbai*
E-mail: *prakashraj06@rediffmail.com*

ABSTRACT

Photofission cross-section of ^{232}Th, ^{238}U and ^{237}Np was measured using bremsstrahlung radiation energy 7.4-9.0 MeV by employing Lexan polycarbonate film as Solid State Nuclear Track Detector (SSNTD). The photon intensity from the Microtron accelerator at a distance of 15 cm from the bremsstrahlung converter (tantalum target) facility was estimated to be 10^{10} photons s^{-1} using the code EGS-4. In this paper, Photofission cross-section of ^{232}Th, ^{238}U and ^{237}Np was calculated with the help of fission fragment angular distribution measurement was discussed. The present experimental measurements were compared with the measurement available in EXFOR Library.

Keywords: Fission bremsstrahlung; ^{232}Th, ^{238}U and ^{237}Np; Angular distributions; fission yields; SSNTD; Photofission cross-sections; Photon difference method.

Pacs No.: 25.85.jg

1. INTRODUCTION

Fission fragment angular distribution of even-even and odd-mass targets induced by the bremsstrahlung radiation near fission barrier is one of the promising areas to obtain information on the effect of target ground state spin on transition state distributions. From the excitation energy dependence of the angular distributions, it is possible to characterize the saddle states in terms of energy and quantum numbers J and K of the nuclei [1]. In view of this, systematic studies have been carried out on the measurements of fission fragment angular distribution of bremsstrahlung radiation induced fission of ^{232}Th, ^{238}U and ^{237}Np in the energy range of 7.4 MeV to 9.2 MeV obtained from Microtron at Mangalore University. From the present study, the information about the effect of ground state spin of target nuclei and energy dependent cross-section near threshold has been obtained.

In the present work, photo fission cross-section of ^{232}Th, ^{238}U and ^{237}Np evaluated with the help of fission fragment angular distribution measurements by using bremsstrahlung radiation from 7.4 MeV to 9.0 MeV have been carried out by employing high efficiency SSNTD technique.

2. EXPERIMENTAL METHOD

To study the ^{232}Th system, a 2 mg/cm^2 thick thorium foil is used as a target. The estimation of the ^{232}Th target has been done by weight method. The target is self supporting and can be mounted on a frame. For the ^{238}U, the target was made by depositing 200 µg/cm^2 thick ^{238}U isotope on an aluminium backing. Amount of electrodeposited ^{238}U target on aluminum was determined by X-ray diffraction method. For the ^{237}Np system, the target was made by depositing 150 µg/cm^2 thick ^{237}Np isotope on an aluminium backing. The thickness of electrodeposited ^{237}Np target was measured by alpha spectrometric technique. The fission chamber used in the experiment was 8 cm high and had a diameter of 8cm. The fission fragments were detected using Lexan polycarbonate films placed at different angles with respect to the incident beam direction and the target was irradiated for 5 hrs. After irradiation the Lexan films were cut into equal strips and etched in 6 N NaOH solution for about 1 hr to develop the fission tracks [2]. These etched Lexan films were washed, dried and analyzed under optical microscope with a magnification of 400 X for the tracks formed during irradiation. During Photo fission experiments the possible contamination of the bremsstrahlung beam with secondary electron and neutrons were estimated using EGS-4 code and it was found to be negligible [3].

3. RESULTS AND DISCUSSION

The photo fission cross-section of ^{232}Th, ^{238}U and ^{237}Np measured in the energy range 7.4 to 9.0 MeV is shown in the Fig. 1, 2 and 3. The fission yield for particular electron end point energy is related to fission cross-section $\sigma(E)$ and bremsstrahlung intensity $N(E, E_{max})$ as follows:

$$Y(E_{max}) = C \int_{0}^{E_{max}} \sigma(E) \, N(E, E_{max}) \, dE \tag{1}$$

In the above equation, the 'C' depends on a function of the irradiated photon flux and the irradiated fissile sample volume. The number of fissile nuclei is included in "N"

and the energy dependence is provided by *"dE"*. The resulting fission cross-section is obtained using photon difference method or iterative method or minimum directional discrepancy. In the present work, photon difference method has been employed to calculate the Photofission cross-section. The Photofission yield obtained from the angular distribution measurements in the energy range as a function of Bremsstrahlung end point energies ranging from 7.4 to 9.2 MeV. In the case of ^{232}Th (γ, f) reaction Katz *et al.*, [4] measured Photofission threshold and it was found to be 5.2 MeV. The Photofission yield obtained in the energy range 7.4-9.2 MeV is a contribution of Bremsstrahlung radiation from threshold to end point energy. In the present investigation, measurement was not done below 7.4 MeV and up to threshold. We have compared our results of the yield against Bremsstrahlung end point energy with the results of Rabotnov et al., [5] Photofission cross-section and it was multiplied with the thin Bremsstrahlung irradiator (calculated by using EGS-4 code) [6] was used to obtain the Photofission yield in the energy range 7.4 MeV to threshold. The Photofission yield obtained from Rabotnov *et al.* [5] is folded as discussed by Arruda Neto *et al.* [7]. The total yield curve in the energy range 5.2-9.2 MeV is unfolded using the photon difference method [8]. The resulting Photofission cross-sections in the entire energy range from 5.2 to 9.2 MeV is shown in Fig. 1. The resulting Photofission cross-section of ^{232}Th of the present investigation was compared with the experimental data obtained from EXFOR library [9].

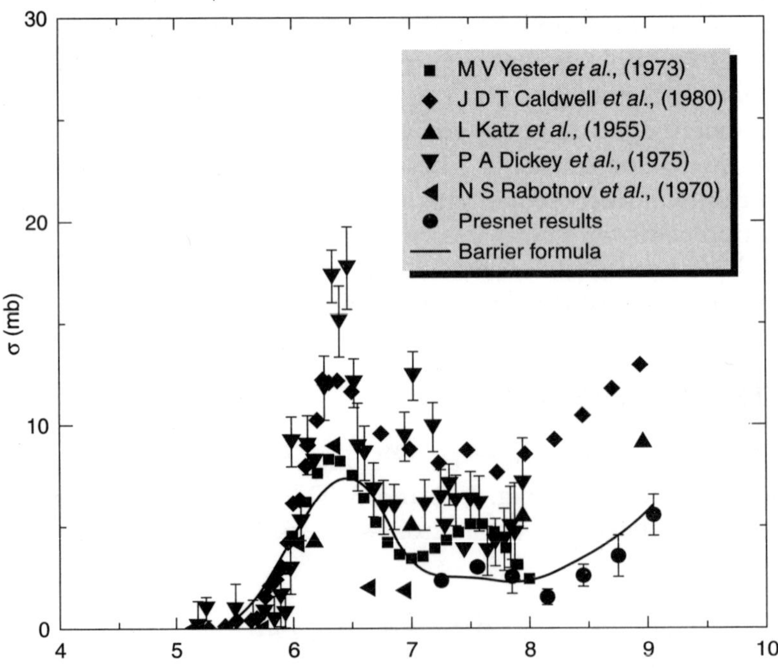

Fig. 1 *Photo fission cross-section of ^{232}Th as a function of photon energy*

From the various analysis carried out, it is clear that for ^{232}Th exhibits almost the same patterns in all investigation. In the present investigation, the resulting Photofission cross-section lies systematically below compared with the data reported earlier. It is expected that the discrepancy in the Bremsstrahlung spectra arises from the difference in the thickness of the target as well as the photon source used in the reported data.

In case of ^{238}U, the Photofission cross-section was evaluated similar to the ^{232}Th. The Photofission yield has been obtained from the measurement of fission fragment angular distribution of ^{238}U in the energy range 7.4 MeV to 9.0 MeV. The Katz *et al.,* [4] found the Photofission threshold of ^{238}U to be 4.6 MeV. The Photofission yield of bremsstrahlung end point energy is the contribution of photons from threshold to end point energy. To evaluate the resulting Photofission cross-section, it is necessary to have the knowledge about the photon spectrum. In the present investigation, measurement data is available in the energy range 7.4 MeV to 9.0 MeV. But below 7.4 MeV to threshold data is not available. Data from the studies carried out by Rabotnov *et al.,* [5] was used for Photofission cross-section measurements for the energy range 7.4 MeV up to threshold and were multiplied with bremsstrahlung spectrum N (E, E_{max}) obtained by EGS-4 code, [7] to obtain the yield curve. The yield curve of the present measurement is in good agreement with the Rabotnov *et al.,* [5] Photofission yield. The total yield curve is unfolded using photon difference method [8]. The measured Photofission cross-section of the present investigation is compared with the results obtained from the EXFOR Library [9] as shown in the Fig. 2. From the figure it was clear that present investigation shows cross-sections almost same pattern as reported earlier.

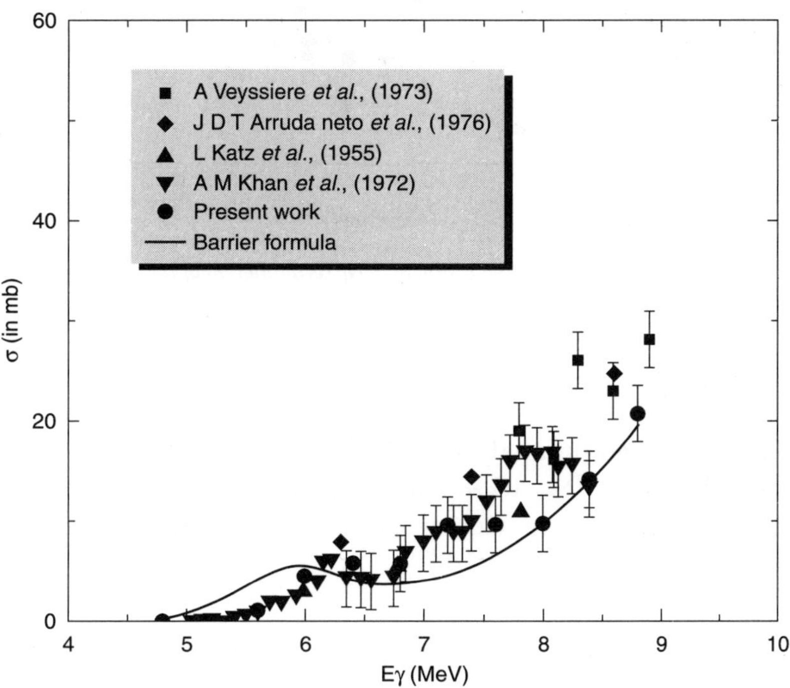

Fig. 2 *Photo fission cross-section of ^{238}U as a function of photon energy.*

The resulting Photofission cross-section of ^{237}Np was evaluated in a similar way to that for the cross-section measurements in the case of ^{232}Th and ^{238}U. Katz *et al.* [4] measured the Photofission threshold for ^{237}Np and found it to be equal to 4.74 MeV. We used the data of Zhuchko *et al.*, [10] for energies below 7.4 MeV to the threshold for the calculation of the Photofission cross-section of ^{237}Np. The Photofission cross-section of Zhuchko *et al.*, [10] is multiplied with the bremsstrahlung spectrum (N (E, E_{max})), obtained using the EGS code [7], to get the Photofission yield. The Photofission yield measured in the present work was folded with the Photofission yield of Zhuchko *et al.*, [10] to obtain a total yield curve. The total curve unfolded by the photon difference method [8]. The resulting Photofission cross-section of ^{237}Np is compared with the results obtained from EXFOR Library is shown in the Fig. 3. From the figures it was concluded that the present experimental results are in good agreement with the results of Katz *et al.*, [11].

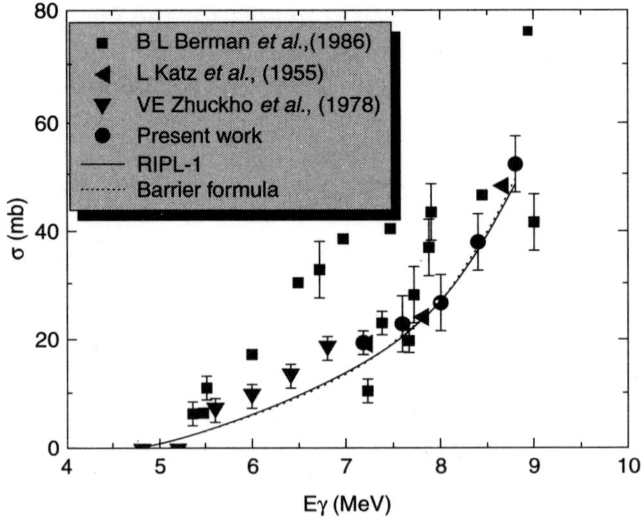

Fig. 3 *Photo fission cross-section of ^{237}Np as a function of photon energy.*

4. CONCLUSION

On the basis of the foregoing analysis it is concluded that the present experimental results of ^{232}Th, ^{238}U and ^{237}Np were compared with the results taken from EXFOR library, consistent with the experiment carried out using Bremsstrahlung gamma rays.

Acknowledgements

We thank Dr S Kailas and Dr R.K Choudhary for their keen interest and various suggestions in the present work. Our thanks are also due to the personnel of Microtron for their help and cooperation during the course of this work. One of the authors, H.G Rajprakash, was thankful to the Board of Research in Nuclear Sciences (BRNS), Department of Atomic Energy, Mumbai, India for providing the financial support.

References

1. Smirenkin G N and Soldatov A S, Phys. Of Atom. Nuclei **59** (1996) 203
2. Fleischer R L *et al.*, Nuclear Tracks in Solids: Principles and Applications. University of California press, Berkeley, 1975.
3. Umakath D *et al.*, Proceedings of National Symposium of Radiation Physics -14, pp 482-485, Gurunanak Dev University, 2001
4. Katz L, Baerg A P and Brown F, Second U.N. Int. Conf. on the Peaceful Uses of Atomic Energy, Vol. **15** (1958) 188
5. Rabotnov N S *et al.*, Sov.J. Nucl. Phys. **11** (1970) 285
6. Nelson W R, Hirayanma H and Rogers D W O, The EGS code system, SLAC-Report-265 (1985)
7. Arruda Neto J D *Tet al.*, Electrofission and Photofission of 238U in the energy range 6-60 MeV, Phys. Rev. C14 (1980) 1499
8. Katz L and Cameron a g w, **The** solution of x-ray activation curves for photonuclear cross-sections, can. j. phys. **29** (1951) 518
9. www.nndc.bnl.gov.in for EXFOR files
10. Zhuchko V.E *et al.*, Study of the Probability of Near-Threshold Fission of the Isotopes of Th, U, Np, Pu, and Am by Bremsstrahlung, Sov. J. Nucl. Phys. **28** (1978) 602
11. Katz, L.*et al.*, 1955. Photofission of U^{238}. Phys. Rev. 99, 98-106.

Projectile Dependence of Inter-shell and Intra-shell Vacancy Transfer Contribution to the L Subshell X-ray Intensity Ratios in Elements; $57 \leq Z \leq 74$

Rohit Thakkar[1], N. Singh[2], P.S.S. Sandhu[2] and K.L. Allawadhi[3,*]

[1]*Ghubaya College of Engineering and Technology, Jalalabad(w), Punjab, India*
[2]*Nuclear Science Laboratories, Physics Department, Punjabi University, Patiala, India*
[3]*Swami Vivekanand Institute of Engineering and Technology, Banur Dist. Patiala Punjab, India*
*E-mail: *allawadhi@yahoo.co.in*

ABSTRACT

The subshell X-ray Intensity Ratios play a very significant role in interpretation of PIXE experiments for non-destructive testing, medical research and trace elemental analysis etc. The accurate measurement of these parameters is, therefore, essential for the correct interpretations of the experimental data. In case of *L* & higher shells, the observed Intensity ratios are not expected to be the intrinsic due to presence of faster intra-shell and the inter-shell transitions. In this work, the contribution of these post-ionization vacancy transfer processes to the *L* subshell intensity ratios and the projectiles dependence of their contribution have been investigated for X-rays induced by protons, deuterons and He ions. It has been observed that both, the intra-shell and inter-shell vacancy transfer processes contribute towards the Intensity ratios, although the quantum of variation induced by these processes depends on both nature and energy of these projectiles.

Keywords: *L* X-ray Emission, Coster-Kronig transitions, Intensity ratios, Inner shell Ionization.

Pacs No.: 25.85.jg

1. INTRODUCTION

The accurate determination of the X-ray intensity ratios plays a very critical role in non-destructive testing of materials, medical research and trace elemental analysis using PIXE techniques. The intensity ratios depend strongly on the subshell vacancy distribution and radiative decay rates. When the vacancies are produced by irradiation of the target of an element by charged particles, the vacancy distribution among the subshells depends upon the mechanisms of excitation of the L subshell electrons and the nature and energy of the incident particles. In case of L & higher shells, the probable shifting of vacancies from the lower subshells (intra-shell)/shells (inter-shell) significantly alter the primary vacancy distribution, thereby changing the 'to be observed' X-ray spectra.

The contribution of the vacancy shifting processes towards the experimental observations has been controlled by some workers in limited cases using techniques like selective ionization of higher subshells. But still, very little work has been done for a detailed insight into the contribution of the vacancies initiated after the primary interaction processes leading to X-ray emission.

In the previous work with photons as projectiles [1-3], attempt has been made by researchers to study contribution/enhancement of some emission parameters by the Coster-Kronig transitions and they have reported a considerable alteration of these parameters in relation with these intra-shell transitions. Our earlier study [4-6], with protons and deuterons as projectiles also indicates a significant contribution of intra-shell transitions to the Subshell Intensity Ratios and X-ray production cross-sections, which depends strongly on incident projectile energy.

In the present work, intra-shell and inter-shell contributions to $I(L_\alpha)/I(L_l)$, $I(L_\alpha)/I(L_\beta)$ and $I(L_\alpha)/I(L_\gamma)$ intensity ratios for X-rays induced by protons, deuterons and He ions have been estimated in elements with $57 \le Z \le 74$. The projectile nature dependence of the contribution of these vacancy rearrangement processes has also been studied.

2. DETERMINATION OF INTENSITY RATIOS

The L subshell X-ray intensities determined by the procedure as explained earlier [4] were used to evaluate the Intensity ratios for the elements $_{57}$La, $_{58}$Ce, $_{59}$Pr, $_{60}$Nd, $_{64}$Gd, $_{66}$Dy, $_{68}$Er, $_{71}$Lu, $_{73}$Ta, $_{74}$W at incident projectile energies ranging from 100 keV to 10 MeV. The intensity ratios $I(L_\alpha)/I(L_l)$, $I(L_\alpha)/I(L_\beta)$ and $I(L_\alpha)/I(L_\gamma)$ have been evaluated for X-ray emission induced by protons, deuterons and He ions. The contributions of the intra-shell (Coster Kronig) and inter-shell (K-to-L_i) transfer processes to the L subshell X-ray Intensity ratios were estimated by applying cut-off for both of these processes separately while evaluating the ratios.

For determination of the intensity ratios, the ECPSSR (Particle Energy Loss (E), Coulomb Deflection (C), Polarization and Binding energy changes of the electrons in a Perturbed Stationary State (PSS) and Relativistic effects (R)) theory based values of ionization cross-sections for K shell and the L_i subshells evaluated for protons and He ions by Cohen and Harrigan [7] and for deuterons by Cohen [8] have been used. For the Coster-Kronig transition probabilities and the sub shell fluorescence yields for different L subshells, the most recent values for $Z \ge 64$ by Campbell's [9] were used. While for elements with $Z < 64$, the values by Krause *et al.* [10] have been taken. The values for the decay rates for different X-ray lines have been taken from the data of Scofield [11]. The K to L_i subshell vacancy transfer probabilities, have been taken from the tables of Rao *et al.* [12].

3. RESULTS AND DISCUSSION

For all the cases under investigation, the L subshell X-ray intensity ratios $I(L_\alpha)/I(L_\beta)$ and $I(L_\alpha)/I(L_\gamma)$ have been found to be energy dependent, while, the intensity ratio $I(L_\alpha)/I(L_l)$ have been observed to be independent of the projectile energy. In general, the intensity ratios $I(L_\alpha)/I(L_\beta)$ and $I(L_\alpha)/I(L_\gamma)$ show strong energy dependence mainly in the lower half of the chosen energy domain, while in most of the cases, they exhibit projectile energy independence for the upper half of energy range chosen in the present work. Among the three projectiles used in the present investigations, the intensity ratios $I(L_\alpha)/I(L_\beta)$ and $I(L_\alpha)/I(L_\gamma)$, in general, are found to be highest for the X-rays induced by protons followed by deuteron and He ion induced X-ray emission respectively.

The maximum contribution of intra-shell transfer to the intensity ratios, in general, is highest for He ions as projectiles followed by deuterons and protons as is evident from Fig. 1(a) for a particular case of $_{74}W$. The contribution of the K to L_i transitions to the L subshell X-ray intensity ratios $I(L_\alpha)/I(L_\beta)$ and $I(L_\alpha)/I(L_\gamma)$ shows variation with the projectile nature and the incident energy in all the elements under investigation; although no regular trend could be inferred upon Fig. 1(b). The inter-K-to-L-shell transfer of vacancies is not found to contribute much for all the L subshell X-ray intensity ratios.

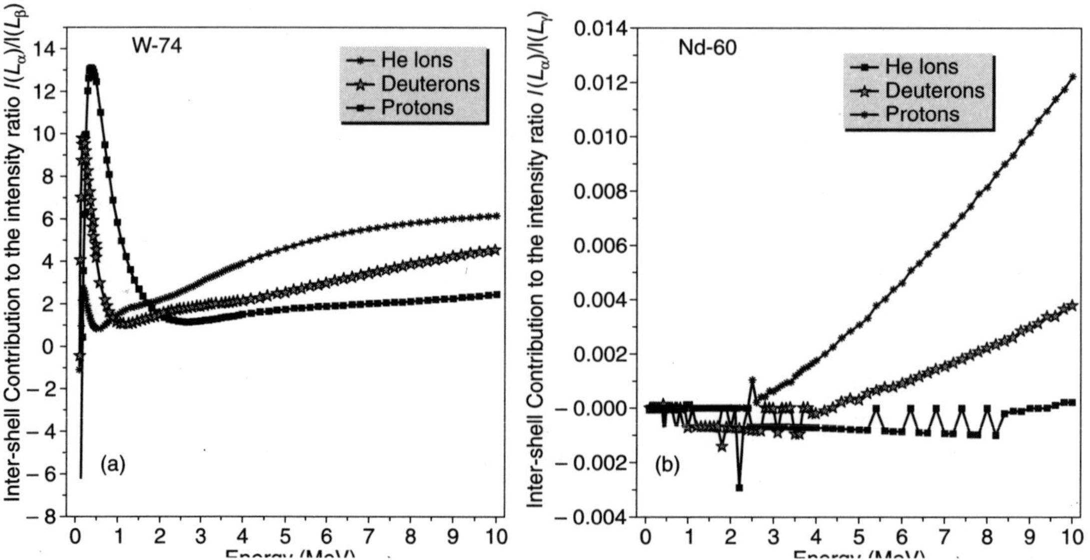

Fig. 1 *Plots of contribution of C K and inter (K-to-Li)-shell transfer processes to L subshell intersity rations Vs He/deuteron/proton energy.*

In nutshell, it may be inferred that the processes initiated after creation of primary vacancies (i.e. intra-shell & inter-shell transfer), whether taken individually or jointly, contribute differently to the intensity ratios for protons, deuterons and He ions, showing dependence on both nature and energy of incident projectile. Further, there are some specific energies for all the three projectiles under investigation, where the contribution of these secondary processes is found to be minimum.

References

1. K.L. Allawadhi, B.S. Sood, Raj Mittal, N. Singh and J.K. Sharma, X-Ray Spectrometry **25**; 233 (1996)
2. E. Oz, N. Ekinci, M. Ertugrul and Y. Sahin, X-Ray Spectrometry **32**; 153 (2003)
3. Mehmet Ertugrul, Appl. Rad. Isot. 57 (1); **63** (2002)
4. Rohit Thakkar, Babita Sharma and K.L. Allawadhi, Radiat. Phys. Chem. **75**; 1482 (2006)
5. Rohit Thakkar, N. Singh and K.L. Allawadhi, Asian J. Chem. **18**, 3283 (2006).
6. Rohit Thakkar, N. Singh and K.L. Allawadhi, Asian J. Chem. **21**, S254 (2009)
7. D.D. Cohen and M. Harrigan, At. Data Nucl. Data Tables 33 (2); 255 (1985)
8. D.D. Cohen, At. Data Nucl. Data Tables **41** (2); 287 (1989)
9. J.L. Campbell, At. Data Nucl. Data Tables **85**; 291 (2003)
11. M.O. Krause, C.W. Nester, C.J. Sparks Jr. and E. Ricci,, Oak Ridge National Laboratoty Report, ORNL - **5399** (1978)
12. J.H. Scofield, At. Data Nucl. Data Tables **14**; 121 (1974)
13. P.V. Rao, M.H. Chen and B. Crasemann, Phys. Rev. A **5**; 997 (1972).

Computation of Bremsstrahlung Cross Section in Nd_2O_3 and $BaTiO_3$

R.S. Madhukeswara[1] H.C. Manjunatha[2*] and B. Rudraswamy[3]

[1]*Government First Grade College, Pandavapura, Mandya district, Karnataka, India*
[2]*Government College for Women, Kolar, Karnataka, India*
[3]*Department of Physics, Bangalore University, Bangalore, Karnataka*
E-mail: *manjunathhc@rediffmail.com

ABSTRACT

A numerical method is used to evaluate External Bremsstrahlung (EB) cross section in paramagnetic substance like Nd_2O_3 and ferroelectric substance like $BaTiO_3$ using tabulated data given for elements. Modified atomic number defined for compound is used to evaluate cross section. The spectral distribution of external Bremsstrahlung excited by $^{90}Sr \rightarrow {}^{90}Y^+ + e^-$ beta particles in these compounds have been evaluated using the computed cross section. The theoretical spectral distribution of Nd_2O_3 is compared with the experimental spectrum measured using 3.8 cm × 3.8 cm NaI(Tl) crystal. The measured spectra show fairly good agreement with the theory at low energy end of spectrum and some deviation (less than 15%) at higher energy end of spectrum.

Keywords: Bremsstrahlung cross section, Ferroelectric, Paramgnetic, NaI detector.

Pacs No.: 23.40.-s

1. INTRODUCTION

Accurate theory for external Bremsstrahlung (EB) has been developed by Tseng-Pratt [1] using the self consistent coulomb field wave function. Seltzer-Berger [2] extended Tseng-Pratt theory for EB produced not only in the field of atomic nucleus but also in the field of an atomic electron and evaluated EB cross section data for various elements. Markowicz *et al.* [3] proposed a new expression for EB intensity and modified atomic number (Z_{mod}) for a compound. Shivaramu [4] estimated the effective atomic number (Z_{eff}) for compounds from the measured EB yields and found that it agrees with Z_{mod}. Bethe-Heitler [5] developed a theory for EB produced in thick target. Manjunatha et al [6 & 7] estimated EB cross section for Bones and compounds used radiation detector. Most of the experimental work on EB spectral study of beta has been carried out with metal as a thick target but with compound as a thick target is lacking. The aim of the present work is to compute the accurate external Bremsstrahlung cross section in in paramagnetic substance like Nd_2O_3 and ferroelectric substance like $BaTiO_3$. The computed cross section was used to calculate the spectral distribution of external Bremsstrahlung excited by $^{90}Sr \rightarrow {}^{90}Y^+ + e^-$ beta particles in Nd_2O_3 and $BaTiO_3$ and it is compared with the experimental spectrum measured using 3.8 cm × 3.8 cm NaI(Tl) crystal.

2. PRESENT WORK

Markowicz *et al.* [3] proposed a new expression to take into account the self absorption of Bremsstrahlung and electron back scattering and to obtain the accurate description of the Bremsstrahlung process

$$I = Const \; \frac{\Delta E}{E_y} \; Z_{mod} \; (E_0 - E_y) \; [1 - f] \tag{1}$$

Here,

$$Z_{mod} = \frac{\sum\limits_{i}^{l} \dfrac{W_i Z_i^2}{A_i}}{\sum\limits_{i}^{l} \dfrac{W_i Z_i}{A_i}} \tag{2}$$

E_γ and E_0 are emitted photon energy and incident electron energy respectively. A_i, W_i and Z_i are atomic weight, weight fraction and atomic number of the ith element in a compound. f is a function of E_0, E_γ and composition (For pure elements $f = 0$). l denotes the number of elements in the compound. *Const* in the equation (1) refers constant. In the present work, it has been evaluated Z_{mod} using Markowicz's equation (2). The estimated Z_{mod} for Neodium Oxide (Nd_2O_3) and Barium Titanate ($BaTiO_4$) are 51.33 and 37.37 respectively. The six elements whose atomic numbers adjacent to that of Nd_2O_3 chosen are Cd, In, Sn, Sb, Te, I and their Z values are 48, 49, 50, 51, 52, 53 respectively. Similarly, six adjacent elements are considered for other two compounds to evaluate EB cross section. The EB cross section for these compounds is evaluated using Lagrange's interpolation technique, Seltzer-Berger [2] theoretical EB cross section data given for elements and the evaluated results of Z_{mod} using the following expression

$$\sigma_{Z_{mod}} = \Sigma \left(\frac{\prod\limits_{Z' \neq Z} (Z_{mod} - Z)}{\prod\limits_{z \neq Z} (z - Z)} \right) \sigma_z \tag{3}$$

where lower case z is the atomic number of the element of known EB cross section σ_z adjacent to the modified atomic number (Z_{mod}) of the compound whose EB cross section $^\sigma Z_{mod}$ is desired and upper case Z are atomic numbers of other elements of known EB cross section adjacent to Z_{mod}. Seltzer *et.al* [2] defined total scaled EB energy weighted cross section as $(\beta^2/Z^2)\, k\, (d\sigma/dk)$, where $\beta = k/T$, k and T are being photon and electron energies in MeV respectively and $(d\sigma/dk)$ is EB cross section per unit energy. The number $n\,(T, k)$ of EB photons of energy k when all of the incident electron energy T completely absorbed in thick target is given by Bethe and Heitler [5] is

$$n\,(T,\,k) = N \int_{1+k}^{T} \left(\frac{\sigma\,(E,\,k)}{(-dE/dx)} \right) dE \qquad (4)$$

where $\sigma\,(E,\,k)$ is EB cross section at photon energy k and electron energy E, N is the number of atoms per unit volume of target and E is the energy of an electron available for an interaction with nucleus of the thick target after it undergoes a loss of energy per unit length $(-dE/dx)$. For a beta emitter with end point energy T_{max}, spectral distribution of EB photons $[S\,(k)]$ is given by

$$S\,(k) = \frac{\displaystyle\int_{T}^{T_{max}} n\,(T,\,k)\,P(T)\,dT}{\displaystyle\int_{T}^{T_{max}} P(T)\,dT} \qquad (5)$$

where $P(T)$ is the beta spectrum of $^{90}Sr \rightarrow {}^{90}Y^+ + e^-$. Evaluated results of $\sigma\,(E, k)$ of eq. (3), tabulated values of $(-dE/dx)$ of Seltzer-Berger [2] was used to get theoretical spectral distribution. The beta source $^{90}Sr \rightarrow {}^{90}Y^+ + e$ was obtained from Bhabha Atomic Research Centre, Bombay, India. The details of experimental arrangement are as explained elsewhere [8]. A 3.8 cm × 3.8 cm NaI (Tl) crystal detector mounted on photomultiplier was coupled to a PC based sophisticated 16 k multi channel analyzer. Target compound such as Nd_2O_3 in the fine powder form were filled in Perspex planchet of 1 cm diameter. The thickness of these compounds was so chosen to stop all the beta particles. The source was placed in a Perspex stand at a distance of 12.5 cm above the face of the detector. The target compound was placed between the detector and the source. A Perspex sheet with thickness sufficient to stop all beta particles is placed on the top of the target compound and with source in position, the spectrum EB + IB + BG was taken. Here IB and BG are internal Bremsstrahlung and background, respectively. The Perspex was then placed below the target compound and the spectrum IB + BG was recorded for the same time. The difference in the two spectra gives Raw EB spectrum. Data were accumulated each time for 12 h. The observed pulse height distribution is the original photon spectrum folded by the response function of the detector system. Hence observed pulse height distribution has been unfolded using the method of Lidden-starfelt [9].

3. RESULTS AND DISCUSSIONS

The estimated EB cross section (Barn/MeV) estimated for various k and T is shown in Figs. 1 and 2. The theoretical EB cross sectional data estimated is found to be higher for Nd$_2$O$_3$ than BaTiO$_4$ at all emitted photon and incident electron energies, which is in accordance with Z^2 dependence of EB cross section. The unfolded measured spectrum obtained for Nd$_2$O$_3$ along with the evaluated theoretical spectrum is shown in Fig. 3 for the energy range 200 – 200 keV. The main contribution to the error in the measured spectra comes from counting statistics. This error is estimated to be varies between 4% at low photon energy, to about 10% at high energy. It is evident from Fig. 3 that the experimental points show closer agreement with theory in the low energy range than in the higher energy end of the spectrum. The positive deviation of experimental point with theory is found to be less than 4.8% at 200 keV, less than 8% at 800 keV, less than 12.3% at 1200 keV and less than 14.5% at 180 keV.

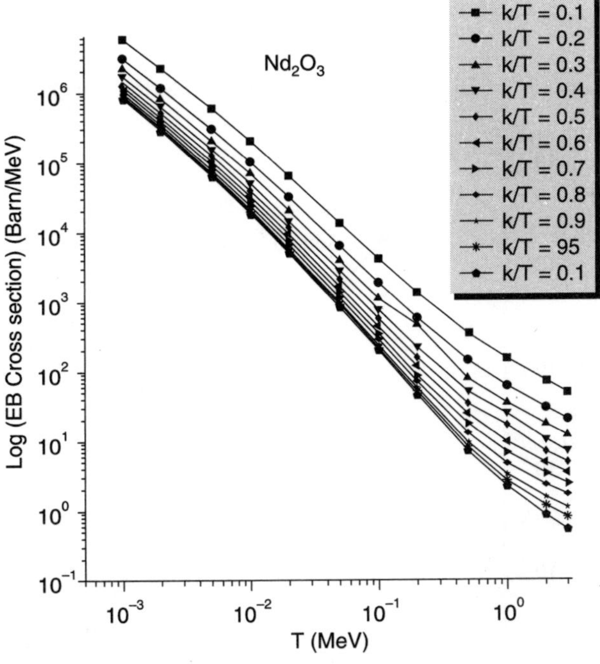

Fig. 1 *Variation of EB cross section with incident electron energy (T) for Nd$_2$O$_3$*

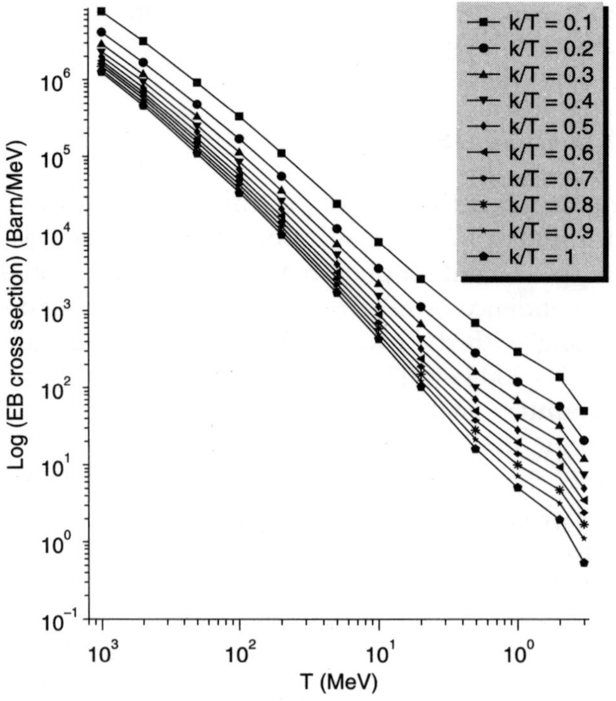

Fig. 2   *Variation of EB cross section with incident electron energy (T) for BaTiO$_4$*

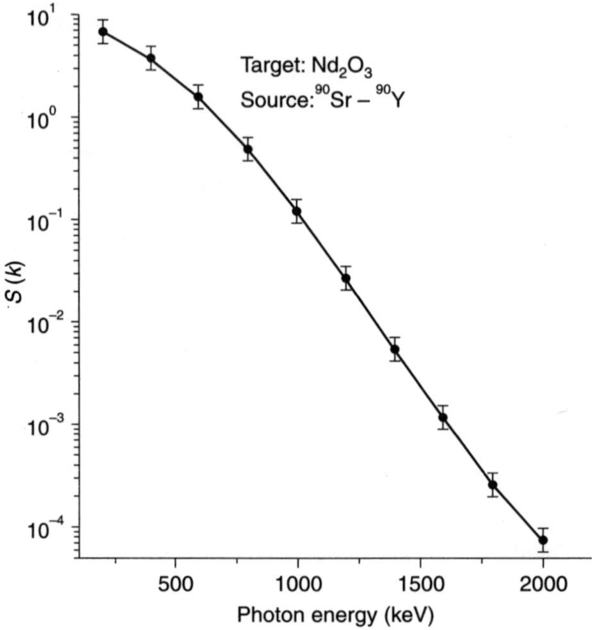

Fig. 3   *Unfolded measured EB spectrum (Circle) with the theoretical distribution (Line) for Nd$_2$O$_3$*

4. CONCLUSION

In conclusion, experimental spectrum show fairly good agreement at low energy end of spectrum and some deviation (less than 14.5%) at higher energy end of spectrum with the theory. Theory can be improved for thick target ferroelectric and paramagnetic compounds by including various 'solid-state effects', namely, multiple scattering, absorption of photons, energy loss of incident electrons and their secondaries and backscattering processes that are inherently present while the Bremsstrahlung photons are emitted from thick targets under bombardment of electrons.

References

1. H.K. Tseng and R.H.Pratt, Phys.Rev.A **3**, 100 (1971)
2. Stephen Seltzer M and Martin. Berger J, At. Data Nucl. Data tables **35**, 345 (1986).
3. Markowicz A.A and VanGriken R.E, Anal. Chem. 56, 2049 (1984).
4. Shivaramu, J. Appl. Phys., **68** (1), 1225 (1990).
5. H.Bethe and W.Heitler, Proc.R.Soc.London, Ser. A **14**, 83 (1934)
6. H.C. Manjunatha, B. Rudraswamy, Appl. Rad. Isot. **65** (4), 397 (2007).
7. H.C. Manjunatha, B. Rudraswamy, Rad. Meas. **42** (2), 251 (2007)
8. H.C. Manjunatha, B. Rudraswamy, Nucl. Instr. and Meth. A **619**, 326 (2010)
9. K. Lidden, N. Starfelt, Phys. Rev. **97**, 419 (1955)

Dose and Residual Activity Estimation for 30 MeV DAE Medical Cyclotron, Kolkata

Sujoy Chatterjee[1*], Tapas Bandyopadhyay[2] and R. Ravishankar[2]

[1]TLD Unit, RPAD, BARC, [2]Health Physics Unit, HPD, BARC
Variable Energy Cyclotron Centre, 1/AF, Bidhan Nagar, Kolkata
E-mail: *sujoy@vecc.gov.in

ABSTRACT

This study has been carried out for the estimation of neutron and photon dose and residual activity for the major radioisotopes for 30 MeV DAE medical Cyclotron which is going to come up soon in Kolkata, India. This will help for radiation protection purposes during normal working of the machine and also for planning its decommissioning at some later stage.

Keywords: Medical cyclotron, Dose, Residual activity Ionization.

1. INTRODUCTION

Medical Cyclotrons are being used for producing various radio-isotopes for diagnostic and therapeutic purpose. The radio-nuclides having very short half-lives and specific decay characteristics are required generally for this purpose. Moreover, these machines are installed in densely populated area. Radiation protection surveillance is of importance from the day of its inception keeping in mind as low as reasonably achievable radiation exposure for radiation workers as well as general public. The planning for decommission of the machine also to be considered during the design stage. The future perspectives of these machines are being considered to extend their applications in research and development in alternate fields such as material science studies due to the possibility of high beam intensity delivery (500 µA or more). The upcoming DAE medical cyclotron (30 MeV proton beam) at Kolkata will cater the need for radiopharmaceuticals of our country and also to provide beam lines for various research and development. The radiation environment around these cyclotron facilities need to be studied very carefully due to high intense proton beams and variety of target materials used in such facilities. An attempt has been made to estimate the dose rate due to prompt secondary neutrons and electromagnetic photons around the target materials like Fe, Cu and Ta of different thicknesses using Monte-Carlo simulation with FLUKA code (Version 2006.3b) [1, 2]. Also an estimation of residual induced radioactivity in materials commonly used for building purposes, such as Concrete have been carried out.

2. MONTE CARLO SIMULATION

A pencil beam of proton is considered at one end of the target having energy of 30 MeV bombarding on thick Cu, Fe and Ta targets. Estimation of neutrons has been done considering continuous energy group cross section data above energy threshold of 19.6 MeV. Below this threshold energy, neutrons are treated by multi-group algorithm. FLUKA has 72 groups for neutron energies starting from 1.00×10^{-11} MeV to 19.6 MeV. Photon interaction cross sections are based on up to date data from EPDL97 database. The variation of the neutron spectral fluence, at distances of 2.0 cm from the Cu and Fe targets, for target thickness of 1.0 cm has been estimated. 2×10^8 number of histories was initiated for the simulation. The variations of photon and neutron dose rates due to prompt photons and neutrons as well as photons from induced activity, at different distances from the farthest edge of the proton irradiated target have been obtained. The dose rates are estimated for 500 m*A of beam intensity of proton for DAE Medical Cyclotron, Kolkata. A computer subroutine along with the Monte Carlo code was used to obtain equivalent dose rate evaluation base on conversion factor recommended by ICRP [3]. The residual induced activities simulated in surrounding Concrete materials due to prompt radiations have been estimated using FLUKA Code when Cu target has been bombarded with 30 MeV protons.

3. RESULTS AND DISCUSSION

Figures 1 and 2 show the spectral distribution for neutrons which shows a local maximum around 800 keV for Fe and 1.65 MeV for Cu target. Fig. 3 and 4 shows the spectral distribution of photons for Cu and Fe target bombarded with 30 MeV protons. The variation of the dose-rates shows exponential reduction with distance that has been shown in Fig. 5 and 6 for neutron and photons from Cu target of 1 cm thickness. Induced activity is mainly dominated for ^{24}Na, ^{28}Al and ^{56}Fe in Concrete for the case of Cu target irradiation with accelerated proton beam.

The estimated radioisotopes are found to be in order with those reported in literature for proton and neutron induced activities for the materials considered [4, 5, and 6]. Residual activity produced has been shown in Table 1.

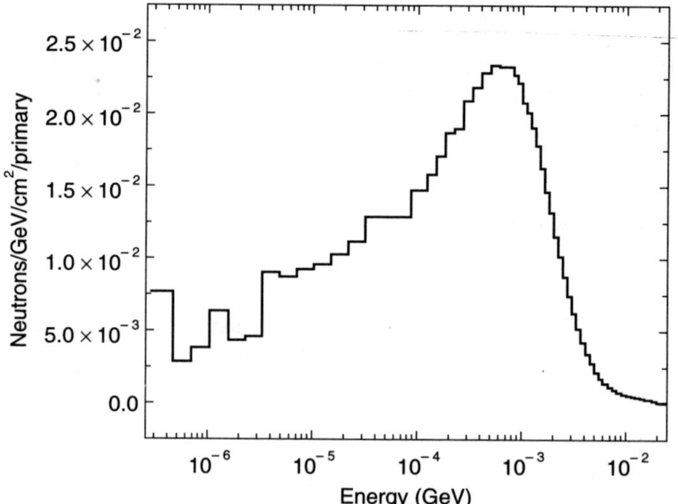

Fig. 1   *Shows the spectral distribution of neutrons from 1 cm thick Cu target bombarded with 30 MeV protons.*

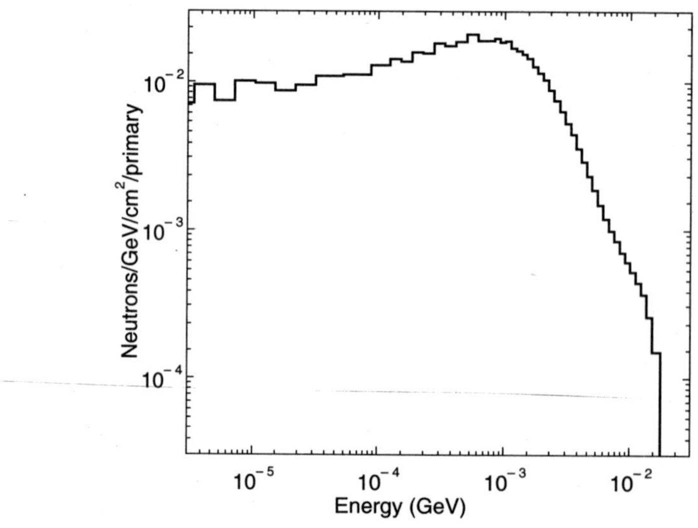

Fig. 2   *Shows the spectral distribution of neutrons from 1 cm thick Fe target bombarded with 30 MeV protons.*

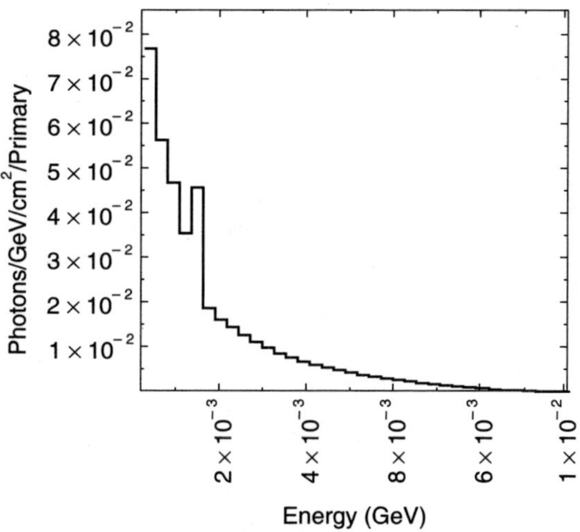

Fig. 3   *Shows the spectral variation of photons from 1 cm thick Cu bombarded with 30 MeV Protons*

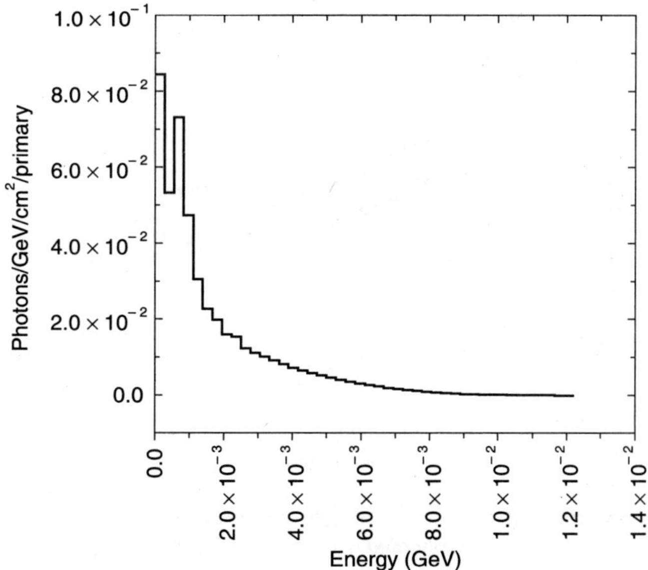

Fig. 4   *Shows the spectral variation of photons from 1 cm thick Fe bombarded with 30 MeV protons*

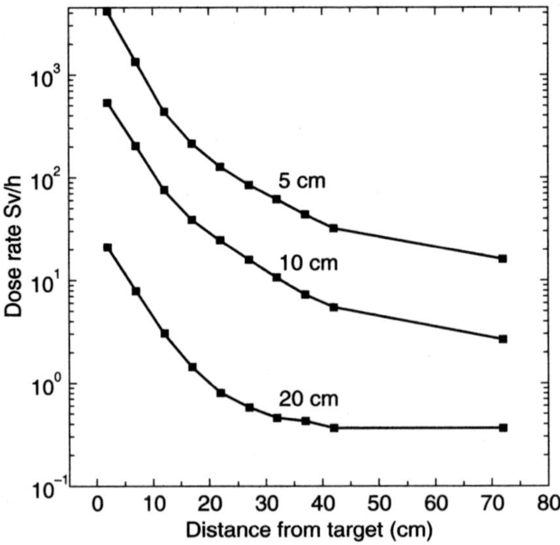

Fig. 5 *Shows the photon dose rate at different distances from 1 cm thick Cu target bombarded with 30 MeV proton*

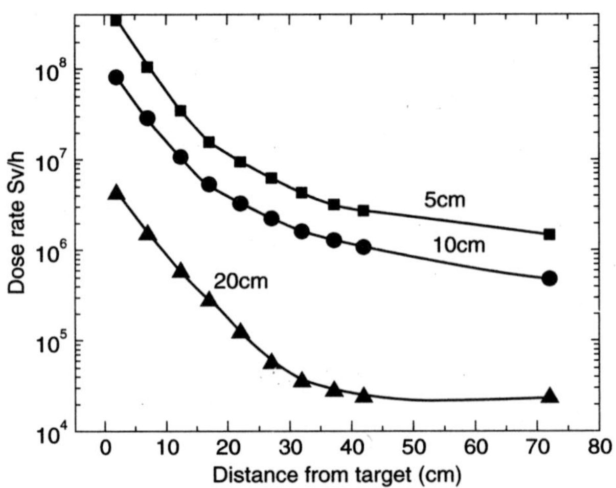

Fig. 6 *Shows the neutron dose rate at diffe-rent distances from 1 cm thick Cu target bombarded with 30 MeV proton*

Table 1 *Shows the major residual nuclei produced in concrete when Cu target is bombarded with 30 MeV protons.*

Residual nuclei produced	No of atoms/cm^3/proton
^{24}Na	1.4858×10^{-07}
^{28}Al	1.7886×10^{-06}
^{56}Fe	4.2724×10^{-07}

Acknowledgements

Authors are thankful to Dr. R.K. Bhandari, Director, VECC, Dr. D.N. Sharma, Associate Director, HSEG, Dr. P.K. Sarkar, Head, HPD, and Dr. Y.S. Mayya, Head, RPAD of BARC for their constant encouragement during this study.

References

1. A. Fasso, A. Ferrari, J. Ranft, and P.R. Sala, "FLUKA: a multi-particle transport code", CERN-2005-10 (2005), INFN/TC_05/11, SLAC-R-773.

2. A. Fasso`, A. Ferrari, S. Roesler, P.R. Sala, G. Battistoni, F. Cerutti, E. Gadioli, M.V. Garzelli, F. Ballarini, A. Ottolenghi, A. Empl and J. Ranft,. "The physics models of FLUKA: status and recent developments", Computing in High Energy and Nuclear Physics 2003 Conference (CHEP2003), La Jolla, CA, USA, March 24-28, 2003, (paper MOMT005), eConf C0303241 (2003), arXiv:hep-ph/0306267.

3. ICRP Publication 74–Dose Conversion Factors (1995).

4. NCRP Report No. 144–Radiation Protection for Particle Accelerator Facilities (2003).

5. DAE Medical Cyclotron Safety report, VECC, Kolkata.

6. Private Communication with Project Manager, DAE Medical Cyclotron, Kolkata.

Production Cross-sections of Radio Nuclides from Fe-54 and Fe-56

P.K. Saran[1], Maitreyee Nandy[2], P.K. Sarkar[3] and Sneh Lata Goyal[1,*]

[1]*Department of Applied Physics, Guru Jambheshwar University of Science & Technology.,
Hisar, Haryana, India*
[2]*Saha Institute of Nuclear Physics, Kolkata, India*
[3]*Health Physics Division, Bhabha Atomic Research Centre, Mumbai, India*
E-mail: *goyalsneh@yahoo.com

ABSTRACT

Neutron induced reaction cross-sections of ^{54}Fe and ^{56}Fe have been estimated for the production of different radio nuclides viz. ^{53}Fe, ^{54}Mn, ^{55}Fe and ^{56}Mn from 1 – 50 MeV neutrons. The excitation functions of these reactions are calculated using the codes EMPIRE-2.19, TALYS-1.0 and ALICE-91. The codes account for the major nuclear reaction mechanisms, including direct, pre-equilibrium and compound nuclear ones. The excitation functions of these isotopes have been compared graphically along with the available experimental data and the results are more or less agreeing up to which energy the experimental data are available.

Keywords: Evaporation model, Pre-equilibrium emission, Exciton model, Geometry dependent hybrid model, Hauser–feshbach theory

Pacs No.: 25.40-h; 28.20-V

1. INTRODUCTION

Activation cross-sections for the production of radio nuclides are of interest for testing nuclear reaction models. Furthermore, in the case of structural materials of a fission reactors, accelerators and accelerator driven systems, the data are important for the estimation of neutron multiplication, nuclear heating, nuclear transmutation and radiation damage effects. Nuclear cross-section data are obtained mostly by nuclear physics experiments and also by nuclear theory. Calculations based on the statistical compound nucleus model, on pre-equilibrium decay models and on direct reaction models are increasingly used to assess cross-sections for nuclear reactions which are difficult to measure.

The present work deals with the evaluation of excitation functions for the production of radio nuclides by $(n, 2n)$, & (n, p) reactions for 1 to 50 MeV neutrons on ^{54}Fe and ^{56}Fe using the computer codes EMPIRE-2.19, TALYS-1.0 and ALICE-91.

2. NUCLEAR REACTION MODEL CALCULATIONS

The EMPIRE-2.19 code [1] accounts for the major nuclear reaction mechanism, including direct, pre-equilibrium (PEQ) and compound nuclear reactions. The secondary compound nuclei (CN) are formed due to subsequent particle emission. The only difference is that although the first CN is initially excited to the unique (incident channel compatible) energy, the secondary CNs are created with excitation energies that spread over the available energy interval. The transmission coefficients are estimated from the optical model subroutine SCAT2. Binding energies are determined using masses recommended by Audi *et al.* [2] whenever available, otherwise theoretical predictions of Moller and Nix [3] are used. The code uses several models to calculate PEQ emissions and the statistical Hauser–Feshbach theory to describe the compound nuclear emissions.

In TALYS [4] code direct reactions are calculated using any one of spherical optical model, DWBA, rotational or vibrational coupled channel analysis and giant resonances. Two component exciton model estimates the PEQ particle emission and the angular distribution of these PEQ particles is determined using Kalbach systematics. Compound nuclear emission is calculated in the framework of Hauser-Feshbach formalism in competition to fission.

For comparison, the cross section values were also estimated using the computer code ALICE-91 given by Blann [5-7]. This code accounts for pre-compound and compound/statistical calculations in general framework of the Weisskopf Ewing evaporation model and hybrid/geometry-dependent hybrid model (GDHM) for pre-compound decay. Here, the computations have been performed using GDHM model. According to this model, the nucleus has a density distribution which can affect PEQ decay in two ways. First, the nucleon mean free path is expected to be longer (on average about a factor of two) in the diffuse nuclear surface. Secondly, in a local density approximation, there is a limit to the hole depth. These two changes were incorporated into the geometry-dependent hybrid model. In the present work we have used optical model inverse cross-section and Fermi gas level density options. The level density parameter $a = A/9$ which is default option of the code. The parameters have been chosen such as to select an evaporation calculation without fission. For pre-compound decay, the initial exciton number is taken as $n = 3$. The binding energies and Q-values used in the present work were all based on the experimental masses. The ALICE code includes experimental masses in block data.

3. RESULTS AND DISCUSSION

In this study the production cross-sections of ^{53}Fe, ^{54}Mn, ^{55}Fe & ^{56}Mn were calculated using the nuclear reaction model codes EMPIRE-2.19, TALYS-1.0 and ALICE-91 for incident neutron energies in the range 1 – 50 MeV. The code EMPIRE was used with different PEQ mechanisms like multistep direct (MSD), multistep compound (MSC) and Hybrid Monte-Carlo Simulation (HMS) approach to the emission of nucleons. With different PEQ models different level density formalisms were used viz.

1. PEQ: MSD+MSC, level density (LEVDEN = 0): EMPIRE specific level densities i.e. BCS + Fermi gas with deformation dependent collective effects, adjusted to experimental "a" values and to discrete levels (EMPIRE – MSD + MSC 0).
2. PEQ: MSD+MSC, level density (LEVDEN = 1): Fermi gas with deformation dependent collective effects and "a" parameters derived from the shell model (EMPIRE-MSD+MSC 1).
3. PEQ: HMS, level density (LEVDEN = 0): EMPIRE specific level densities i.e. BCS + Fermi gas with deformation dependent collective effects, adjusted to experimental "a" values and to discrete levels (EMPIRE – HMS 0).
4. PEQ: HMS, level density (LEVDEN = 1): Fermi gas with deformation dependent collective effects and "a" parameters derived from the shell model (EMPIRE – HMS 1).

The excitation functions of ^{53}Fe & ^{54}Mn from n + ^{54}Fe reaction and of ^{55}Fe & ^{56}Mn from n + ^{56}Fe reaction have been shown in Figs. 1-4 along with the available experimental data [8]. For the reactions under consideration the experimental data are available up to about 20 MeV neutron energy. The cross-sections computed using codes EMPIRE, TALYS and ALICE are more or less agreeing with the experimental data up to which energy the experimental data are available. Beyond that region the HMS model in EMPIRE gives the results which are more near to the results with TALYS and ALICE codes with both the level density options used. The MSD + MSC model incorporating EMPIRE specific level density i.e. BCS + Fermi gas level density with deformation dependent collective effects (EMPIRE–MSD + MSC 0) yields higher cross-sections for the production of ^{54}Mn and ^{56}Mn for neutron energies above 20 MeV and for the production of ^{53}Fe and ^{55}Fe for neutron energies above 35 MeV as compared to the results of other codes, whereas, MSD + MSC models using Fermi gas level density with deformation dependent collective effects (EMPIRE-MSD + MSC 1) gives consistently low production cross-sections as compared to TALYS and ALICE codes in all the reactions under study.

The ALICE code for the production of ^{53}Fe and ^{55}Fe by $(n, 2n)$ reaction gives the results which are far away from the results of EMPIRE and TALYS codes. This difference in results may be due to the difference in the evaporation models used. A strong influence of PEQ mechanism on the production cross-sections is observed by the difference in the results of EMPIRE code with two PEQ models used.

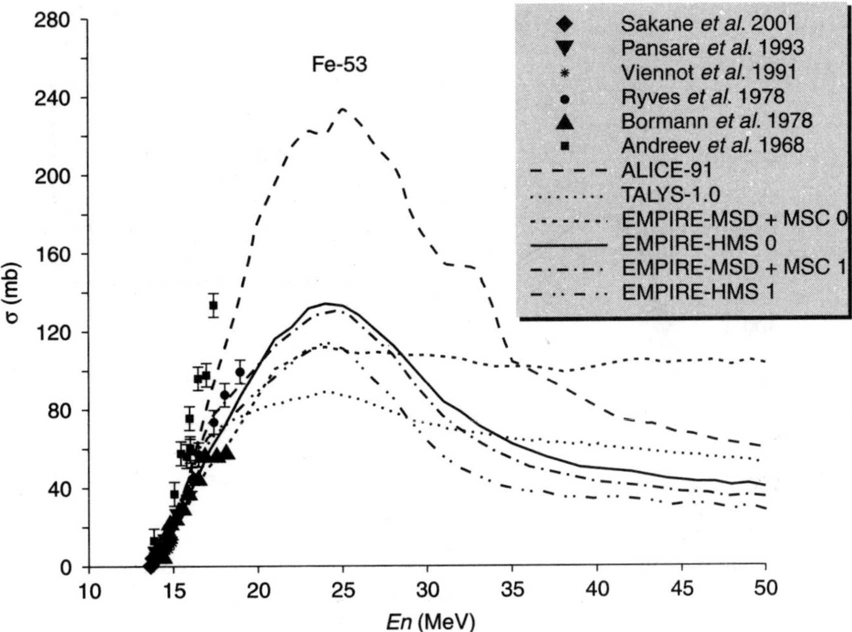

Fig. 1   *Excitation function of Fe-53 from (n + Fe-54)*

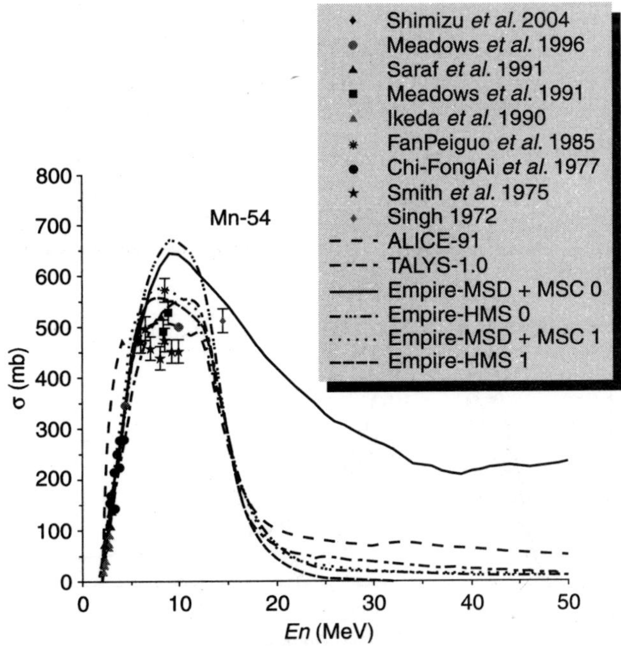

Fig. 2   *Excitation function of Mn-54 from (n + Fe-54)*

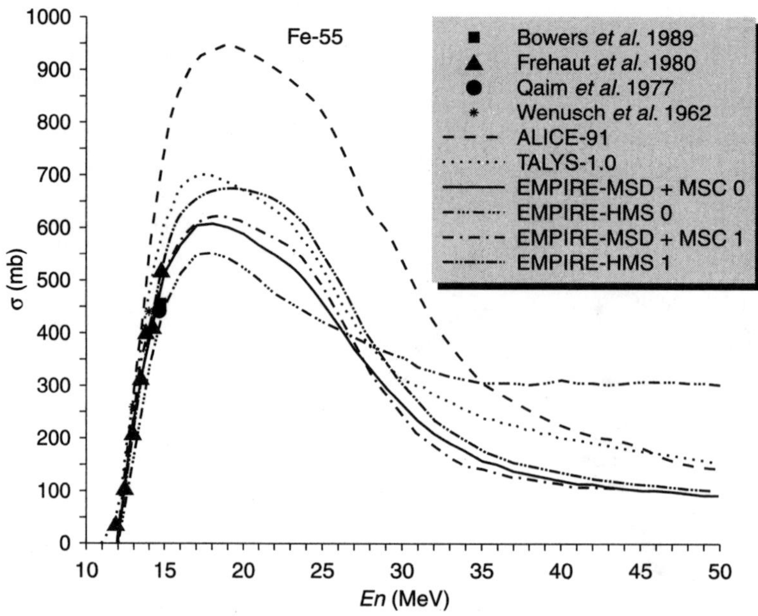

Fig. 3 *Excitation function of Fe-55 from (n + Fe-56)*

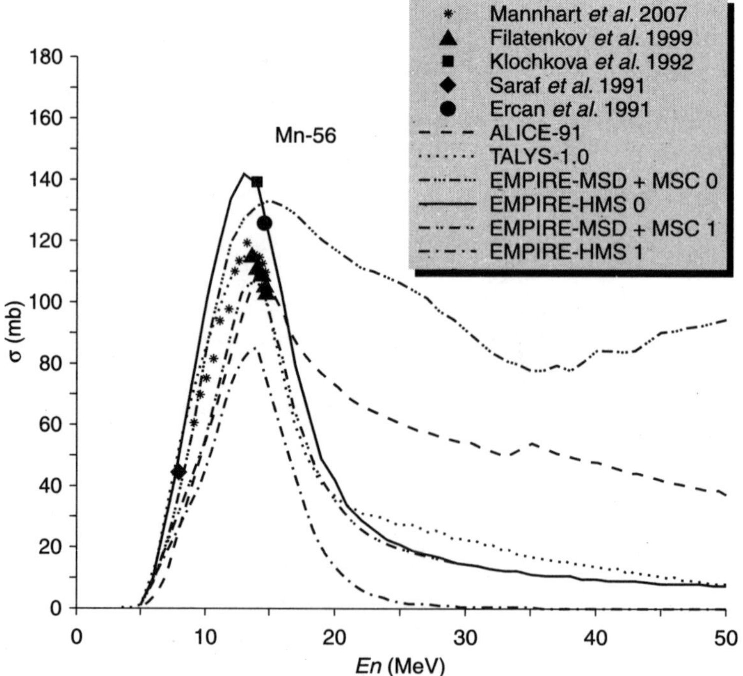

Fig. 4 *Excitation function of Mn-56 from (n + Fe-56)*

Acknowledgements

Financial support from DAE-BRNS, BARC, Mumbai by giving major research project is thankfully acknowledged.

References

1. M Herman EMPIRE nuclear reaction model code (2.19 Lodi) (IAEA, Vienna, Austria) (22 March, 2005)

2. G Audi, A H Wapstra Nucl. Phys. **A595** 409 (1995)

3. P Moller, J R Nix At. Data Nucl. Data Tables **26** 165 (1981)

4. A J Koning S Hilaire and M C Duijvestjjn TALYS: Comprehensive nuclear reaction modeling. In Proceedings of the International Conference on Nuclear Data for Science and Technology—ND2004 AIP vol. 769 (eds) R C Haight, M B Chadwick, T Kawano and P Talou, Sep. 26–Oct.1, 2004(Santa Fe, USA), p 1154–1159 (2005)

5. M Blann *Recent Progress and Current Status of Preequilibrium Reaction Theories and Computer Code ALICE* UCRL-JC10905 Z (Lawrence Livermore National Laboratory, California, USA) (1991)

6. M Blann *Code ALICE/85/300* (Lawrence Livermore National Laboratory, Report NO. UCID-20169) (1985)

7. M Blann and H K Vonach *Phys. Rev.* **C28** 1475 (1983)

8. www.nndc.bnl.gov/exfor/exfor00.htm Experimental Nuclear Reaction Data (EXFOR/CSISRS) Database version of October 11, 2010

Estimation of Secondary Particles from 200 MeV/u ^{12}C Ions Bombarding a Water Phantom using FLUKA

Sunil C.[1*], Wissmann F.[2] and P.K. Sarkar[1]

[1]*Health Physics Division, Bhabha Atomic Research Centre, India*
[2]*PTB, Bundesalle 110, 38116 Braunschweig, Germany*
E-mail: *sunilc@barc.gov.in*

ABSTRACT

The secondary particle fluence and ambient dose equivalent has been estimated from 200 MeV/u ^{12}C beam incident on a water phantom using the FLUKA Monte Carlo code and compared with the experimental results carried out at GSI. The results indicate that charged particles contribute more to the dose equivalent in the forward direction while neutron dose equivalent in seen to dominate in the lateral directions

Keywords: Heavy ion, Secondary particles, Dose estimation

Pacs no.: 87.53.Bn, 87.53.Bn, 87.15.ak

1. INTRODUCTION

Particle radiation therapy is being increasingly put to use in several countries. Carbon ion therapy facilities are operational in HIMAC and HIBMC Japan while such facilities are planned in CNAO Italy, Heidelberg Germany and Gunma University Japan. There is a growing concern of latent secondary cancer risks produced as a result of secondary particle interaction [1] outside the treatment volume. With typical beam energies of few 100 MeV/u, charged particles are found to dominate in the forward direction while neutron particles are dominant at backward angles. While there are several double differential yield measurements reported for common target materials at projectile energies similar to that used in ion therapy, very few such measurements have been carried out from a water phantom simulating a human body bombarded by the projectiles as would happen in a beam therapy scenario. Recently micro dosimetric technique has been applied to investigate the charged and neutral particle dose using a TEPC based detector PI-DOS (PTB-Inflight DOSemeter) developed by PTB [2], while

WENDI neutron dose equivalent monitor was used to measure the neutron dose equivalent resulting from a water phantom irradiated by 200 MeV/u 12C projectiles [3]. The double differential energy distributions of several particles emitted from a water phantom irradiated by 200 MeV/u 12C ions have been measured experimentally using TOF technique at GSI [4].

In this work, the experimental setup of TEPC detector PI-DOS is simulated to estimate the secondary particle yields and the dose equivalents as is measured by it. The simulations are carried out using the FLUKA Monte Carlo code [5]. The neutron fluence obtained from the simulation is compared with the TOF measurements carried out at GSI.

2. SIMULATION

FLUKA [5] Monte Carlo code is used to model the geometry and simulate the secondary particles produced from the beam-phantom interaction. The phantom is considered to be a circular water column of dimensions 18 cm height and 15 cm diameter. The beam was incident on the circular surface of the cylinder centered at the height of the cylinder. The detector center is kept at 2.0 m distance from the center of the water phantom. In this simulation, the detector is taken to be the innermost sphere and scoring is done inside this volume at 0°, 5°, 10°, 20°, 30° and 90° with respect to the beam direction. For calculations at different angles, the beam direction was changed while keeping the detector setup constant. The DPMJET and RQMD models in FLUKA were invoked for heavy ion reactions with explicit transport of heavy charged particles. USRTRACK track length estimators were used to the score the fluence inside the detector volumes and the deq99c user sub routine was used to fold the fluence with the ICRP fluence to dose conversion coefficients through AUXSCORE card. The USRBIN estimator was used to obtain the fluence and energy absorption gradients in rectangular three dimension plots. Energy cutoffs were not applied while scoring. Differential energy spectra were estimated using three different bin sizes for all particles to obtain better statistics. Thus, particles of energy from 20 MeV to 120 MeV were scored in 20 bins of 5 MeV width, while energy ranging from 120 MeV to 240 MeV were scored in a bin width of 10 MeV and particles above 240 MeV were scored with bin size of 50 MeV width. In addition, a separate scoring estimator was also utilized to obtain the full energy range and to fold it with the conversion coefficients. This ensures no loss of information in the estimated dose values. The spread in beam energy and beam dimensions are considered to be negligible and are not taken into consideration in the present calculations. Five separate runs with different random number seeds with cumulative number of histories of 10^8 were used to estimate the results and the associated uncertainties.

3. RESULTS AND DISCUSSION

The neutron dose equivalent measured using the πDOS detector and by using the WENDI neutron dose monitor and the results obtained from the FLUKA calculations are shown in Fig. 1 [6].

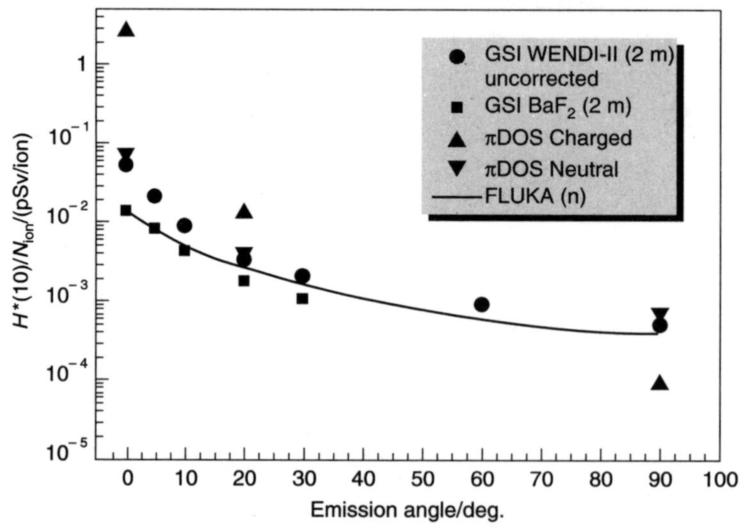

Fig. 1 *The neutron dose equivalent obtained from FLUKA calculations, WENDI measurements and the neutral dose measured by the πDOS developed by PTB.*

It can be seen the result from FLUKA calculations and the measurements agree very well at 20°, 30° and 90°. The results, however, disagree for forward angles smaller than 20°. There, the calculation agrees much better with BaF$_2$ data measured at GSI [4]. This might be explained by the increased fluence of the charged particles at forward angles (mainly protons and He isotopes), which produce secondary neutrons in the surroundings of πDOS leading to an increase in the measured dose rate in the neutral mode. The experimental results from πDOS plotted in figure 1 have been corrected for the charged particle signals. Alpha particles, tritons and 3 He particles are found to have higher energies when compared to the light charged particles. From Fig. 2, it can be seen that the yield of light charged particle like proton and deuterons fall less rapidly with increasing angles when compared to the heavier charged particles such as alpha particles,^{3}He and tritons. However, the yields of heavier charged particles are higher when compared to the light charged particles in the forward directions.

The higher energy and yield of alpha particles in the forward direction may influence the response of the detectors at these angles. The energies of these particles are such that they can travel 2.0 meters in air and yet retain sufficient energy to produce events in the counters. In the case of WENDI, the thin layer of high Z material

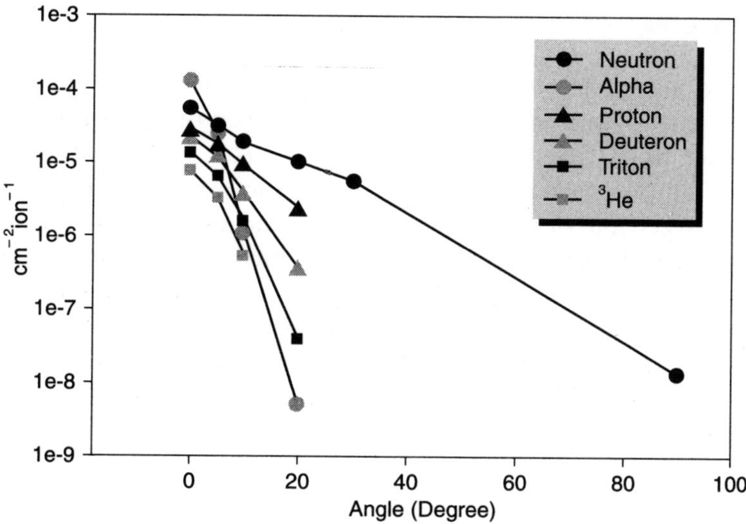

Fig. 2 *The angular distribution yields of secondary particles obtained by FLUKA calculation.*

used to obtain response for high energy neutrons will cut off such particles of higher energies. Thus, the differences in WENDI results are smaller than the PI-DOS counters. The effect of charged particles on the response of these instruments will have to be further investigated and quantified for an accurate and precise correction. Such a correction in the TEPC readings is expected to reduce the difference in the forward directions. The neutron spectra obtained from the present FLUKA calculations are also compared with the experimentally measured values from GSI. The result so obtained at 0° is shown in Fig. 3. Here it can be seen that the spectrum at 0° calculated by FLUKA agree with the experimental measurements. However, there are discrepancies observed at other angles.

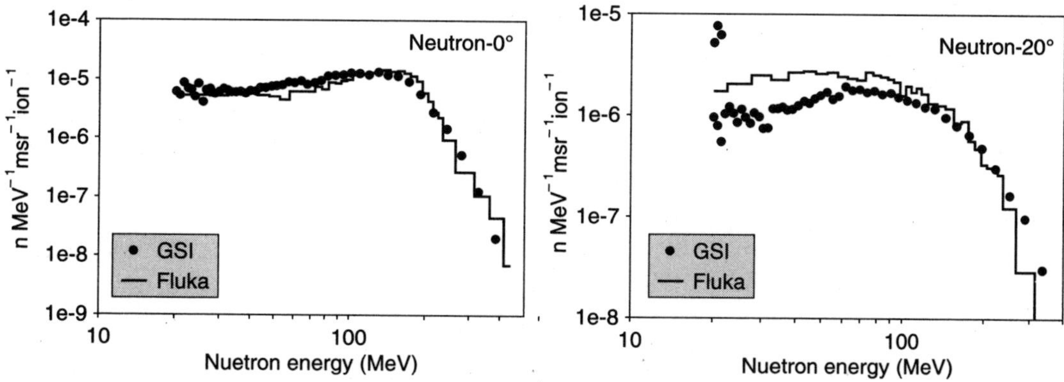

Fig. 3 *The neutron spectra obtained at 0° compared with the experimental data. The closed circles are data obtained from the GSI experiment while the lines are FLUKA simulations.*

4. CONCLUSION

The neutron dose equivalent estimated using the FLUKA Monte Carlo code is in agreement with the measurements carried out using the tissue equivalent proportional counter developed at PTB and the WENDI neutron dose equivalent monitor at 20°, 30° and 90°. The results however disagree in the extreme forward angles such as 0° and 20°. The charged particle contribution is found to be substantial in the forward direction as compared to the backward angles, with alpha particles dominating the yield. The high energy and yield of alpha particles can result in events in the detectors. This has a direct bearing on the outcome of the neutral particle dose measured by TEPC and WENDI detectors and hence needs to be further investigated. The neutron fluence and the total dose are found to be highly forward peaked. The neutron energy spectra calculated by FLUKA are found to agree with the measured spectra at 0° while there are discrepancies at other angles.

References

1. P. J. Taddei, *et al, Phys. Med. Biol.,* **53** 2131-2147, (2008)
2. F Wissmann., *et al Radiat. Prot. Dosim.* **110** 347-349, (2004)
3. H. K. Iwase, *et al., Radiat. Prot. Dosim.* **126** 615–618, (2007)
4. K Gunzert-Marx, *et al New Journal of Physics* **10**, 075003, (2008)
5. A Fasso, *et al* FLUKA: a multi-particle transport code, CERN-2005-10. (2005)
6. F. Wissmann *et al., Radiat Environ Biophys.* **49**, 331–336, (2010)

Status of In-house Development of Compact Electrostatic Deuteron Accelerator for Neutron Generation

B.K. Das[*1], A. Shyam[1], R. Das[1], and A.D.P. Rao[2]

[1]*Energetics and Electromagnetics Division, Bhabha Atomic Research Centre,*
Autonagar, Visakhapatnam, India
[2]*Department of Nuclear Physics, Andhra University, Visakhapatnam, India*
E-mail: *dasbabu31@gmail.com*

ABSTRACT

In recent years, due to specific features of compact neutron generators, their demand in elemental analysis and imaging of the illicit materials has been increased in scientific community. Compact in size, controlled operation and radiation safety like features of neutron generator are suitable for research work with illicit materials. An accelerator based neutron generator can be operated in steady mode as well as in pulse mode. The main embodiment of this type of generator includes ion source, ion acceleration system, target etc. Such type of neutron generator is under development in our laboratory. In this paper, various physics and technical issues related to the important components of this generator and the progress at our laboratory is discussed.

Keywords: Neutron generator, ion source, high voltage vacuum insulation

Pacs No.: 29.25.Dz, 07.77.Ka, 84.70.+p

1. INTRODUCTION

In the growing scenario of terrorism throughout the world, restriction of the movements of the illicit materials in air, sea and road routes become very important. Imaging of the cargo container, luggage bags is not enough but it becomes essential to do elemental analysis of all the containers. For this purpose, a Non-Destructive Testing is required which can detect the elements inside the container without doing any physical change to the present objects. Prompt Gamma Activation Analysis (PGAA) in the form of Thermal Neutron Analysis (TNA) and Fast Neutron Analysis (FNA) [1, 2] are promising techniques in which, the nature of the element as well as the quantity present inside the container can be detected. For this purpose, production of neutrons plays an important role. Compact neutron generator has several advantages over other type of neutron sources like nuclear reactors and radioisotope sources. In case of neutron generator, there is not any safety criticality, can be operated in steady and pulse mode, no radiation hazard when switched off and easily transportable. We are developing one compact neutron generator in our laboratory. The neutrons can be produced from D-D or D-T reactions. For both the reactions high-energy deuterium ions are required. The schematic diagram of a typical electrostatic compact neutron generator is shown in Fig. 1. The various embodiments of this generator are described in next sections.

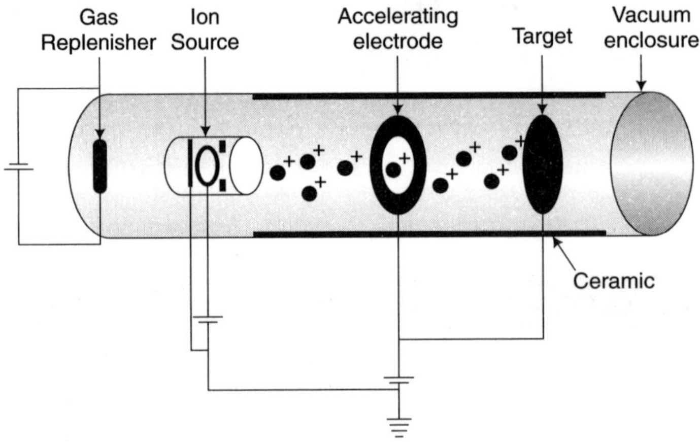

Fig. 1 *Schematic diagram of the accelerator*

2. FEATURES OF THE NEUTRON GENERATOR

The main features of the neutron generator includes ion source-that creates deuterium ions, ion acceleration optics - that accelerates the ions up to the required energy and the target where the neutrons are generated due to nuclear interaction of the high energetic deuterium ions and the neutral deuterium which is present in the metal target.

2.1 Ion Source

Formation and extraction of ions are important issues. Penning ion source [3] is suitable because of simple structure, easy instrumentation, filament less operation and long life. One compact size penning ion source was developed for this purpose. The detail description of this source has been described elsewhere [4]. Plasma was created in between two cathode plates and one anode placed in between them. Deuterium plasma was created in pressure range of 1×10^{-4} torr to 5×10^{-5} torr, at a potential difference of 2 KV between the cathodes and the anode and a magnetic field of 500 gauss along the axis produced by one permanent magnet. This ion source acts as self-extracted ion source. The ions produced in the plasma region come out through one aperture made in one of the cathode plate known as the plasma electrode. The extracted ion current was measured by one faraday cup. The variation of the extracted ion current with respect to the deuterium pressure is shown Fig. 2. At a distance of 70 mm from the extraction aperture we had measured a deuteron current of 60 μA. This much of deuteron current is sufficient to produce a neutron flux of 10^9 n/s from D-T reactions.

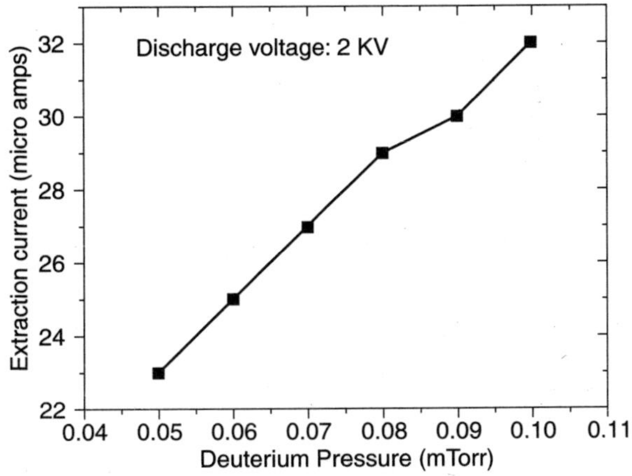

Fig. 2 *Extraction deuteron current*

2.2 Acceleration of Ions

Once the ions are extracted from the ion source, it is required to accelerate them to the required energy. For acceleration of the ions from the extraction aperture, one single electrode is placed at a gap of ~70 mm. The high voltage breakdown between the grounded plasma electrode and the acceleration electrode is important. The high voltage breakdown in vacuum gap depends upon various factors like, the gap distance, the area, geometry, material and surface condition of the electrode, gas pressure and the nature of the gas. The relation between the breakdown voltage, the gap length and the effective surface area are given in equations (1) and (2) respectively

$$V_{\text{max}} = kd^n \tag{1}$$

Where V_{max} is the maximum breakdown voltage, k is a constant that depends upon the electrode material and gap length. The value of k is given as 312 [5], d is the gap length and n is a constant that depends upon the pressure range. For pressure of 1×10^{-4} torr, $n = 0.11$ [6].

$$V_{\text{max}} = k \left[\frac{A_{eff}}{\text{cm}^2} \right]^n \tag{2}$$

Where V_{max} is the maximum breakdown voltage, k and n are material and geometrical constants. The values are $k = 342$ KV and $n = 0.16$ [7].

Considering these entire factors, electrodes have been made out of stainless steel. The surface has been elcetropolished with roughness of < 1 µm. The high voltage strength of the electrode arrangement was examined for high voltage up to 80 KV.

3. CONCLUSION

The compact deuterium accelerator has been developed. This accelerator can generate few hundred microamperes of deuterium ion with ion energy up to 80 KeV. The ion extraction, acceleration and vacuum insulation tests were performed. In future experiments, we will endeavour for production of neutrons by using different targets like deuteriated, tritiated solid targets or pure reactive targets like titanium. As the neutron yield from D-T reaction is 100 times higher than D-D reaction, this generator in its present form can produce neutron flux in the order of 10^9 n/sec from D-T reactions.

References

1. A. Buffer, Radiation Physics and Chemistry **71** (2004) 853–861
2. G. Vourvopoulos, NIMB, **89** (1994), 388-393
3. *Ion Sources*, Huashun Zhang, Springer, Berlin, 1999
4. B. K. Das and Anurag Shyam, Review of Scientific Instruments, **79**, 12, 123305, (2008)
5. Kustom R. L., *J. Appl. Phys.*, **41**, 3256-3268 (1970)
6. Akira Yamamoto, Akhihiro Maki and Yutaka Maniwa, *Japanese J. of Appl. Phys.* Vol 16, No. **2** (1977), 343-354
7. U Sch⇐mann and M Kurrat, 20th International Symposium on Discharges and Electrical Insulation in Vacuum, Tours, France, June 30 – July **5**, 2002

Construction of Beamline Radiation Shielding Hutches for Indus-2 Synchrotron Radiation Source

Sanjay Chouksey[1*], G. Haridas Nair[2], V.G. Sathe[1], Vishal Dhamgya[1], M. Jaganath[2], A.K. Sinha[1], G.S. Lodha[1] and Gurnam Singh[1]

[1]*Raja Ramanna Centre for Advanced Technology, Indore (MP), India,*
[2]*Bhabha Atomic Research Centre, Mumbai, India*
E-mail: *chouksey@rrcat.gov.in*

ABSTRACT

Indus-2 is an electron synchrotron radiation source. The Synchrotron Radiation (SR) photons are transported to the experimental station through twenty-six beamlines. The radiation environment around the beamline mainly comprises of Bremsstrahlung Radiation (BR) and Sychrotron Radiation (SR). Radiation shielded hutches are required for the protection for personnel against these radiation hazards. Standardization of the hutch design and fabrication process was done. This paper discusses hutch layout design, standardization of fabrication, shielding philosophy, material selection, quality checks, installation and subsequent optimization based on radiation measurements.

Keywords: Indus-2, Synchrotron radiation beam lines, Beam line hutches, Bremsstrahlung, Ozone, Modular design, Shielding

1. INTRODUCTION

Indus-2 is a 2.5 GeV electron synchrotron radiation source (SRS)[1], which provides a spectrum of photons from Infrared to X-ray region. The Synchrotron Radiation (SR) photons are transported to the experimental station through appropriate beamlines. Twenty-six beamlines are planned in Indus-2. The SRS is housed inside a 1.5 m thick concrete shielding wall which has penetrations for installation of beamlines to tap the SR from 0°, 5° & 10° ports of each bending magnet. The radiation environment around the beamline comprises of Bremsstrahlung Radiation (BR) and Synchrotron Radiation (SR)[2]. This can be either direct or scattered. When SR beam is brought out in a beamline, shielded hutch (encloses the entire beamline) is required for the protection for personnel against the radiation hazards. Main objective of hutch design is to keep radiation levels in working areas within stipulated limits set by Atomic Energy Regulatory Board (AERB) under all machine operation conditions. Despite having large variation in the requirements of different beam lines, individual components for hutches of SR beamlines have been standardized to a large extent to facilitate quick design and installation of the beamline hutches.

2. LAYOUT DESIGN

Basic configuration of the hutch layout has divergence geometry, starting from the ring wall towards an experimental station. The general rule is that the dividing line for adjacent beamlines is the bisector between two beamline axes. After several iterations, a layout was finalized to accommodate whole beamline in two hutches. The first hutch is called "Optics Hutch" placed closed to the shielding wall which houses optical components of the beam line. The second hutch is called "Experimental Hutch" which houses experimental station. Longer beamlines have an "Intermediate Hutch" between the two Fig. 1. Flexibility, compactness, accessibility are some of the important design features of the hutches. Wherever there is a severe space limitation, a common sliding door is provided in the optics hutch wall to allow easy access during maintenance.

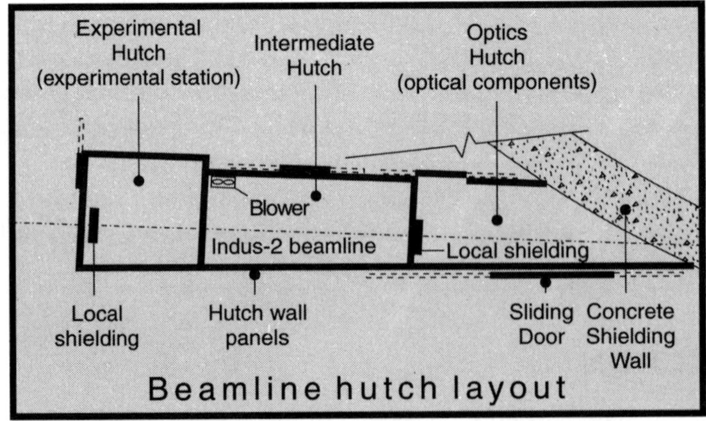

Fig. 1 *Typical beamline hutch layout on Indus-2 SRS*

3. DESIGN PHILOSOPHY

The radiation shielding is designed in such a way that outside the shield, dose rate is less than 1 µSv/h for normally accessible area. This is based on the shielding philosophy recommended by Atomic Energy Regulatory Board[3].

4. ASSESSMENT OF THE SHIELD THICKNESS

In order to determine the shielding requirement for both SR and BR shielding calculations [4] and preliminary experiments were also performed. LiF Thermo Luminescent Dosimeters (TLD) sandwiched between copper absorbers were used in the direct SR beam at the end of a front end. The SR and BR passing through a thin beryllium window was allowed to fall on the TLD-Cu absorber stack. Fig. 2 shows the absorbed dose measured using LiF TLD embedded in copper absorbers normalized to stored beam current. The dose rate with no absorber gives the direct SR dose rate normalized to beam current for the experimental condition, 7.5 mA @ 2.0 GeV. It can be seen that at a thickness of 3 mm the SR is totally getting absorbed and the dose rate shown beyond is due to bremsstrahlung x-rays. Improvement in vacuum in the ring is expected to reduce the BR dose rate. An equivalent thickness of lead shielding for the sidewalls adopted is – 1 mm/2 mm lead sndwitched between Mild steel(MS) panels, perpendicular walls – 3 mm lead in MS panels and roof – 3 mm aluminum sheet. The direct gas BR is stopped by local shielding of 100 mm thick interlocking lead bricks each, behind Double Crystal Monochromator (DCM) in the optics hutch and at the end of the experimental hutch. BR stop thickness is obtained based on the calculations suggested by G.Tromba[5] on gas bremsstrahlung.

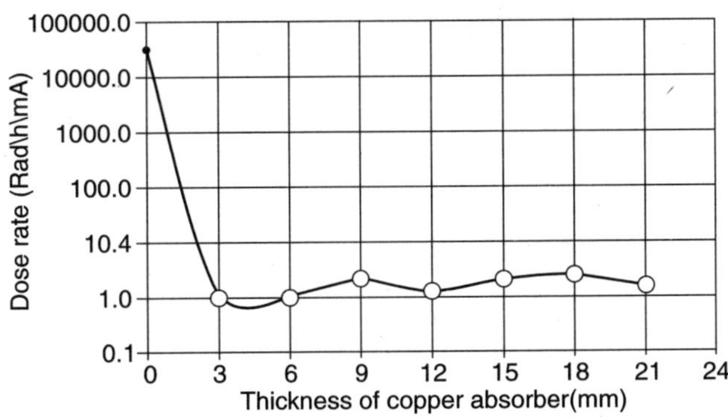

Fig. 2 *Absorbed dose profile in copper absorbers measured with LiF TLDs*

5. FABRICATION

The main characteristic of the standard hutch is its modular design. The shielding hutches are fabricated out of steel/lead sheet panel in modular construction with a provision for augmentation of shield based on prevailing radiation levels. Lead sheet of required thickness is sandwiched between both the panels. The hutches are built with full height 2.5 meter × 1.25 meter panels. A few special panels are included to match the geometry of a given hutch. The material used in the construction of the outer and inner panels and frame supports are of cold rolled closed annealed (CRCA) steel sheet. Horizontal and vertical structural members are fabricated out of 3 mm thick CRCA steel sheets. The floor structure members are grouted on to the floor by M 10 anchor bolts. Outer panel is fabricated out of a 1.2 mm CRCA steel sheet. They are folded at all edges with holes at desired locations for mounting on vertical and horizontal members. Using various tools, jigs, fixtures and dies during manufacturing, ensure interchangeability of structural members. CNC CO2 LASER cutting and bending of sheet metal results in perfect form of sections. The roof of the hutches are removable, to allow the installation of the heavy beamline equipments using overhead crane. CRCA steel panels and structures are powder coated for surface protection against rusting. Manually operated sliding doors with safety interlocks are provided for personnel and material access Fig. 3.

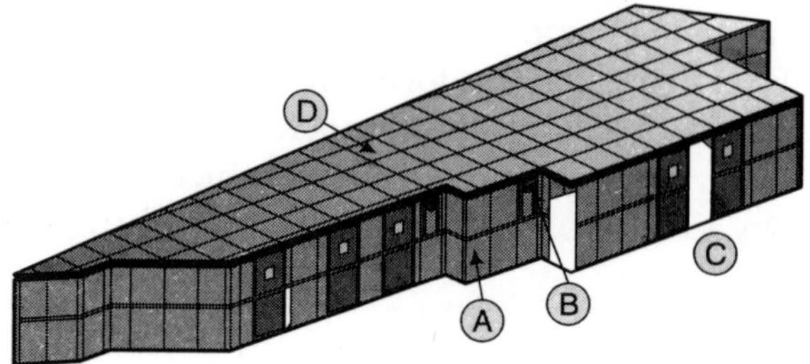

Fig. 3 *The schematic view of a radiation shielding hutch in Indus-2. (A) Hutch panel (B) Exhaust fans (C) Sliding doors (D) Roof panels*

6. QUALITY ASSURANCE

Adopting good engineering practices from design till installation ensured radiation tightness at the joints. Qualification of surface coating was done as per relevant ASTM standards. Dimensional check, density check, visual inspection and finally radiometry test were carried out to ensure the quality of lead bricks.

7. SERVICES

Instrument quality compressed air and chilled water are delivered to each hutch. Roof mounted exhaust fans are provided to take care of the ventilation due to ozone production. Provisions are made for lighting fixtures inside the hutches. Two tier perforated cable trays, one for power cables and another for signal cables, are provided on top of a hutch.

8. CONCLUSION

Designing proper radiation shielding hutches in a confined area and its installation with severe constraints was a challenge as this work was of its first kind in our country. Installation for six SR beamline hutches has already been completed. Effectiveness of the radiation shielding has been validated by radiation survey done by Health Physicists.

Acknowledgements

Authors express sincere thanks to Dr. P.D. Gupta, Director, RRCAT, for guidance and encouragement in the work. Thanks are due to Shri Jafar Ali for preparing AUTOCAD drawings. The authors are grateful to Dr. S.K. Deb for his support in implementation of shielding arrangement.

References

1. Technical report of synchrotron radiation source, Indus-2 (Pub Sep 1998)
2. N. Ipe *et al.* -Guide to beamline radiation shielding design at the advanced photon source (APS report/ANL/APS/TB_7, SLAC TN 93_5)
3. Safety Directive no. 2, 1991, Atomic Energy Regulatory Board, India (1991)
4. Haridas. G *et al.* Technical note on radiation shielding at synchrotron radiation beamline (Bending magnet source) hutches of Indus-2 SRS (April 2010)
5. G. Tromba *et.al*, Nulcl.Inst. Methods A 292 700-705 (1990).

Development of an Accelerator based Intense 14-MeV Neutron Generator for Fusion Neutronics Experiments at IPR

Sudhirsinh Vala[1*], Shrichand Jakhar[1], Mitul Abhangi[1], Rajnikant Makwana[1], C.V.S. Rao[1] and T.K. Basu[2]

[1]Institute for Plasma Research, Bhat, Gandhingar, India
[2]Raja Ramanna Fellow of DAE
E-mail: *Sudhir@ipr.res.in

ABSTRACT

An accelerator based 14-MeV neutron generator for neutronics studies related to fusion reactor is under development at the Institute for Plasma Research. Neutrons are generated from the nuclear reaction $T(d, n)\alpha$ by bombarding a solid tritium target with deuterons accelerated up to 300 keV in an electrostatic accelerator. The 14-MeV neutron generator consists of an ECR ion source, beam extraction system, accelerating column, vacuum pumping system, beam profile monitor, beam steerer, faraday cup, a high-voltage power supply and an isolation transformer. The ECR ion source, beam extraction system and the required power supply units are kept on the high voltage deck, which is at 300 KV floating potential. The input electric power to the entire power supply units is provided through an isolation transformer. Vacuum of the order of 10–7 mbar is maintained inside the beam line by using a 550 LPS turbo molecular pump. This paper describes the status of 14-MeV neutron generator. With a deuteron beam current of 1 mA, the expected neutron yield is 1011 n/s.

Keywords: Accelerator, 14-MeV neutron generator, ECR ion source

Pacs No.: B-03

1. INTRODUCTION

Neutronics play an important role in the nuclear technology of D-T fusion reactor. This is due to the fact that 80% of the primary energy generated in the burning plasma is released to the surrounding blanket region as kinetic energy of the 14-MeV neutrons.

In order to study neutronics of fusion reactor blankets, a program is underway at the Institute for Plasma Research using 14-MeV neutron source. The 14-MeV neutron generator will be used to develop techniques to measure nuclear responses in the blanket assembly, such as tritium breeding, neutron and gamma ray heating, shielding effect etc. and compare them with the calculated values. This paper describes the main components of 14-MeV neutron generator and the result of beam current measurement.

2. 14-MEV NEUTRON GENERATOR

14-MeV neutrons are generated from the nuclear reaction ^{3}H (d, n) ^{4}He by bombarding a solid tritium target with deuterons accelerated up to 300 keV in an electrostatic accelerator. With a deuteron beam current of 1 mA the expected neutron yield is 10^{11} n/s. The partly assembled experimental set up of 14-MeV neutron generator is shown in Fig. 1. The neutron generator is composed of six separate sections: ion source, bending magnet, acceleration tube, vacuum system, beam diagnostic system and target assembly. The ion source, extraction system, the 90° bending magnet and its high voltage power supply are kept on a high voltage deck, which is at 300 KV floating with respect to the ground potential. The input electric power to the entire power supply unit is provided through an isolation transformer. The beam diagnostic equipments, vacuum pump and the target are at ground potential. The 14-MeV neutron generator design parameters are given in Table 1.

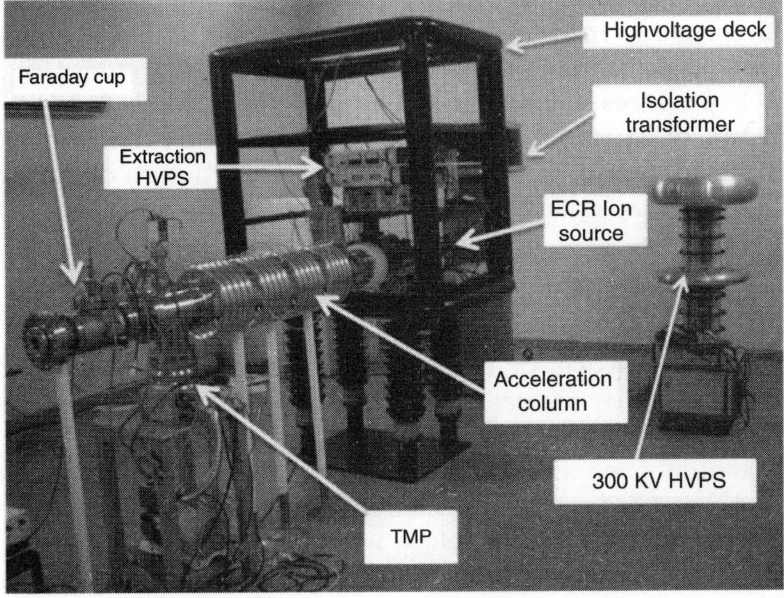

Fig. 1 *Partly assembled 14-MeV neutron generator*

Table 1 *Design parameters of 14-MeV neutron generator*

Particular	Parameter
Beam Energy maximum	300 KeV
Type of Ion Source	ECR Ion source
D^+ Beam Current at the target	1 mA
Beam Spot size	~15 mm
Tritium Target Activity	20 Ci
Target Diameter	35 mm
Neutron yield	10^{11} n/s

2.1 Ion Source

The deuterium plasma is produced in an ECR ion source and it is capable of delivering 300 W power at 2.45 GHz frequency. The advantage of the ECR ion source is that it does not have any filament so a stable ion beam can be produced for longer duration. The deuterium plasma is confined in a region where there is superposition of an axial magnetic field and radial magnetic field of multi-pole magnet [1]. Deuterium ion beam is extracted from the ECR ion source at an extraction voltage of 20 kV then it is focused by an einzel lens. The extraction and focusing systems are important parts of the neutron generator. They have been designed using SIMION 8 ion optics code [2]. The design of the focusing system is optimized to obtain maximum ion beam current.

2.2 90° Bending Magnet

A 90° double focusing magnet is used to separate D+ from molecular ions. The bending magnet is a C-Shaped 90-degree, double focusing, DC dipole electromagnet. It produces a maximum magnetic field of 0.2 T, which corresponds to a Larmor radius of 24 cm [3]. As the magnet would be placed and operated on a high voltage platform, special care has been taken in the design of the magnet to have minimum weight, good acceptance and moderate mass resolution. The magnet and the power supply are air-cooled.

2.3 Acceleration Tube

The acceleration tube consists of four sections of vacuum extensions with shaped electrodes fitted inside. They are designed for ultrahigh vacuum and each section of the accelerating column is rated for 75 kV in air and 200 kV in SF_6. A series of shielding electrodes are mounted on the inside of the tube to prevent the inner ceramics surface from ion bombardments and contaminations. A multiple series of resistor string in air is used to distribute the electrostatic potential uniformly along the tube. The holding voltage of tube is 300 KV in air. The high voltage power supply unit is a commercial Crockroft & Walton type with a maximum current of 10 mA.

2.4 Vacuum System

One 550 l/s turbo-molecular pump with oil free scroll pump is connected to the exit of the accelerating tube to achieve 10^{-7} mbar in the acceleration tube. The exhausts of vacuum pump will be swept into a tritium monitoring system.

2.5 Beam Diagnostic System

After 300 kV acceleration column, a beam steerer, beam profile monitor and faraday cup is used for ion beam diagnostic. Beam steerer is used to steer the accelerated ion beam in X- & Y-direction perpendicular to beam axis. Beam Profile Monitor is diagnostic equipment that measures the intensity distribution and the position of the ion beam. Faraday cup is used to measure the ion beam current.

2.5.1 *Beam Steerer*

The beam steerer is kept at ground potential and is placed just after the 300 keV Accelerating tube. It is four rectangular plate electrodes assembly. Two parallel plates are required to steer the D+ ion beam in X-direction and remaining two are for steering the beam in Y-direction perpendicular to previous direction [4]. All the four rectangular plate electrodes are connected to four MHV ceramic feed-through, which are connected to 5 kV DC power supply placed outside.

2.5.2 *Beam Profile Monitor*

The Beam profile monitor (BPM) is a device that measures the intensity distribution and position of a beam of charge particle. A single wire formed in to a 45° helix is rotated about the axis of the helix in vacuum at frequency of about 18 cps. It sweeps across the beam in two orthogonal directions in every cycle. The secondary electron current released from the wire as it intercepts the beam is a measure of the beam intensity at every instant. The two resulting signals are then displayed on an oscilloscope. The BPM has a nominal 6.0" (15 cm) housing and a molybdenum beam entrance aperture with a diameter of 2.75" (6.98 cm). The cross-sectional diameter of the molybdenum scanning wire is 1.5 mm. The BPM is ultra high vacuum compatible. A magnetic coupling is used to transfer motion into the vacuum system for the rotating scanning wire [5]. There are no sliding seals.

2.5.3 *Faraday Cup*

The purpose of the faraday cup is to measure the accelerated deuterium beam current accurately. The cup is made of Tantalum having 25 mm diameter and operated through electro pneumatic system. The cone is electrically insulated from the cooling system and the cooling is accomplished by radiation to a surrounding cooling coil.

It intercepts the beam during it's IN position and ion current striking on it is measured by a log amplifier connected through BNC feed through. Secondary electrons are induced when beam strikes on the cup. A negatively biased suppressor electrode placed near the cup suppresses the secondary electrons [6].

2.6 Target Assembly

The Tritium target is used to produce the 14-MeV neutrons from $T(D, n)$ ^{4}He fusion reaction. The Tritium loaded on titanium fixed on a 45 mm diameter and 1 mm thick copper substrate is used as a target. The target is placed inside a target holder which is coupled with a drift tube at the end of the accelerator. When D+ beam strikes the target and produce heat which is extracted by a closed loop chilled water system.

3. CONCLUSION

The testing of the ECR ion source on high voltage deck as well as testing of the required high voltage power supply (up to 60 kV) and isolation transformer have been completed. The design of extraction system was done by using SIMION-8 ion optic code. The testing of extraction system is complete and an ion beam current of 746 µA at 18 kV could be achieved and it will be further optimized for maximum ion beam current. Primary testing of beam line components like Acceleration column, Beam Steerer, Beam profile monitor, and Faraday cup have also been completed. Design of the 90° bending magnet has been done by using OPERA-3D and its engineering design is being finalized. The experiments for the diagnostic low energy ion beam are in progress.

References

1. M. Schlapp, R. Trassl, M. liehr and E. Salzborn, IEEE (Particle accelerator conference, 1997), Vol 3, Page: 2708-2710
2. SIMION Version 8.0 user manual
3. U. Lakshminarayana, Manu Bajpai, Sudhirsinh Vala, Shrichand Jakhar, Mukti Ranjan Jana, C.V.S. Rao, Subrata Das, Ritesh Malik, and S.S. Prabhu proceeding in ISARP-11
4. NEC Electrostatic ion Beam steerer user manual
5. NEC BPM 83 user manual
6. NEC FC 18 user manual

Energy Dispersive XRF Method to Study the Chemical Shift in X-ray Energies

Kamaldip Kaur and Raj Mittal

Nuclear Science Laboratories, Physics Department, Punjabi University, Patiala
E-mail: *polu_gre@yahoo.co.in, rmsingla@yahoo.com*

ABSTRACT

Energy dispersive X-ray fluorescence set up comprising of low power X-ray tube and Si (PIN) detector has been used for chemical shift measurements in X-ray energies following a method that involves the division of measurement time into sub-divisions and in each sub-division XRF spectrum of two compounds of an element are, in turn, obtained. A polynomial fit to the determined photo-peak energies from the spectra removes the statistical fluctuations of the data. The results are corrected afterwards to remove the non-statistical errors. The method was used for chemical shift measurements of K x-ray energies of Fe, Ca and K in Fe and $FeSO_4.7\,H_2O$; $CaCO_3$ and $CaSO_4.2H_2O$ and KBr and KNO_3 compounds. The results show energy shifts 2.74 ± 0.40 eV, 0.98 ± 0.68 eV and 1.95 ± 0.74 eV respectively for Fe, Ca and K.

Keywords: XRF, Chemical shift, $FeSO_4$

1. INTRODUCTION

Generally, in energy measurements, the fluctuations occur in the experimental data due to counting statistics and non-statistical effects. The statistical errors can be reduced either by increasing the counting rate or by increasing the counting time. The non-statistical error in position of X-ray emission line is due to fluctuations caused by the environmental factors, electric apparatus instability and some unpredictable factors. The energy dispersive X-ray spectrometers with energy resolution in the range 150 eV – 250 eV at 5.9 keV are not suitable to measure the chemical shifts (~ a few eV) [1] of characteristic X-ray emission line. Xiao et al [2] has introduced an EDXRF method for measurements of chemical shift after theoretically exploring its possibility. The method copes with the problem of non-statistical errors and has been used to measure energy shifts in a limited number of cases. The work presented in this paper involves a similar study of chemical shifts using low-power X-ray tube source for excitation of samples and Si (PIN) detector having resolution of 210 eV at 5.9 keV for the detection of emitted X-ray lines.

2. PROCEDURE

According to the method, the measurement time was divided into a number of sub-divisions (2N). The sample with specified chemical state was made the reference material and was measured in the odd sub-divisions; the other sample in which the chemical state is to be determined is measured in the even sub-divisions. The line position energies from the spectra in odd subdivisions E1's and in even subdivisions E2's, were separately applied polynomial fits to have fitted energies E1' and E2'. The fitted results give the fluctuations caused only by the non-statistical effects. The fluctuation free results E1″ and E2″ were evaluated as

$$E1'' = E1 - ((E1' + E2')/2) + E0 \tag{1}$$

$$E2'' = E2 - ((E1' + E2')/2) + E0 \tag{2}$$

$$E0 = \sum_{n=1}^{N} \frac{E1 + E2}{2N} \tag{3}$$

So, the corrected results ($E1''$ and $E2''$) are free from the non-statistical errors. The mean peak positions and measurement error in $E1''$ and $E2''$ data were calculated using the relations:

$$\overline{E} = \sum_{n=1}^{N} \frac{E_n}{N} \tag{4}$$

$$S_E = \sqrt{\frac{1}{N-1} \sum_{n=1}^{N} (E_n - \overline{E})^2} \tag{5}$$

$$\sigma_E = \frac{S_E}{\sqrt{N}} \tag{6}$$

The $\overline{E}$'s of each of the two sets of data, subject to the limitations of the apparatus, yield the net difference in line positions $E1$ and $E2$ of two samples that is because of the chemical shift.

3. EXPERIMENT

The experiment was performed using the low-power Neptune X-ray tube set up and the Si (PIN) detector (AMPTEK model XR-100CR) Fig. 1. Photons from the X-ray tube were incident on the target. The fluorescent X-rays produced by the target elements were detected in the Si (PIN) detector which is at 90° to the source in a single reflection geometry. One of the compound/metal was selected as sample 1 while the other as sample 2. Time of the experiment was subdivided into even subdivisions. Half of the time of each subdivision was used for collecting the spectrum and the other half for saving it and changing the sample. The anode voltage of the tube was set just above the K edge energy of sample element for exciting its K X-rays. The filament current was set to reduce the dead time of counting system and the detector gain was set to increase the precision of the measurements.

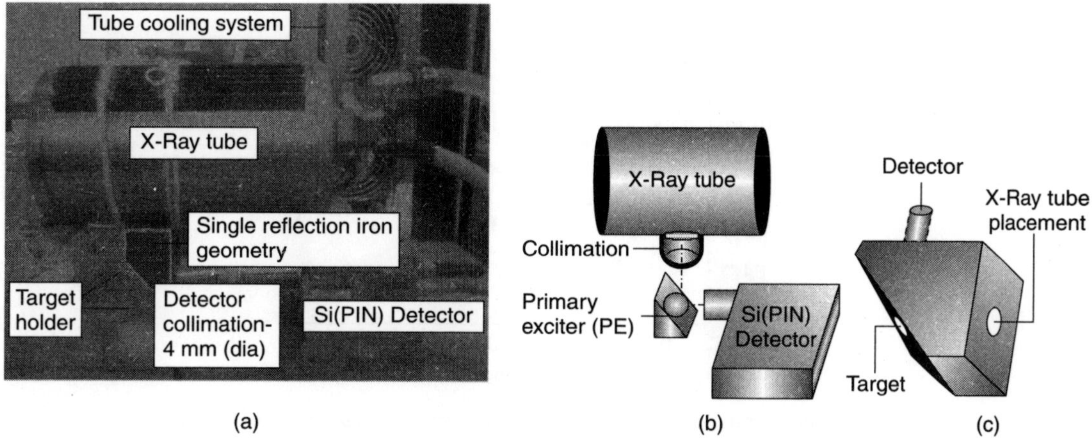

Fig. 1 *(a) A view of single reflection iron geometry with X-ray tube as photon source, target and Si (PIN) detector. (b) the schematic arrangement of tube, geometry and detector.(c) geometry from single iron piece.*

4. RESULTS AND DISCUSSION

The typical spectra obtained in a set of even and odd sub divisions for Fe and $FeSO_4.7H_2O$, are shown in Fig. 2.

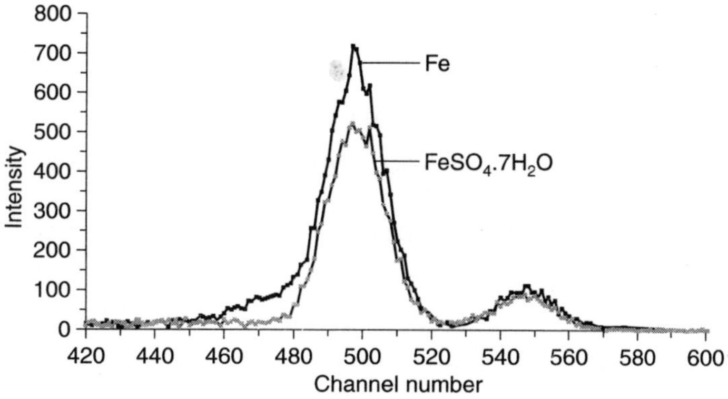

Fig. 2 *Typical spectra of Fe and $FeSO_4.7H_2O$ in an even and an odd time sub division.*

The peak positions (*E1* and *E2*) of the samples were calculated in each time slot. The peak energies versus time slots were plotted. The plotted points showed fluctuations and were applied polynomial fits. Fig. 3 and 4 show typical plots of the energy values verses time slots and the polynomial fits to the energy values for Fe and FeSO$_4$ samples respectively. After applying the polynomial fits, the fitted results (*E1′* and *E2′*) for both the samples were obtained and calculations (equations (1) and (2)) were performed on the values to remove the non statistical errors.

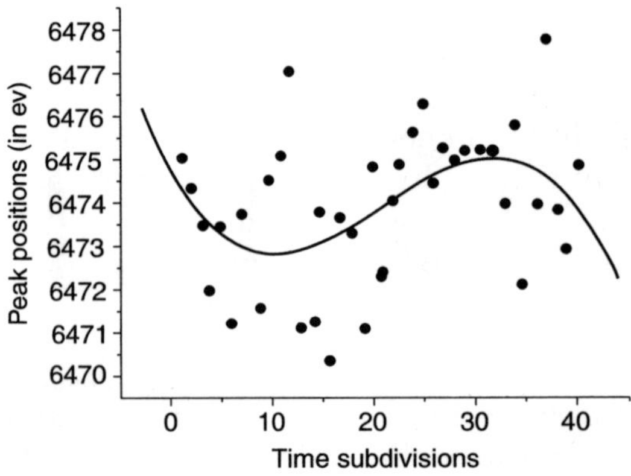

Fig. 3   *Plot of peak positions of Fe with time slots and its polynomial fitting.*

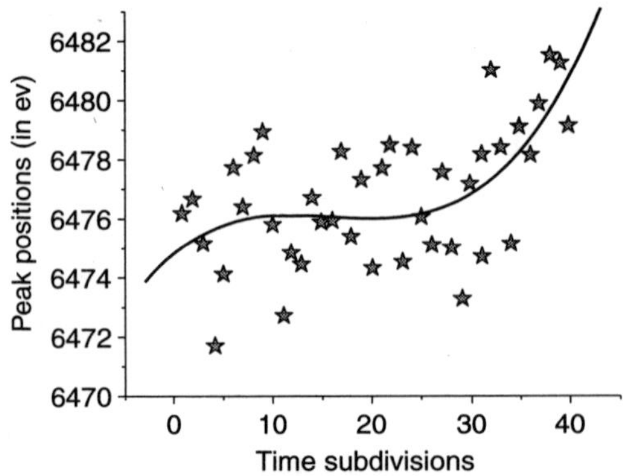

Fig. 4   *Plot of peak positions of FeSO$_4$ with time slots and its polynomial fitting.*

The mean peak positions and measurement error S_E and σ_E from *E1″* and *E2″* data calculated using the relations (4), (5) and (6) are being listed in Table. 1 along with the combined error in the result as:

$$\sigma = \sqrt{\sigma_1^2 + \sigma_2^2} \tag{7}$$

Table 1 *Chemical shifts of different compounds with error.*

Compound/Metal	Mean peak position, $\overline{E}$; *(in eV)*	Measurement error, σ_E *(in eV)*	Chemical shift with error *(in eV)*
Fe & $FeSO_4.7H_2O$	6473.96 & 6476.70	0.26 & 0.31	2.74 ± 0.40
$CaCO_3$ & $CaSO_4.2H_2O$	3708.52 & 3707.53	0.63 & 0.26	0.99 ± 0.68
KBr & KNO_3	3282.48 & 3280.53	0.70 & 0.22	1.95 ± 0.73

5. CONCLUSION

The measured chemical shifts in energy states of elements Fe, Ca and K in their different compounds with an energy dispersive system using a statistical data-processing method after exploiting the limitations of the detector gives results below its resolution limit, which is an encouraging work in this direction.

Acknowledgments

The financial assistance from BRNS, Government of India in the form of project grant (ref no. 2007/37/6/BRNS) for this work is highly acknowledged.

References

1. Grieken R.E. Van and Markowicz A. *Handbook of X-ray Spectrometry*, Marcel Dekker Inc., New York (2002).
2. Xiao Y. *et al.*, *The Japan Society for Analytical Chemistry*, **14**, (1998).

M X-ray Production Cross Section Measurements with Photons from Low Power X-ray Tube in Elements Th and U

Sheenu Gupta, V.K. Mittal and Raj Mittal[*]

Nuclear Science Laboratories, Physics Department, Punjabi University, Patiala, India

E-mail: *rmsingla@yahoo.com

ABSTRACT

M-lines X-ray fluorescence (XRF) cross-sections have been measured using low power X-ray tube and Si PIN detector in a single reflection geometrical set-up by selective excitation of M-shell of experimental target by adjusting the tube anode voltage. In the single reflection set-up, the continuous Bremsstrahlung excites the M-shell vacancies. Therefore, a procedure has been followed to find the weighted average energy and total intensity of the Bremsstrahlung that excites the target. A fair agreement of measured values with the existing one supports the procedure.

Keywords: X-ray fluorescence cross section, Bremsstrahlung radiation and Excitation and De-excitation of inner-shell vacancies.

Pacs No.: 32.70.Jz, 32.30.Rj

1. INTRODUCTION

Applications of low energy fluorescence x-rays (XRF) in various fields has raised the importance of data on M X-ray fluorescence cross sections. Total M-shell XRF cross sections have been evaluated theoretically and measured experimentally by a number of workers for some high Z elements [1-4]. As reported by Sharma et al [5], most of the work done is at 5.959 keV from Fe-55 source, the data on individual M line cross section is very scarce because of the complexities involved in the M-shell X-ray spectrum. The use of radioactive sources imposes limitations on intensity of incident photons. To get rid of this limitation, presently, an attempt has been made to find M line XRF cross sections using a low power X-ray tube along with a Si PIN detector in compact single reflection geometry. The selective production of only M-shell vacancies in Th and U has been done with Bremsstrahlung radiation from the tube operated at 6 and 6.5 kV for Th and U respectively. Some innovative efforts were made to find the weighted average energy of incident radiation above the M-edge. The details are as given below.

2. EXPERIMENTAL SET-UP

The experimental arrangement consists of a 90° single reflection geometry, made from a small iron cube of dimensions 5 cm × 5 cm × 5 cm, along with a low power (100 W) x-ray tube with Rh anode as a photon source and a Peltier cooled Si PIN detector having resolution of ~210 eV at 5.959 keV to record the x-rays. The target is placed at 45° to both the incident and emitted radiation. The arrangement of tube, targets and detector is shown in schematic Fig. 1.

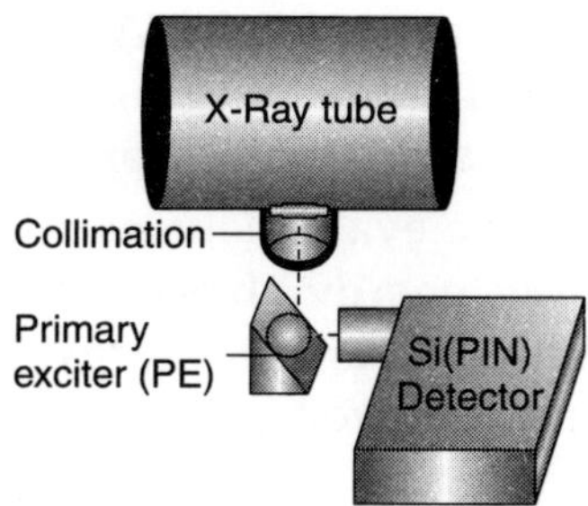

Fig. 1 *X-ray path from X-ray tube to detector, distances and collimation sizes are not according to the scale.*

The experimental targets in metallic form are exposed to tube photons in the geometry and the resulting fluorescent M X-rays and scattered Bremstrahlung Fig. 2 are recorded in Si PIN detector.

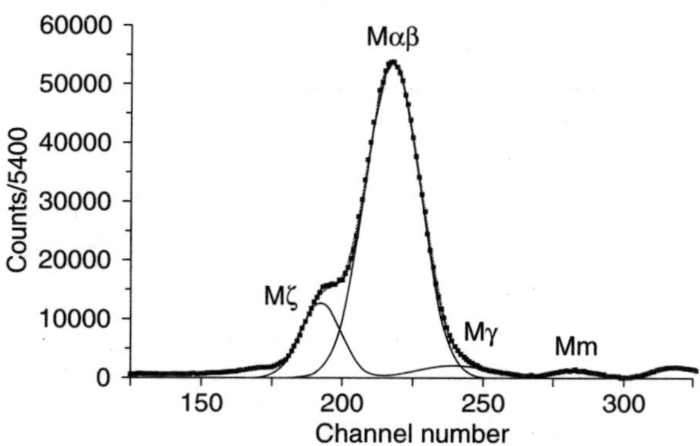

Fig. 2 *Typical spectrum of Th with background subtraction.*

Tube anode voltages are such that these do not excite the Fe of geometry but all the M sub-shells of both the targets. Filament current is set at 0.9mA to keep the dead time of detector less than 1%. The tube radiation scattered from the target are accounted by recording the spectra with equivalent Al targets of Th and U. The five M sub-shells give rise to a large number of M X-ray lines those appear under different peaks due to limited resolution of the detector. The experiment is run for sufficient time to obtain the statistics of counts <1% under the various peaks of the background subtracted net spectrum.

The production cross-section in b/atom for a group of M X-rays, $\sigma_{Mg}(g = \xi, \alpha\beta, \gamma \,\&\, m)$ is given as

$$\sigma_{Mg} = \frac{N_{Mg} \; \text{Mol} \cdot wt \cdot}{nt \; \beta \; (i, \; M_g) \; \varepsilon_{Mg}} \tag{1}$$

where N_{Mg}, count rate under the M_g X-ray peak; Mol. Wt., molecular weight of the target; n, Avogadro's number; t_M, thickness of target in gm/cm^2 and $\beta(i, M_g)$, self absorption correction factor of the target at incident energy i and emitted M_g X-ray energy.

The calculations of σ_{Mg} involve the evaluations of $\beta(i, M_g)$ at incident and M_g X-ray group energies and ε at M_g X-ray energy those are evaluated from using XCOM software [6]. To find the weighted average energy E_i of incident tube photons between the $M 5$ edge energy and tube voltage, an innovative procedure is followed. For this, a spectrum of borax by placing its target in place of experimental target is taken at the same operating voltages. The region of borax spectrum, $M 5$ edge energy of the experimental target-energy corresponding to the tube anode voltage, is divided into strips (t), each strip of energy width 200 eV. Numbers of X-rays under each strip are accounted. At the mean energy of each strip, total Klien-Nishina scattering and elastic scattering cross-sections are evaluated [6]. Dividing the sum of cross-sections with the corresponding number of strip counts, incident intensity in each energy interval is computed. Then using the intensity I_t and mean energy E_t of each energy interval, average incident energy E_i is found as

$$E_i = \frac{\Sigma_t E_t I_t}{\Sigma_t I_t} \tag{2}$$

The weighted X-ray energy of M_g group is calculated from level energies [7] and intensities [8] by grouping the M X-ray lines falling in the regions of intense M lines $\pm$ detector resolution.

The efficiency of detector in the M X-ray region is found by recording the spectrum of each low Z element, S, K, Ca, Ti, V and Cr whose K X-ray energies lie in energy range of M X-rays of elements of interest in the same set-up. The correction for scattered tube photons from the target is applied by subtracting the borax spectrum form it. Using the formulation (1), the efficiency of detector for each target is calculated as,

$$\varepsilon_K = \frac{N_K M_K}{n t_K \beta_K \sigma_K}$$

(3)

where the different parameters have the same meaning as defined above. σ_K, K X-ray production cross sections are found from the code KCSPIF [9]. With known efficiency at M X-ray lines, other required parameters and the counts under the respective M X-ray photo-peaks, M X-ray cross sections are evaluated. The error in measured cross-section is the quadratic sum of errors in different parameters used in the evaluation of cross-section as evaluation of peak areas ($\leq$ 1%), efficiency measurements ($\leq$ 3%), absorption correction factor ($\leq$ 5%) and target thickness measurements ($\leq$ 4%). The evaluated cross-sections along with uncertainties are compared with the existing experimental and theoretical values in the Table 1.

Table 1 *Measured M-subshell X-ray fluorescence cross-sections in b/atom along with the existing experimental and theoretical values.*

Element	$\sigma_{M\alpha\beta}$	$\sigma_{M\gamma}$	σ_{Mm}
Th	6890 ± 689	410 ± 41	117 ± 12
	7286 ± 510[5], 6997[8], 6778[10]	401 ± 60[5], 396.3[8], 367.5[10]	198 ± 19[5], 215[8], 198.7[10]
U	6950 ± 695	554 ± 55	89 ± 9
	7761 ± 543[5], 8042[8], 7947[10]	484 ± 48[5], 512[8], 451.9[10]	215 ± 21[5], 237.8[8], 235.3[10]

3. RESULTS AND DISCUSSION

Since M_ξ line falls in the region of scattered Rh L-lines and atmospheric fluorescent Ar K-line, so it is cumbersome to find out the exact peak area, therefore, cross-sections are evaluated only for $\alpha\beta$, γ & m groups. For both the elements, the measured cross-sections within experimental uncertainties are in agreement with earlier experimental and theoretical values. For U, the results are little bit on the lower side as compared to the earlier quoted values. The fair agreement of measured values with the existing one supports the procedure.

Acknowledgments

The authors duly acknowledge Prof. Lodha, Head, X-ray optics section, RRCAT, Indore for supporting this work. The financial assistance from BRNS, Govt. of India in the form of project grant (ref. no. 2007/37/6/BRNS) for this work is highly acknowledged.

References

1. Garg, R.R., Singh, S., Mehta, D., Kumar, S., Shahi, J.S., Garg, M.L., Singh, N., Mangal, P.C., Trehan, P.N., X-ray Spectrom. **20** 91 (1991).

2. Puri, S., Mehta, D., Chand, B., Singh, N., Hubbel, J.H., Trehan, P.N., Nucl. Instrum. Methods B **73** 319 (1993).

3. Shatendra, K., Singh, N., Mittal, R., Allawadhi, K.L., Sood, B.S., X-ray Spectrom. **14** 195 (1985).

4. Mann, K.S., Singh, N., Mittal, R., Allawadhi, K.L., Sood, B.S., J. Phys. B **23** 2497 (1990).

5. Sharma M., Sharma V., Kumar S., Puri S. and Singh N., Rad. Phys. Chem. **75** 1503 (2006).

6. Berger M J and Hubbell J H Computer code XCOM, version 3.1, National Bureau of Standards, Gaithersburg, MD, USA, available from http://physics.nist.gov/(1999).

7. Storm E and Israel H I Nucl. Data tables A **7** 565 (1970).

8. Chen M.H. and Crasemann B., Phys. Rev. A **30** 170 (1984).

9. Bansal M. and Mittal R. Rad. Phys. Chem. **79** 583 (2010).

10. Chen M.H. and Crasemann B., Phys. Rev. A **21** 449(1983).

Studies of ^{16}N in Primary Coolant and its Correlation with Reactor Power in CIRUS

Anilkumar S.[1*], P. Sumanth[2], S.K. Prasad[1], V.M. Thakur[1],
Amit K. Verma[1], R.K.B. Yadav[1], D.A.R. Babu[1],
G. Bharadwaj[2] and D.N. Sharma[1]

[1]Radiations Safety Systems Division, Bhabha Atomic Research Centre, Mumbai, India
[2]Research Reactor Maintenance Division, Bhabha Atomic Research Centre, Mumbai, India
E-mail: *anilk@barc.gov.in

ABSTRACT

The significant contribution to external radiation field in the nuclear reactor is from the high energy gamma photos from ^{16}N. This short lived radionuclide is produced in the reactor by the neutron activation of ^{16}O in coolant water and its production rate is proportional to the fast neutron flux. The paper discusses the details of experiments carried out to study the concentration of ^{16}N in primary coolant water line of CIRUS reactor using gamma spectrometry techniques. The data obtained from our studies showed very good correlation between the ^{16}N content in the coolant water and the reactor power. The study also emphasized the feasibility of using ^{16}N monitoring as a reactor power monitor.

Keywords: ^{16}N, Gamma spectrometry, Reactor power.

1. INTRODUCTION

^{16}N radionuclide is formed in the reactor due to the neutron activation reaction of oxygen present in coolant water by ^{16}O (n, p) ^{16}N reaction. ^{16}N is a short-lived beta active radionuclide with a half life of 7.13sec. The main radiological concern of this radionuclide is its high energy photons of 6.13 MeV (67%) and 7.11 MeV (4.9%) [1]. This radionuclide is regarded as the major contributor for the coolant activity and radiation levels thereof. The issue of coolant activation and transport of coolant activity to the turbine and balance of plant is of great concern in boiling water reactors [2]. In swimming pool type reactors the radiation field on pool surface is mainly due to the high energy photons from ^{16}N where coolant and moderator is light water. The production of this nuclide is expected to be proportional to the power of the reactor and the concentration of this nuclide in the coolant water can be correlated to the reactor power. The ^{16}N activity measurement at the nuclear plant is also considered for monitoring potential steam generator leakages in nuclear power plants. The possibility of utilizing the ^{16}N activity measurements in coolant water for monitoring the reactor power is investigated [3]. A preliminary study has been carried out in the primary coolant (PCW) line of CIRUS reactor. For the present experiment 3″ × 3″ NaI (Tl) and 2″ × 2″ BGO detectors were used for the gamma spectrometric measurements at various power levels of the reactor. The paper describes the details of the measurements carried out for the gamma spectrometric monitoring of ^{16}N in CIRUS reactor.

2.   MATERIALS AND METHODS

The primary coolant water (PCW) taken through sampling lines in failed fuel detection (FFD) room of CIRUS reactor was monitored for ^{16}N using NaI(Tl) and BGO based gamma spectrometry system. The measurements were carried out at reactor power levels of 0.4 MW, 4 MW, 10 MW and 20 MW. The NaI(Tl) detector used in the experiment is an integral assembly with built in HV, Preamplifier and amplifier. The amplifier output from the detector assembly is analysed using USB based MCA system coupled to a lap top computer. A similar spectrometry set up was used for the BGO detector. The detectors were shielded using circular lead shield with a thickness of 1.5″ mounted on moving platform during measurement. The energy calibration of the spectrometer for high energy photons were carried out using 662 keV (^{137}Cs), 1173 keV, 1332 keV (^{60}Co) and 4438 keV (Am-Be) gamma energies. The geometry for the water sampling system consists of spiraled copper tubes of 12 turns with a diameter of 12.5 cm. The copper tube is having a diameter (ID) of 4.3 mm and thickness of 1mm. The spiral geometry was preferred to a well in tank geometry as results of measurements using this geometry showed stagnancy in water at bottom leading to a very high background. The detector was placed inside the spiral assembly co axially. A parametric study of background vs delay time was carried out to optimize the delay time of measurement. The height of the spiral assembly is 7.56 cm matching with the height of the NaI (Tl) detector. The spectral data was acquired for a counting period of 300 sec. The spectral data were taken using both NaI (Tl) and BGO detectors for various power levels. Repeated measurements were taken for each power level.

The offline analysis of the gamma ray spectra was carried out to estimate the intensity of ^{16}N peak in the spectrum. Because of the unavoidable presence of the escape peaks in the spectrum due to high energies, an energy region of 4.5 MeV to 7.5 MeV is fixed for estimating the count rate due to ^{16}N. The background spectrum was also acquired for 300 sec. using both the detectors. Typical background counts of 58 counts and 1026 counts were obtained in 300 sec for NaI (Tl) and BGO detectors respectively. The complete analysis of the spectrum shows significant activity due to ^{64}Cu (12.7 h), ^{28}Al (2.24 min), ^{24}Na (14.95 h) and ^{56}Mn (2.58 h) are observed in addition to ^{16}N. The contribution due to fission γ is negligible as the system was located away from the core. The gamma ray spectra recorded in the ^{16}N ROI for different power levels for NaI (Tl) detector is shown in Fig. 1. The net counts in the ROI are taken after subtracting the background counts. The details of the count rates measured in the peak regions of ^{16}N gamma energies for NaI (Tl) and BGO detectors are shown in Table. 1.

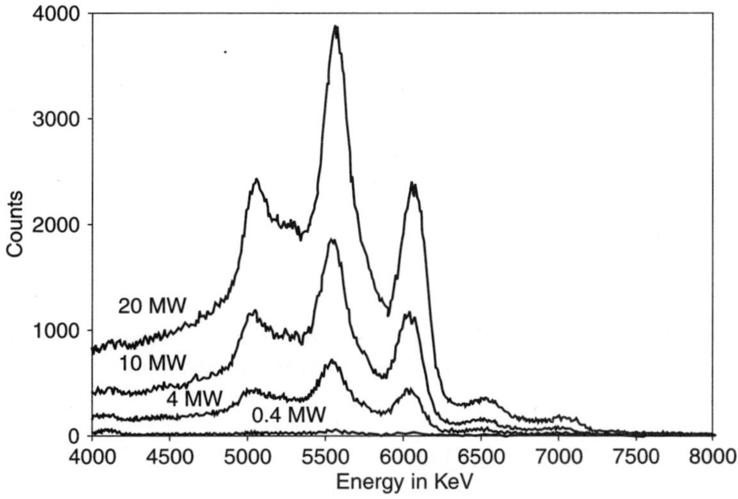

Fig. 1 *The region of ^{16}N in the gamma ray spectra obtained for different reactor power levels using NaI (Tl) detector*

Table 1 *Net corrected count rate in ^{16}N region from NaI (Tl) and BGO based spectrometric measurements*

Reactor Power (MW)	NaI (Tl)	Count rate (Cps) for BGO
0.4	24.3	11.7
4.0	349.1	202.5
10.0	921.6	531.2
20.0	1930.9	1111.6

In order to obtain the actual count rate (CPS) of ^{16}N due to activity in the spiral assembly during the counting period it is required to apply the decay correction for the ^{16}N activity during the transit time in the assembly. The flow rate (cc/min) of PCW water was monitored for all power levels during the spectral data acquisition. The volume of the water in the 12 turn spiral assembly is 68.5 ml. The ^{16}N activity decays during its transit through the spiral coil and continuous fresh activity is injected in to the spiral because of the continuous flow. The transit time of the activity in the spiral geometry for various power levels is calculated from the respective flow rate. Then the correction factors of the count rate are calculated for the transit time at which the ^{16}N activity decays during the acquisition. The corrected net count rates obtained in the ^{16}N ROI of the spectrum taken at various power levels are plotted. The correlation of count rate obtained in gamma ray spectrum for ^{16}N in different power levels using NaI (Tl) detector is shown in Fig. 2.

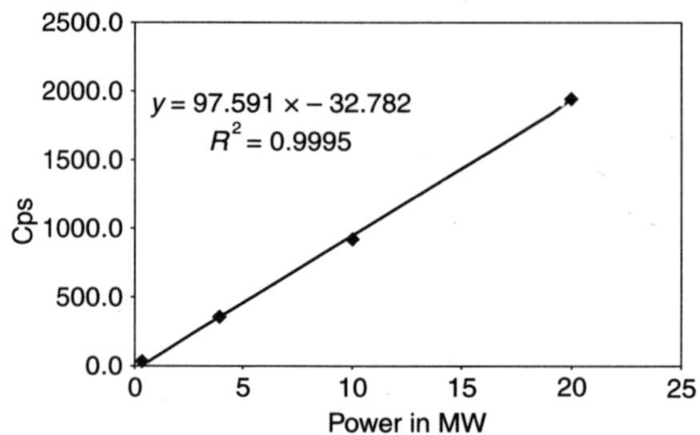

Fig. 2 *The plot showing the correlation of count rate obtained in the ^{16}N region of gamma spectrum for different reactor power levels using NaI (Tl) spectrometer*

3. RESULTS AND DISCUSSION

From the present experiments it is established that the concentration of ^{16}N in the PCW line can be measured using NaI (Tl) or BGO based gamma ray spectrometry system employing a spiral flow arrangement around the detector. The actual activity concentration of the nuclide in the coolant line also can be done using the efficiency factor for the detection of high energy gamma photons from ^{16}N decay. It is difficult to find certified radioactive sources with energies this high to calibrate the detector. The alternate method suggested is to calculate the detector response to photons of this source with the geometry by Monte Carlo methods. But with the fixed detection system in a defined geometry it is possible to correlate the net counts in the ^{16}N region to the reactor power. The count rate data obtained from the ^{16}N ROI shows excellent correlation with the reactor power level as measured by core n-detectors. Relevant correction factors evaluated taking in to account the decay of ^{16}N during the transit through the spiral assembly which enables to use this monitoring system as a reactor power monitor. With proper optimization of geometry, detector (NaI (Tl) or BGO), and measurement set up it is possible to have a reactor power monitor in future reactors.

Acknowledgements

Authors express their sincere gratitude to Dr.A.K. Ghosh, Director HS & E Group and Shri V.K. Raina, Director Reactor Group for their constant encouragement through out the work.

References

1. Hu-Xia Shi, Bo-Xian Chen, Ti-Zhu Li and Di Yun, *Applied Radiation and Isotopes*, **57**, 517 (2002)
2. B. Fishcer, M. Smolinski and J. Buongiorno, *Nuclear Technology*, **147** 269 (2004)
3. W.A. Jester, T. Daubenspeck, S. Pandey and X.Xu , *J. Radioanalytical and Nuclear Chemistry* **244** 453 (2000)

Use of (*n, xn*) Reactions to Enhance High Energy Neutron Response in a Modified Bonner Sphere

Biju K.[1,*], S.P. Tripathy[2], Sunil C.[2] and P.K. Sarkar[2]

[1]*Radiological Physics and Advisory Division, Bhabha Atomic Research Centre, Mumbai, India*

[2]*Health Physics Division, Bhabha Atomic Research Centre, Mumbai, India*

E-mail: *bijuk@barc.gov.in*

ABSTRACT

Conventional bonner spheres with metallic shells are used to enhance the response for the high energy neutrons using (*n, xn*) reactions. This study provides the enhancement in the neutron response using the low density metallic shells, zirconium and beryllium which has considerably good (*n, 2n*) cross sections. The response with high density metallic shells, lead and tungsten are also studied and compared with zirconium and beryllium. The results of Zr shell over the 9 inch HDPE sphere, at higher energies, seem interesting and can be explored further for the development of low-weight spectrometer whereas Be do not show multiplication that can be used with any reasonable effect.

Keywords: Bonner sphere, Converter, Neutron spectrometer.

1. INTRODUCTION

Neutron dosimetric problems in high energy, high current accelerators primarily for the application of accelerator driven technologies like ADSS [1] necessitates spectrum measurement up to few hundreds of MeV or beyond. High-density polyethylene (HDPE) multisphere or Bonner sphere spectrometer using Li-6 thermoluminescence detector at the center is one among the mostly used systems in a high intensity or pulsed radiation field that takes care of the undesirable dead times or pile up effects.

Response to high energy neutrons does not increase significantly even by enlarging the sphere diameter further, rather it affects the lower energy part of the spectrum due to over-moderation. Therefore, metallic shells are introduced along with the HDPE spheres to enhance the response via (*n, xn*) nuclear processes [2]. Many studies have been reported on the use of the high density metallic shells for the enhancement of the response of Bonners spheres upto high energies. The optimization of thickness of the metallic shells, mainly for the high-density metals, e.g., W, Pb, etc. are obtained using MCNP based calculations [3]. However, the weight of the spectrometer increases significantly with addition of these metallic shells, which sometimes limit the use of these detectors to measure the neutron field at inconvenient places.

In this work, the response is studied for low density metallic shells such as Zr and Be using Monte Carlo simulation code FLUKA [4]. These materials have considerably good $(n, 2n)$ cross-section and, moreover, Be has much lower threshold energy ($\sim$ 3 MeV), which could open up the channel at lower neutron energy leading to enhancement in the overall response. The response with Pb and W are also studied and compared.

2. MONTE CARLO SIMULATIONS

The geometry as simulated using Monte Carlo code FLUKA is shown in Fig. 1. A parallel monoenergetic beam is made to incident normally on the bonner sphere. The radius of the beam is kept equal to the radius of the sphere. Neutron fluence spectrum is scored using track length estimator in a sphere of radius 0.5 cm at the centre of the HDPE sphere.

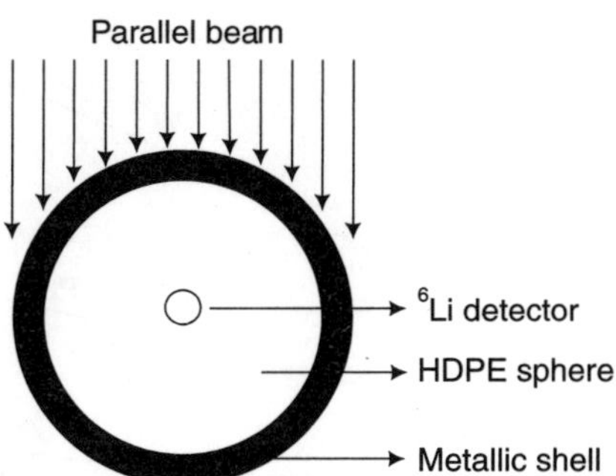

Fig. 1 *The simulated irradiation geometry*

The ^{6}Li detector response was calculated using the formula

$$R = \int \phi\,(E)\,\sigma_{(n,\,t)}\,(E)\,dE * A$$

where $\varphi(E)\,dE$ is the fluence (cm^{-2}) at energy between E and $E + dE$, $\sigma_{(n,\,t)}\,(E)$ is the cross section (cm^2) and A is the area of the neutron beam (cm^2).

Initially a bare HDPE sphere of 9 inch diameter is used to estimate the ^{6}Li(n, α) reaction inside the detector cell volume. Subsequently, 1 cm thick metallic hollow spherical shells of the four chosen materials were used to cover the HDPE sphere and the response was calculated again. The ratio of these two results gives the enhancement in the response.

3. RESULTS AND DISCUSSION

The reaction cross section and the threshold energy values were obtained from the ENDF/B-VII cross section data base [5] and were analyzed to identify the elements that could be used as suitable metallic shells. Table 1 presents the density, $(n, 2n)$ reaction cross section and the threshold energy for W, Pb, Zr, and Be.

Table 1 *Useful properties for metallic shell selection.*

	Threshold energy (MeV)	$\sigma\ (n, 2n)$ (barn)	Density (g cm^{-3})
^{182}W	9.0	2.02	19.25
^{206}Pb	9.0	2.23	11.34
^{96}Zr	12.5	1.25	6.52
^{9}Be	3.0	0.56	1.85

Of these W and Pb have been well studied and are used in various configurations of extended dose monitors and Bonner spheres. Use of Zr and Be have not been reported in the literature. In the present study, the response of ^{6}Li detector is compared for these four different cover materials. Fig. 2 shows the enhancement in the response of ^{6}Li detector, for the HDPE sphere covered with a 1cm metallic shell to that of the bare HDPE sphere, in the neutron energy range of 50-200 MeV.

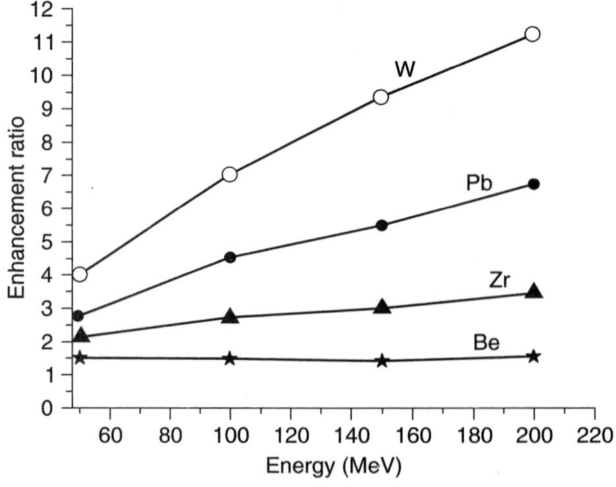

Fig. 2 *The ratio of ^{6}Li response with 1 cm shell (W, Pb, Zr and Be) on a 9" HDPE to Li-6 response with 9 inch HDPE sphere as a function of neutron energy.*

From Fig. 2, it is clear that W and Pb have better enhancement properties when compared to Zr and Be which is primarily due the higher density of these materials containing more number of reactions. Fig. 3 presents the enhancement of Zirconium as a function of shell thickness. At higher energies for larger thickness, say 2 cm, the enhancement ratio for Zr improves. The results for Be is not as promising as one would expect from the lower threshold and the peak cross section. This is due to the fact that the elastic cross section constitutes 2/3 of the total cross section while the $(n, 2n)$ cross section is just 1/3. W and Pb have $(n, 3n)$ channel also opening up at higher threshold energies thereby ensuring a constant multiplication while Fermi breakup is the preferred mode in the case of Be at higher energies. Fig. 4 presents the differential neutron spectra exiting from the metal shell and falling on the moderator shell for 200 MeV incident neutrons.

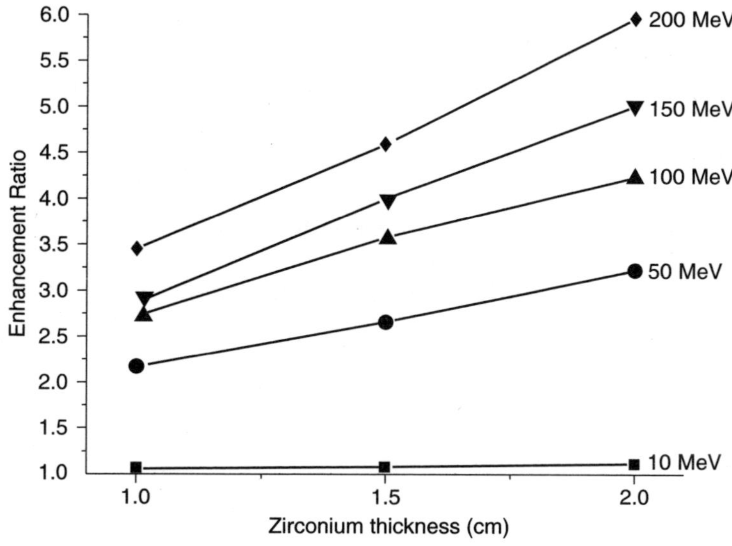

Fig. 3 *The ratio of ^{6}Li response with Zr shell on a 9" HDPE to Li-6 response with 9 inch HDPE sphere as a function of shell thickness.*

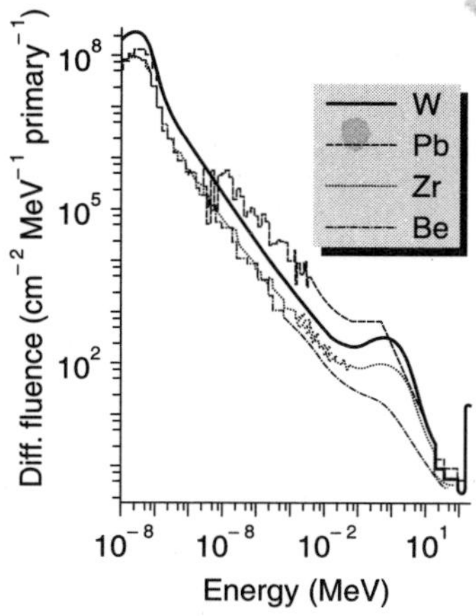

Fig. 4 *Exit spectrum form the metallic shell*

3. CONCLUSION

The results of Zr shell over the 9 inch HDPE sphere, especially at higher energies, seem interesting and can be explored further for the development of low-weight spectrometer. However, results from Be do not show multiplication that can be used with any reasonable effect. Further studies are in progress with different detectors to optimize the enhanced response for high energy neutrons.

References

1. P K *Sarkar Radiat. Meas.* In Press (2010)
2. Birattari, A Ferrari, C Nuccetelli, M Pelliccioni and M. Silari *Nucl. Inst. Meth.* A **297** 250 (1990)
3. S Serre, K Castellani-Coulié, D Paul, and V Lacoste, *IEEE trans. on nucl. science* **56** 3582 (2009)
4. A Fasso, A Ferrari, J Ranft and P Sala Fasso R A *FLUKA: a multi-particle transport code,* CERN-2005-10 (2005)
5. http://www.nndc.bnl.gov/, last accessed 21 October 2010

Absolute Measurement of Neutron Yield for d-t Neutron Generator using Water Activation Technique

Saheb Hussain Md[1*], Shrichand Jakhar[2], Mitul Abhangi[2], Rajnikant Makwana[2], Sudhirsinh Vala[2], A. Gopala Krishna[1], C.V.S. Rao[2] and T.K. Basu[2#]

Jawaharlal Nehru[1] Technological University[2], Kakinada, Andhra Pradesh Institute for Plasma Research, Bhat, Gandhinagar, #Raja Ramanna Fellow of DAE
E-mail: *saheb228@gmail.com

ABSTRACT

In a Burning Plasma Experiment (BPE) like ITER, fusion power is continuously monitored by measuring the neutron flux Neutron activation method using the conventional solid metal foils results in poor time resolution. The activation of water by 14-MeV neutrons where $^{16}O\ (n,\ p)\ N^{16}$ nuclear reaction takes place is the most promising candidate for this purpose. The short half-life of N16 radio nuclide produced in this reaction ($T1/2$ ~7.13 sec) and the threshold energy being high enough (~ 10.4 MeV) makes it a preferred technique. In order to check the feasibility of this technique, an experimental set-up was designed. The experimental set-up consisted of a water pipeline running between the neutron generator and a detector. The detector was a fully shielded and collimated NaI(Tl) spectrometer. The water column was irradiated with the 14-MeV neutrons produced in a sealed neutron generator. Water when irradiated with 14-MeV neutrons produces gamma photons of energies 6.128 MeV (67.06%) and 7.115 MeV (4.94%) from the N16 nucleus which is measured with NaI(Tl) spectrometer. From the measured gamma intensity, the D-T neutron yield is estimated. For this purpose the NaI(Tl) spectrometer is calibrated for its absolute efficiency by MCNP (Monte Carlo N- Particle) calculation.

Keywords: Neutron yield, D-T neutrons, Fusion power.

Pacs No.: 28.20.Fc

1. INTRODUCTION

In foil activation process an accurate measurement of the neutron yield is not possible with good time resolution. So we have considered activation of flowing water based on the ^{16}O (n, p) N^{16} reaction for the accurate estimation of fusion neutron yield. In D-T neutron generators, it is important to know the neutron source yield accurately with good time resolution. When water is irradiated by 14.1 MeV neutrons, radioactive ^{16}N nuclei are produced by the ^{16}O (n, p) N^{16} reactions from oxygen atoms present in water molecules. It can serve as the most promising candidate for this purpose due to its short half-life of 7.13 s and cross section threshold energy of 10.4 MeV which is close to the D-T neutron energy. An experimental setup consisting of water and gamma spectrometer is required for the measurement. The water is irradiated with D-T neurons and two gamma energies 6.128 MeV (67.06%) and 7.115 MeV (4.94%) from ^{16}N are measured with the NaI (Tl) detector. From the measured gamma activities the D-T neutron yield is calculated. Neutron activation performed by using solid sample metals (or) foils, results in poor temporal resolution. The advantage of the neutron activation in flowing water is that the estimation of D-T source neutron yield can be done accurately with good time resolution which is not possible to do by foil activation technique. Moreover, this process has a feature in continuous measurement for long durations. The experimental set up is easily maintained, because the measurement station is located outside of the biological-shield, away from neutron generator. In addition, long-term sensitivity change is expected.

1.1 The Major Tasks Under this Project are as Follows

- Calibration of gamma-ray detector efficiency at 6.13 MeV.
- Design of Shield for gamma (γ)-ray detector.
- Measurement of neutron yield

1.2 Overview of Experimental Setup

2. EXPERIMENTAL SETUP

The experiment was carried out at the Neutronics Laboratory of Institute for Plasma research at Gandhinagar. Fig. 1 shows a schematic of the experimental set up. Water is pumped to the neutron generator irradiation point up to counting station of gamma ray Detector set up point. A spiral part of S.S pipe around the generator is made so that much quantity of water is exposed to neutrons in order to achieve sufficient count rate. Water flowed spirally along the water pipe installed around the neutron generator emission center to have enough long resident time for irradiation. The total length of spiral part was about 250 cm. Stain less steel pipe of 1.20 cm inner diameter was considered for the pumping system throughout. Water was sent from the measurement area to the neutron generation area by a pump. Water was irradiated by the D-T neutrons. The irradiated water returned to the counting station area, and the gamma–ray was measured by NaI (Tl) detector. Water Traveled for long duration after passing through the counting station which was helpful to recycle the same water for a longtime. After the decay of ^{16}N activity, the water was sent to the neutron emission point again [4] [5] [6].

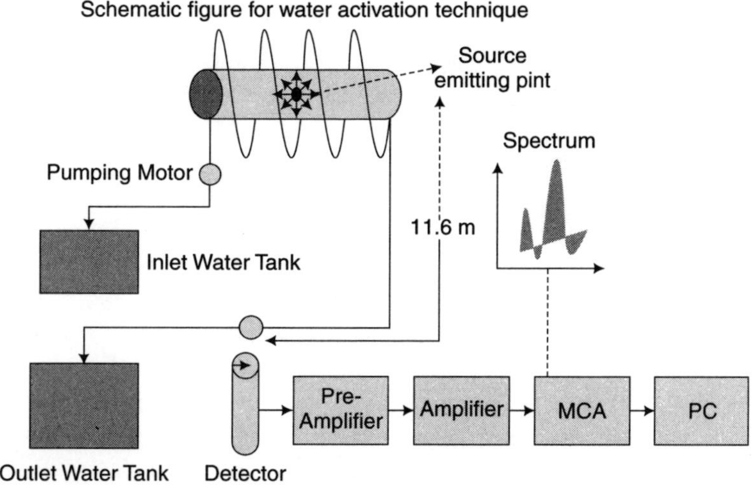

Fig. 1 *Water activation flow diagram*

2.1 Instruments Required

1. D-T Neutron generator
2. NaI (Tl) detector (3″ × 3″)
3. Water Loop System

2.2 Gamma Ray Counting Station Arrangement

In case of ^{16}O (n, p) N^{16} reaction, gamma rays in the range of 6-7 MeV are produced, so the "photoelectric interactions" with detector material are negligible. Compton scattering and pair production are roughly equally possible. Total attenuation co-efficient per unit "material density" is 0.03 cm^2/g. The cross section of the pair production increases with the square of the charge number z of the detector material. Therefore, high –Z materials are suitable for the detection of 6.13 MeV and 7.13 MeV gamma rays. We considered NaI (Tl) inorganic scintillator material because the density of NaI (Tl) is 3.67 g/cc. Fig. 2 shows the Gamma ray counting point and the gamma ray counting station. Expected gamma ray count rate activity of ^{16}N is calculated using the Monte Carlo N-particle code MCNP. For reducing the background counts from natural sources and ^{40}K sufficient lead shielding (8.30 cm Ω) around the NaI (Tl) scintillator was assembled. Generally 1.4 MeV peak is seen from ^{40}K element [1] [2] [3].

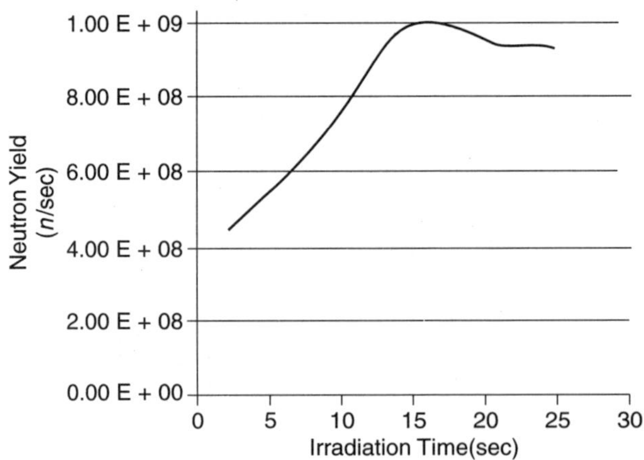

Fig. 2 *Neutron yield measured graph*

Table 1 *Time resolution Result*

Parameters	Flow velocity at 150 cm/s	Flow velocity at 177 cm/s
Inner Diameter of pipe (cm)	1.22	1.22
Kinematic Viscosity(cm^2/s)	0.008	0.008
Reynolds no	22875	26992.5
Friction factor (f)	0.025	0.024
Friction Velocity (U_t) cm/s	8.50	9.82
k_t (Dispersion Coefficient)	52.37	60.53
$\Delta w/v$	0.16	0.14
Δt (Time Resolution) ms	160.34	146.08

Table 2 *Neutron Yield Experimental Calibration*

Irradiation Time (s)	Total Counts	Counting Time (s)	Counts/s	Neutro Yield Measured (neutrons/s)
2	619	6.36	97	4.30×10^8
5	737	6.01	122	5.40×10^8
10	1013	5.97	169	7.60×10^8
15	1425	5.95	239	1.0×10^9
20	2939	13.75	213	9.50×10^8
25	4297	20.48	209	9.30×10^8

3. CONCLUSION

The neutron yield has been measured from the activity of N^{16} at two different water flow velocities, 177 and 150 cm/s. NaI (Tl) detector was arranged at a distance of 1295 cm away from the irradiation point. The neutron generator was operated for various time periods like 2, 5, 10, 15 sec etc. Errors are due to (1) Anisotropic emission of neutrons, (2) Detector Efficiency at 6.13 MeV Peak was calculated using MCNP. In this measurement, accuracy of the ^{16}O (n, p) N^{16} reaction cross section value is an important issue. The emitted neutron spectrum is dominated by 14 MeV neutrons. So the reaction rate is determined using the cross section in the range of 14–15 MeV. In the literature, the experimental values of the cross section is scattered from 30 to 50 mb in this energy range. The evaluated values have 5% difference among different nuclear data libraries. Neutron flux also varies from one spiral to other due to the varying distance from the source point. Some of the errors can be reduced but most of them can't be neglected. Background counting can be reduced by providing sufficient Lead shielding around the NaI (Tl) Detector.

References

1. Radiation Detection and Measurement by Glenn F. knoll, Wiley, 3rd Edition, Publication Date: (2000).

2. Measurement and Detection of Radiation. Tsoulfanidis Nicholas, Taylor and Francis, 2nd Edition, (1995)

3. Introductory nuclear physics by Krane, Kenneth s, & David, John Wiley & Sons, 2nd Edition, (1988)

4. Practical Gamma-ray spectrometry by Gordan.R, Publisher: John Wiley & Sons, Ltd, 2nd Edition (2008.)

5. Absolute measurement of D–T neutron Flux with a monitor using activation of Flowing water Y. Uno, J. Kaneko, (2001)

6. Activation System, Neutron Activation by JAERI (Japan Atomic Energy Research Institute, Mukoyama, Publication Date 2001-Jul-10.

Gamma Background Rejection in C_4F_{10} Superheated Droplet Detector by Pulse Analysis

Mala Das[1,*], Susnata Seth[1], Satyajit Saha[2] and Pijushpani Bhattacharjee[1]

[1]*Astroparticle Physics & Cosmology Division, Saha Institute of Nuclear Physics, Kolkata*
[2]*Applied Nuclear Physics Division, Saha Institute of Nuclear Physics, Kolkata*
E-mail: *mala.das@saha.ac.in*

ABSTRACT

Superheated droplet detector is one of the demanding detectors both for neutron dosimetry and cold dark matter search experiment due to its unique discrimination property between higher and lower ionizing radiations such as neutron & gamma rays. In this paper, the gamma response of superheated droplet detector with C_4F_{10} as the sensitive liquid in soft gel matrix is presented and the effect of energy deposition as a function of temperature of the detector is investigated. The power spectrum shows that the energy released during the bubble formation process via the pulse formation by condenser microphone does not depend on the temperature of the detector. This information is very useful in determining the gamma background when the detector is in a multiple radiation fields.

Keywords: Droplet, Gamma rays, Neurons, WIMPs, Pulse

Pacs No.: 29.40.–n; 95.35.+d

1. INTRODUCTION

Superheated droplet detector consisting of drops of superheated liquid is one of the demanding detectors for neutrons and WIMPs dark matter search experiment [1, 2]. Since its discovery in 1979, the detector is mainly used in neutron dosimetry. For neutron dosimeter, the type of the detector liquid used is the refrigerant liquid of low boiling point such as R12 (CCl_2F_2; b.p. – 29.8°C), R114 ($C_2Cl_2F_4$; b.p. 3.7°C), R134a ($C_2H_2F_4$; b.p. – 26°C) etc.

C_4F_{10} (b.p. – 1.7°C) is the liquid used for WIMPs dark matter search experiment by PICASSO (Project In Canada to Search for Supersymmetric Objects) collaboration at 2 km deep underground SNO Lab, Canada [3]. WIMP induced nuclear recoils are similar to neutron induced nuclear recoils and therefore have identical technique of detection. By operating the experiment at underground, using liquids of different boiling points and compositions, by varying the operating conditions, and employing different rejection techniques, neutrons and gamma rays and other lower ionizing radiations can be rejected which are the leading backgrounds for WIMP dark matter search experiment. PICASSO experiments consists of 32 detectors, each of 4.5 L polymer based hard matrix holding millions of droplets of about 200 micron size with total 2.6 Kg superheated liquids. In the present work, the detector was fabricated in viscous gel medium with C_4F_{10} as the sensitive liquid and gamma sensitivities of the detector with subsequent rejection technique has been studied. Drop size distribution of C_4F_{10} has also been measured. This work is a part of the *R & D* activities related to next generation large mass low background PICASSO detector for dark matter search experiment.

2. PRESENT WORK

The responses of C_4F_{10} detector in presence of ^{137}Cs gamma source was measured by varying the temperature of the detector in the range of 30°C to 55°C. Cs gamma source was obtained at Dept. of Physics, Bose Institute, Kolkata. The volume of the unpurified detector, fabricated at SINP, is 40 ml and it is fabricated under high pressure unlike the present PICASSO detector fabrication which is at low temperature. The detector is placed inside the water bath by controlling the temperature of the water by a temperature controller (Metravi, DTC-200) with precision ± 1°C. The acoustic pulse formed due to bubble nucleation was measured by a condenser microphone, which subsequently converts the acoustic signal to the electrical signal. Trace of each pulse has been recorded in dizitised storage oscilloscope (model no. MSO7032A, Agilent Technologies) and power (*P*) has been calculated from the amplitudes of the pulses [4]. Power, *P* is defined as $\mathrm{Log}_{10}\left(\sum_i |V_i|^2\right)$ where V_i is the pulse amplitude in volt of the digitized acoustic pulse at the ith time bin, and the summation extends over the duration of the signal [4]. Differential power (*P*) distribution of the pulses has been calculated from the trace measurement at different temperatures for ^{137}Cs gamma ray induced events. Drop size distribution in the detector has been measured using the microscope with PC based software.

3. RESULTS

Few results of the various measurements are shown in Fig. 1 through Fig. 4. Fig. 1 shows C_4F_{10} drop size distribution in soft gel matrix used for the present measurement. Observed response of the detector in presence of ^{137}Cs source is presented in Fig. 2.

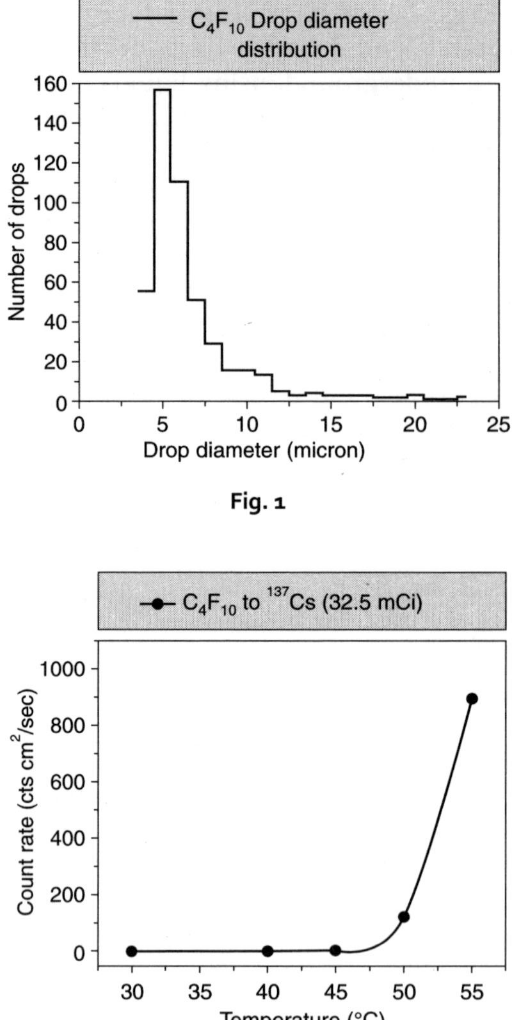

Fig. 1

Fig. 2

Fig. 3 contains the typical pulse output from condenser microphone due to gamma induced event, as recorded in CRO. The duration and amplitude of the pulse depends on the type of the source (neutrons, gamma rays etc) of bubble nucleation. The calculated power distribution is shown in Fig. 4.

4.   DISCUSSIONS AND CONCLUSIONS

The measurement on the droplet size distribution in the detector as displayed in Fig. 1 shows a peak around 5 micron diameter. In the present experiment, the threshold

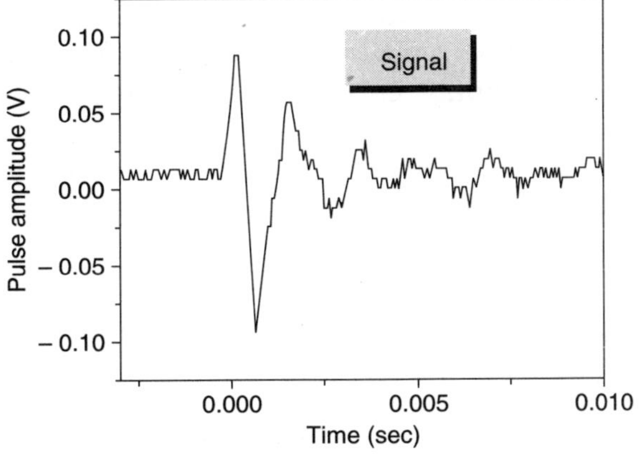

Fig. 3

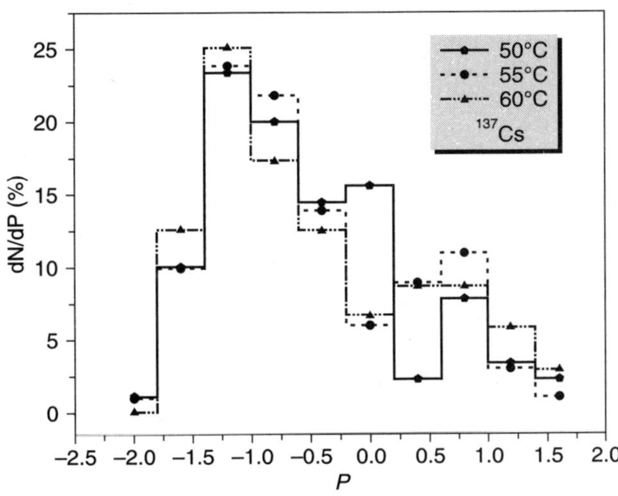

Fig. 4

temperature (T_{th}) of gamma sensitivity of the detector fabricated at the unpurified condition arises in the range 45°C < T_{th} ≤ 50°C (Fig. 2). It is estimated to be about 55.8°C according to d'Errico (2001) which is the mean value of the boiling point (– 1.7°C) and the critical temperature (113.3°C) of the liquid [5]. It is to be noted in this connection that the gamma sensitive threshold temperature of C$_4$F$_{10}$ detector in purified condition, fabricated at clean room environment, as reported by PICASSO experiment is 55°C [3]. Fig. 3 shows that the signal rises at the beginning of the bubble formation and then slowly decreases with many undulations corresponding to the vibrations of the oscillating bubble. From Fig. 4, it is clear that the power distribution does not depend on temperature unlike as observed for neutron induced events. It was observed for neutron induced cases that P depends on temperature and therefore, a temperature dependent power cut is usually applied to reject the neutron induced events [3]. Power distribution encounters the energy released during the bubble nucleation process. For gamma rays, the energy deposition in the detector medium by the electrons occurs almost at the end of their tracks. Therefore, with increases in temperature, although the probability of nucleation increases which causes the increase in count rate but the energy deposition effective to trigger nucleation does not change noticeably.

Therefore, the present observation shows that a temperature independent power cut would be sufficient to reject the gamma induced events for superheated droplet detector employing C_4F_{10} as sensitive liquid, both, for neutron detection in mixed radiation field, and WIMPs search experiment. Further studies in different conditions are going on.

Acknowledgments

Authors want to acknowledge Prof. Viktor Zacek, University de Montreal, Canada and PICASSO collaboration for valuable discussions on the results of the present work.

References

1. R.E. Apfel, US Patent **4143274** (1979).
2. PICASSO collaboration, *Nucl. Instrum. Meths.* A **555**, 184 (2005).
3. PICASSO collaboration, *Phys Lett.* B **682**, 185 (2009).
4. Mala Das, S Seth, S Saha, S Bhattacharya, P Bhattacharjee, *Nucl. Instrum. Meth* A **622**, 196 (2010).
5. F. D'Errico, *Nucl. Instrum. Meth.* B **184**, 229 (2001).

Characterisation of a Si-PIN Diode Detector and its Usage in Gamma Spectroscopy

Abhijit Bisoi[1], A. Nandi[2], M.K. Ray[1], N.A. Rather[1], S. Ray[1], D. Pramanik[3], and M. Saha Sarkar[1*]

[1]Saha Institute of Nuclear Physics, Kolkata, India
[2]Indian Institute of Technology, Kanpur, India
[3]Bengal Engineering and Science University, Shibpur, Howrah, India
E-mail: *maitrayee.sahasarkar@saha.ac.in

ABSTRACT

We have characterised a silicon PIN diode detector and compared its resolution and efficiency with a LEPS detector and tested whether the PIN detector can be used as a detector complementary to a LEPS in nuclear gamma spectroscopy.

Keywords: Si PIN diode, LEPS, Resolution, Efficiency

Pacs No.: 29.30.Kv; 29.40.Wk; 29.25.Rm

1. INTRODUCTION

Silicon PIN diode detectors are usually used for the detection of X-ray and low-energy gamma rays. Depending on the thickness of the Be window, 90% of the incident photons reach the detector at energies ranging from 2 to 3 keV. The detection efficiency is a function of the Si wafer thickness. For a 300 micron thick wafer, efficiency is ~100% at 10 keV, falling to ~1% at 150 keV. These detectors are usually rugged, cheap, available in light, portable configuration with liquid Nitrogen free cooling and can handle high count rate. Small volume planar detectors (Low Energy Photon Spectrometer (LEPS)) are usually sensitive for detection of x-ray and gamma rays below 100 keV. The efficiency of the LEPS detectors starts falling off sharply from around 40 keV. The Silicon PIN diode can be used as a detector suitable for detection of X-rays and/or gamma ray in the energy region where the LEPS detector efficiency decreases rapidly.

In the present work, we have characterised a silicon PIN diode detector and compared its resolution and efficiency with a LEPS detector and tested whether the

PIN detector can be used as a detector complementary to a LEPS in nuclear gamma spectroscopy.

2. EXPERIMENT

The detectors used are, (i) a PIN diode detector (XR-100CR) from AMPTEK Inc. USA, (ii) a small volume planar HPGe (LEPS) from ORTEC with 80 mm^2 active area. The silicon PIN diode detector has an active area of 25 mm^2 (500 μm thick). The detector window is made from beryllium (25 μm thick) and the detector has been thermoelectrically cooled to a very low temperature (around – 45°C) to reduce the leakage current to insignificant level [1]. The radiation sources used are primarily ^{241}Am and ^{133}Ba. For the PIN diode, total 2.5*10^6 counts were collected for the 13.9 keV peak. The peak to background ratio was determined to be 200: 1. A typical energy spectrum of PIN diode detector for ^{241}Am source is shown in Fig. 1.

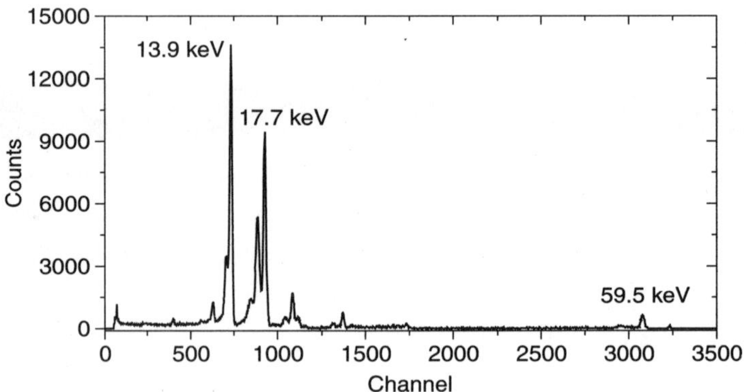

Fig. 1 *Energy spectrum of 241 Am source taken with a PIN diode detector.*

In Pin diode detectors, the electron-hole pairs created by X-Rays, which interact with the back contact of the detector, are collected more slowly than normal events. These events result in incomplete charge collection (Fig. 2) and give rise to spurious peaks. Using the RTD or Rise Time discrimination circuit provided in the detector electronics one can eliminate these peaks (Fig. 2). RTD circuit rejects these pulses due to their long rise times. Similarly, X-ray events that are produced near the edges of these detectors may also result into partial charge collection resulting in a "tail" at the low energy side of an energy peak. An external tantalum collimator was used to eliminate this effect.

3. RESULTS

For characterisation of the detector and to compare it with the LEPS, the energy resolutions as well as the efficiencies of the detectors have been determined. The LEPS efficiency

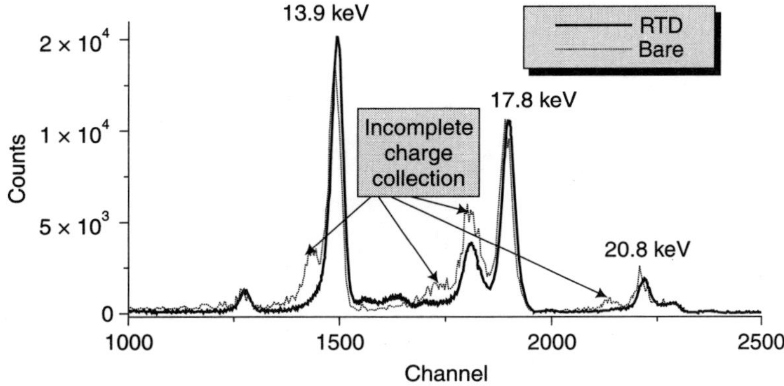

Fig. 2 *Spurious peaks due to incomplete charge collection.*

has been measured till 1408 keV by using ^{152}Eu source also. We have plotted the full width at half maxima (FWHM) of the x-rays peaks to compare energy resolutions (Fig. 3) of these two detectors by using ^{241}Am source. Detection efficiencies of these two detectors (PIN diode & LEPS) with photon energy are shown in Fig. 4. The efficiencies of these two detectors at 13.9 keV of ^{241}Am have been normalised to 1. The comparison of the efficiencies of PIN diode and LEPS is given in Table 1 and Fig. 4.

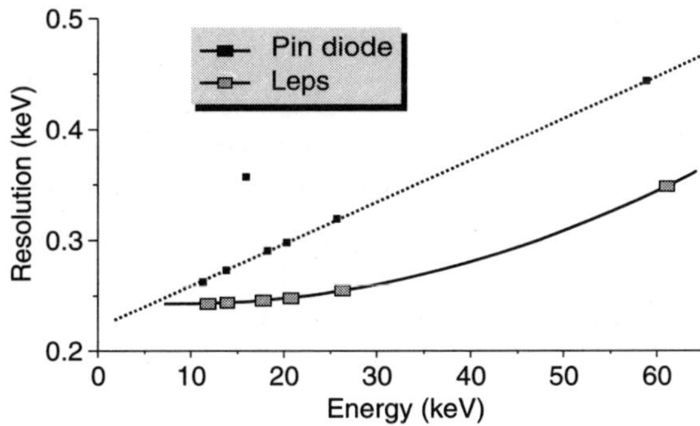

Fig. 3 *Energy resolution of the detectors*

4. DISCUSSION

From Fig. 3, we can see that energy resolution of LEPS is better than PIN diode detector. But there is an indication that at energies below 8 keV, there may be an improvement in PIN resolution. In Fig. 4, for LEPS, peak position of efficiency curve is around 50 keV and below 20 keV it sharply decreases, whereas for PIN diode detector, detection efficiency around 50 keV region is very poor but below 20 keV its detection efficiency improves. Efficiency ratios from Table 1 also indicates that indeed Si-PIN diode detector can be used as complementary to the LEPS.

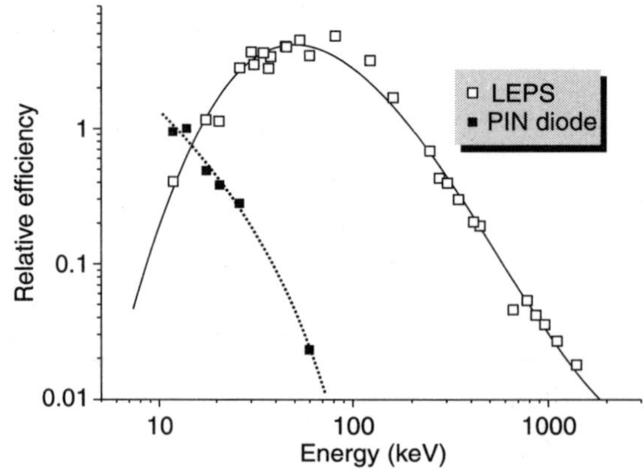

Fig. 4   *Detector efficiency shown as a function of photon energy. Red circles are data points for PIN diode detector and the black squares are that for LEPS. The symbols include the errors. The lines are drawn to guide the eyes.*

Table 1   Comparison of the features of the two detectors.

Type of measurement	Pin diode Detector	LEPS
Efficiency ratio (eff at 11.8 keV)/(eff at 59.5 keV)	44.46	0.13

5.   CONCLUSION

The characteristics of the PIN diode detector show that it can be used as complementary to the LEPS in gamma spectroscopy. Calibrated sources emitting X-rays below 10 keV will be used to characterise Si-Pin diode at lower energies. The improvement of the energy resolution and peak-to-total ratio of this detector by using Rise Time Discriminator (RTD) has been tested. A suitable collimator in front of the detector window has been used to improve the spectrum quality. The usage of Pin diodes as a survey meter will be attempted in future.

Acknowledgements

The authors sincerely thank Prof. Monoranjan Sarkar for allowing them to use the Si-PIN detector and Mr. Pradipta Das for his technical help during the experiment.

Reference

1. G. F. Knoll *Radiation Detection and Measurement* (Wiley India Pvt. Ltd.) (2009).

Comparative Study: Performance of Neutron Survey Meters, Area Monitors with Experimental Validation of Medium Energy Response $n + \gamma$ Survey Meter for 40 and 60 MeV α on Ta

R. Ravishankar[1*], Tapas Bandyopadhyay[1], Sujoy Chatterjee[2], M. Bar[1], Mausumi Sengupta Mitra[1] and Lekha M. Chowdhury[1]

[1]*Health Physics Unit, VECC, HPD, BARC, Kolkata*
[2]*TLD UNIT, RPAD, BARC, VECC, Kolkata*
E-mail: *rravi@vecc.gov.in*

ABSTRACT

The uncompromising quandary in the estimation of dose due to prompt high energy secondary radiations like neutrons and other particles emitted, has prompted to do the experimental and computational studies in case of intermediate energy positive ion cyclotrons like K-130 Cyclotron, Variable Energy Cyclotron Centre, Kolkata. A comparative study of performance of different types neutron and gamma survey meters, area radiation monitors for flux and dose rates have been carried out with experiments using 40 and 60 MeV α on Ta target using K-130 Cyclotron, VECC, Kolkata. Neutron and gamma energy folded spectra have been obtained using BC-501 detectors and dose estimations were done for both n and γ. Estimations of flux and dose have been carried out also using empirical formulations.

Keywords: Neutrons, Neutron survey meters, Energy response.

1. INTRODUCTION

General Neutron Survey Instruments (response normally upto 14 or 20 MeV) and personnel monitoring devices are not available for measuring high energy neutrons above 20 MeV. Computer simulations based on Monte-Carlo methods may be an alternative for the estimation of dose to the persons exposed by chance around those facilities but nevertheless the experimental verification cannot be substituted and are very vital in validating the virtual experiment viz. simulation. Neutron survey instruments are used to detect neutrons with a wide range of energies (thermal to ~14 MeV) and directions. Practically most of the survey instruments are deficient in terms of energy and angular dependence of response to some extent. Mostly two types of neutron survey meters are in use: Andersson-Braun [1] and Leake type [2] [3]. They use BF_3 or ^3He proportional counter located inside a moderator of CH_2 that has a perforated thermal neutron attenuating layer located within it. Such instruments present an ergonomic challenge being heavy and bulky and have caused injuries during radiation surveys. These conventional survey meters though designed for measuring neutrons up to 17 or 20 MeV have been found to have poor response above ~10 MeV which make them insufficient tool in high energy accelerator facilities [4]. Ludlum 2363 Gamma Neutron Survey meter from Ludlum Measurements INC with Proton Recoil Scintillator Los Alamos (PRESCILA) neutron probe is a low weight alternative (~2 Kg) capable of extended energy response (up to 100 MeV), high sensitivity with moderately good gamma discrimination capability. Also it is cable of measuring both neutron and gamma radiation simultaneously with both analogue and digital display.

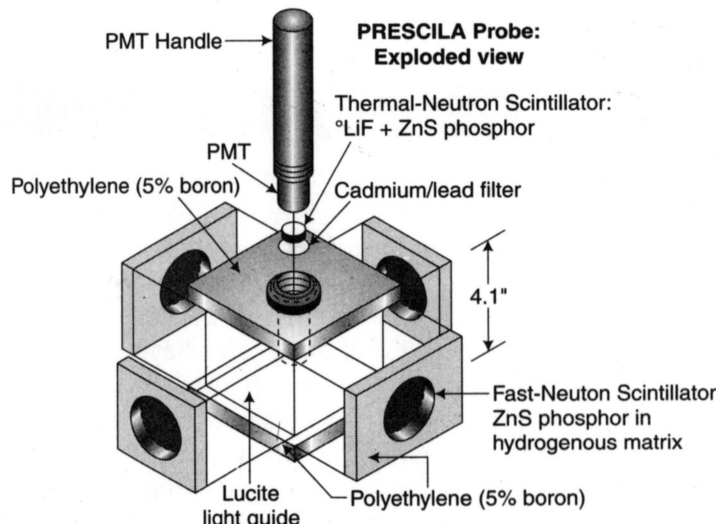

2.   INSTRUMENT FEATURES

The instrument contains an internal energy compensated G-M detector for gamma and 42-41 PRESCILA probe for neutron. An array of ZnS (Ag) based scintillators is located inside and around a Lucite light guide which couples the scintillations to bi-alkali PMT. The use of both thermal and fast scintillators allows the instrument to be used for a wide range of energy spectrum (beyond 20 MeV). The light guide and borated polyethylene frame provide moderation for thermal scintillator element. The inherent pulse height advantage of proton recoils over electron tracks in the scintillator makes it possible to use standard pulse height discriminator instead of any specialized and sophisticated one. The probe is having excellent sensitivity for neutron (350 cpm per mrem/h) and gamma (1050 cpm per mrem/h). The directional response is uniform ($\pm$ 15%) over a wide range of energies. Response linearity has been characterized to over 20 mSv/h. Gamma rejection is effective in gamma fields up to 2 mSv/h. The exploded view of PRESCILA is given below.

3. EXPERIMENT DETAILS

The experiments have been carried out at K-130 VECC with thick Ta target bombarded by 40 and 60 MeV α beams for different beam currents. The range of 40 and 60 MeV α in Ta are 0.013 cm and 0.024 cm respectively. The Ta target was kept in a target holder-collimator arrangement in such a way the secondary neutrons undergo insignificant degradation in energy and number.

Measurements were carried out at 1.2 m from target for neutron dose estimation with Andersson-Braun design type instruments Victoreen 190 N and Studsvik Digipig 2222 A and also with PRESCILA Probe type Ludlum 2363 Gamma Neutron Survey meter for a few angular positions. Also Neutron flux were measured with Area Neutron Monitors (BF_3 based-ECIL Make) for the same set up. The experiments were carried out for different projectile currents (100 nA, 200 nA, 300 nA, 400 nA and 500 nA).

Dose estimations from gross neutron flux measurements obtained from Area Neutron monitors have been analyzed. With BC-501 organic scintillator detector, using Pulse shape discrimination techniques neutron and gamma energy folded spectrums have been generated using unfolding methods for the same setup. The results of dose and flux for both neutron and gamma have been compared. Neutron yields and thereby dose (using suitable flux to dose conversion factors) have been compared with existing empirical formulations in the literature.

Area Gamma monitors have been used to estimate the γ exposure rates at the same setup. Ludlum 2363 was employed to estimate the γ dose also. $CaSO_4$: Dy TLDs have been placed at the same setup, exposed for fixed time and the dose due to γ have been estimated for the same conditions. Results of the experimental data and analysis will be given during presentations.

Prior to experiment all the three neutron survey meters have been tested for dose evaluation with 5 Ci Pu-Be Neutron source at different distances for two angular spacings. The results have been compared with neutron flux measured with area neutron monitors also.

4. RESULTS

The following tables (Tables 1 and 2) show the results of the experiments viz. the neutron dose rates and gamma dose rates. For Area Neutron Monitors (ANM) the dose rates are evaluated from the flux.

Table 1 *40 & 60 MeV alpha on Ta – Sensors kept at 1.2 m at 0 degree and 60 degrees neutron dose rates in mRem/h – dose rates evaluated wherever necessary*

S. No.	Victoreen 190 N	DigiPig 2222 A	Ludlum 2363 : only n part	Evaluated – BC501 A n spec.	Evaluated from ANM flux	Details of Expt
1	2450	550	350	435	300	40 MeV, 100 nA, 0 deg
2	1850	480	340	315	300	40 MeV, 100 nA, 60 deg
3	4500	1410	1050	1100	750	60 MeV, 50 nA, 0 deg
4	4505	1240	980	860	750	60 MeV, 50 nA, 60 deg

Table 2 *40 & 60 MeV alpha on Ta – Sensors kept at 1.2 m at 0 degree and 60 degrees Photon Dose rates in mRem/h – Dose rates evaluated wherever necessary*

S. No.	Automess 6150 AD	Ludlum 2363 only γ	Evaluated – BC501 A γ spect.	$CaSO_4$: Dy TLD. Dose/hr (with 18.5 mm buildup)	Area gamma monitor (ECIL)	Details of expt
1	29	34	32	37	20	40 MeV, 100 nA, 0 deg
2	27	35	31	36	18	40 MeV, 100 nA, 60 deg
3	49	64	76	79	58	60 MeV, 50 nA, 0 deg
4	46	59	62	72	55	60 MeV, 50 nA, 60 deg

5. CONCLUSION

The following conclusions have been made based on the results:

For 60 MeV alpha many reaction channels might have opened and hence more dose with lesser incident primary. Deflector system was not steady and better statistics could not be achieved. It was observed that Vioctoreen 190 N survey-meter has always overestimates the neutron dose. Ludlum 2363 survey-meter appears to have good response above 20 MeV. ECIL Area Radiation Monitors of both Neutron and Gamma have been found to have reasonable response. More experiments being planned with higher energies, varieties of projectiles-targets combinations in future. TLD Dose estimations are to be found in order with others

References

1. Andersson, I.O. and Braun, J.A neutron rem counter with uniform sensitivity from 0.025 eV to 10 MV. In: Proceedings of the Symposium on Neutron Dosimetry, Harwell (1962) (Vienna: IAEA), Vol. **2**, pp. 85–89 (1963).

2. Leake, J.W. A spherical dose equivalent neutron detector. Nucl. Instrum. Methods phys. Res. A **45**, 151–156 (1965).

3. Leake, J.W. An improved spherical dose equivalent neutron detector. Nucl. Instrum. Methods Phys. Res. A **63**, 329–332 (1968).

4. Patterson, H.W. and Thomas, R.H. Accelerator Health Physics. (NY: Academic Press) (1973).

5. National Council on Radiation Protection and Measurements. Radiation protection for particle accelerator facilities. NCRP Report No. 144 (USA: NCRP) (2003).

6. Guo, Z.Y., Allen, P.T., Doucas, G. and Mck Hyder, H.R. Thick target fast neutron yields. Nucl. Instrum. Methods Phys. Res. B **29**, 500 (1987).

7. Clapier, F. and Zaidins, C.S. Neutron dose equivalent rates due to heavy ion beams. Nucl. Instrum. Methods Phys. Res. **217**, 489 (1983).

8. Nandy, M., Bandyopadhyay, T. and Sarkar, P.K. Measurement and analysis of neutron spectra from a thick Ta target bombarded by 7.2 A MeV 16O ions. Phys. .

9. International Commission on Radiological Protection.Conversion coefficients for use in radiological protection against external radiation. ICRP Publication 74. Ann. ICRP 26 (3–4). (Oxford: Elesevier Science) (1996).

10. Instruction Manual Ludlum Model 2363, Ludlum Measurements Inc., Texas.

11. Private communication, Ludlum Measurements Inc., Texas.

12. K-130 Cyclotron Safety Report 2010.

Response of Bubble Detectors and Pocket Dosimeter to Monoenergetic Neutrons from D-D and D-T Reactions

Rupali Pal[1*], Saroj Bishnoi[2], M.P. Chougaonkar[1] and Y.S. Mayya[1]

[1]*Radiological Physics and Advisory Division*
[2]*Laser and Neutron Physics Division*
Bhabha Atomic Research Centre, Trombay, Mumbai, India
E-mail: *rupalirohatgi@rediffmail.com*

ABSTRACT

High energy accelerators are installed in the country for application in medical, material science studies and particle physics. There is need for development of advanced neutron dosimetry systems for high energies. This paper discusses the response of bubble detectors and neutron pocket dosimeters to monoenergetic neutrons produced by D-D (2.45 MeV) and D-T (14.1 MeV) reactions in neutron generator. The response of the detectors can be used for generating a calibration factor, which can be applied when used in reactor and accelerator environment.

Keywords: Monoenergetic neutrons, Neutron dosimetry, Bubble detectors, Pocket dosimeters.

Pacs No.: 87.53.Bn, 29.40.-n

1. INTRODUCTION

High energy accelerator programs are coming up in the country for application in medical, material science studies and particle physics. There is need for development of advanced neutron dosimetry systems for high energies. There are several different types of neutron detectors, both active and passive used for neutron dosimetry and spectrometry. Some of the active detectors are rem counters, Bonner spheres with bare ^{3}He or BF_3 proportional counters, TEPC, LiI (Eu) scintillators, and Si diode based neutron pocket dosimeters. Passive methods include CR-39 track detectors, bubble detectors, activation foils, thermoluminiscent detectors, radioluminiscent glass detectors. This paper discusses the response of bubble detectors and neutron pocket dosimeters to monoenergetic neutrons produced by D-D (2.45 MeV) and D-T (14.1 MeV) reactions in neutron generator. The response of the detectors can be used for generating a calibration factor, which can be applied when used in reactor and accelerator environment.

2.   MATERIAL AND METHODS

2.1   Bubble Detectors

The bubble detector consists of tiny superheated liquid droplets dispersed throughout in a firm elastic polymer (polyacrlamide) [1]. The polymer is contained in a small capped and sealed plastic (plastic detector) test tube. A low boiling point liquid floating above the polymer exerts adequate pressure on it so as to keep it in a radiation insensitive state. The detector is sensitized by removing the cap and discarding the liquid. Upon neutron irradiation visible bubbles are formed in the detector, the number of bubbles being a measure of the neutron dose equivalent. In our experiment we have used BUBBLE DETECTOR-PND (Bubble Detector Personal Neutron Dosimeter) procured from BTI, Canada. This dosimeter has built in compensation for temperature effects on the superheated droplets. This bubble detector responds to fast neutron in the energy region from 100 KeV to 15 MeV.

2.2   Neutron Pocket Dosimeters

Active detectors based on silicon PIN diodes were developed for individual radiation protection purposes in mixed neutron and photon fields. These devices have combination of converters such as $^6Li/^{10}B$ and polyethylene radiators over semiconductor diodes to seek response from slow to fast neutrons. Incident neutrons interact with the converter and produce charged particles that can deposit energy in the semiconductor and cause a signal. Pocket dosimeters used in irradiation to D-D and D-T reactions were procured from, Aloka Co, Japan. (Model PDM-313).

2.3   Neutron Generator Facility

Irradiations were carried out with indigenously built neutron generator facility in BARC. It produces monoenergetic neutrons based on the $D (D, n) \,^3He$ & $D (T, n) \,^4He$ fusion reactions. This is a 300 kV DC electrostatic accelerator (based on Cockcroft & Walton type multiplier) in which the D^+ ion beam current is accelerated and bombarded on a deuterium target. It has demountable target mechanism which makes it easier to switch over between Deuterium or Tritium target. In our experiments we calibrated bubble detectors and neutron pocket dosimeter using D-D (2.45 MeV) and D-T (14.1 MeV) monoenergetic neutrons. Corresponding neutron yield (n/sec) was monitored online with a calibrated ^3He detector.

2.4 Experimental Set-Up

The bubble detectors and neutron pocket dosimeters were exposed to the monoenergetic beam of 2.45 and 14.1 MeV in known neutron yield. The detectors were placed on a semi-circular perspex stand with uniform radius of 25 cm around the beam as shown in Fig. 1. Two bubble detectors and two pocket dosimeters were exposed in free air at three different angles of 0°, 45°, 90° with respect to incident D^+ ion beam to observe the response of the detectors with varying neutron fluence. After irradiation, the bubbles in the bubble detectors were counted in the reader system to calculate cumulative dose. The pocket dosimeter output were in terms of personal dose equivalent (mSv). These experimental values of neutron dose were compared with the values computed analytically. The source strength in 4π space was 4.3×10^6 n/s in D-D reaction and 9.5×10^6 n/s for D-T reaction. The irradiation was for 5 and 10 min for D-D and D-T reactions respectively. The fluence at 25 cm was calculated and multiplied by dose conversion factor (h*(10)) for the respective energies (ICRP 74) to arrive at the reference (delivered) neutron dose equivalent.

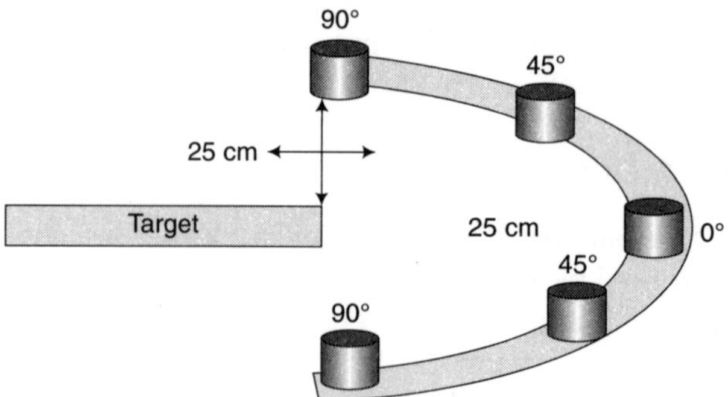

Fig. 1 *Schematic diagram of the experimental setup.*

3. RESULTS AND DISCUSSION

The ratios of measured dose to delivered dose for bubble detectors and pocket dosimeters in D-D and D-T reactions are shown in Table 1. In D-D reaction (2.45 MeV), the ratio in bubble detectors was excellent (unity) showing that response at normal irradiation matches well with calculated values, but in D-T reaction (14.1 MeV), the ratio drops by 43%.

In pocket dosimeters, at 2.45 MeV, the ratio (as seen Col. 3 & 4, Table 1) drops to 0.57 at normal irradiation but there is a fair agreement in 14.1 MeV in the same direction. pocket dosimeters show underestimation of neutron dose at around 2 MeV. The variation in sensitivity of bubble detectors and pocket dosimeters in these neutron energies is in concordance with the observations of Silari et al and D'Errico et al respectively[1, 2].

Table 1 *Ratio of the Measured dose to delivered for bubble detector and pocket dosimeter to monoenergetic neutrons.*

Angles	Ratio of measured dose to reference dose			
	D-D reaction		D-T Reaction	
	Bubble detector	Pocket dosimeter	Bubble detector	Pocket dosimeter
0°	1.00	0.57	0.43	0.98
45	0.77	0.42	0.32	0.90
90	0.68	0.14	0.28	0.49

Graphical representation of the response of bubble detectors and pocket dosimeters to varying neutron fluence with angles is shown in Fig. 2 and Fig. 3 respectively. It is also observed from Fig. 2, as the angle of irradiation increases radially with respect to the target, the ratio in bubble detectors drops to nearly 30% in both energies. As the detectors have cylindrical geometry they have no angular response. Hence, the decrease in response is only due to varying neutron fluence at different angles. the neutron flux is mainly in forward direction and is dependent on the emission angles of neutrons from the source reactions (angular anisotropy)[3]. In case of pocket dosimeter, there is a decrease by nearly 50%. This increase in variation is due to planar nature of the detectors. However, this experiment implies that bubble detectors and pocket dosimeters are sensitive to neutron yield variation.

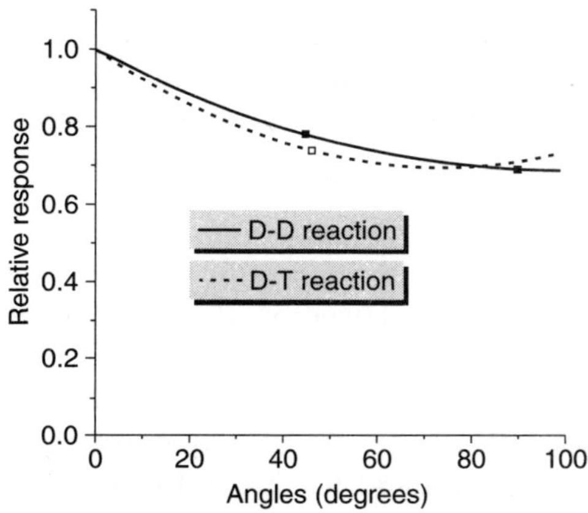

Fig. 2 *Response of bubble detectors at 2.45 MeV (d-d reaction) and 14 MeV (d-t reaction) monoenergetic neutrons.*

4. CONCLUSION

Considering the observations,

(i) The decline in response of the neutron detectors at different angles shows that they are very sensitive to the neutron fluence variation.

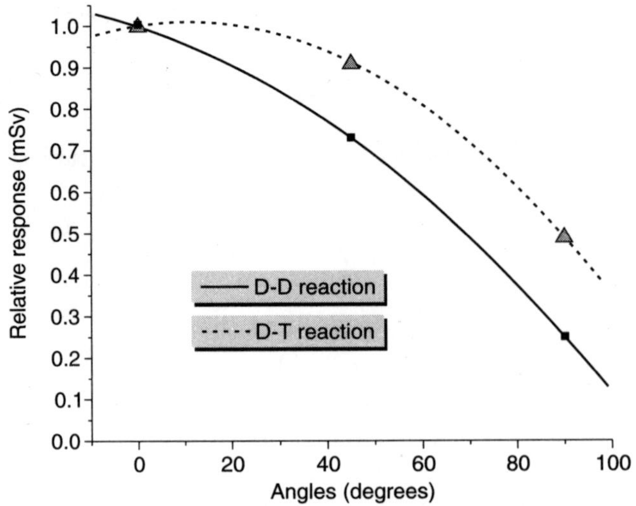

Fig. 3 *Response of neutron pocket dosimeter to 2.45 MeV (d-d reaction) and 14 MeV (d-t reaction) monoenergetic neutrons.*

(ii) The bubble detectors are suitable for measurement of dose in neutron environment around 2 MeV. They show an underestimation in energies above 10 MeV.

(iii) Pocket dosimeters are suited for high energies such as 14 MeV but underestimate dose at around 2 MeV.

(iv) The knowledge of detector response at different neutron energies will enable the user to use these detectors in reactor and accelerator environment with appropriate calibration factor.

Acknowledgements

We are thankful to Dr. S. Kailas, Director Physics Group, BARC and Dr. Amar Sinha, LNPD, BARC for allowing me to carry out the experiment at their facility. We are thankful to Dr. A.K. Ghosh, Director, Health, Safety & Environment Group, and Dr. Y.S. Mayya, Head, RPAD, for his encouragement and support in this work. I am also thankful to Shri Tarun Patel for his help in the experimental setup of the detectors.

References

1. H. Ing, R.A. Noulty and T.D. McLean *Radiation Measurements,* **271** pp 1-11 (1997).
2. M. Silari, Radiation *Protection Dosimetry* **243** pp. 230-244 (2007).
3. Francesco D'Errico, M. Luszik Bhadra, T. Lahaye *Nuclear. Instruments and Methods in Physics Research A* **505** pp 411-414 (2003).

Performance of a Commercially Available Real-time Neutron Spectrometer

Sunil C.[*], S.P. Tripathy, D.S. Joshi, G.S. Sahoo and P.K. Sarkar
Health Physics Division, Bhabha Atomic Research Centre, Mumbai, India
E-mail: *sunilc@barc.gov.in

ABSTRACT

A portable neutron spectrometer was used to measure mono energetic and continuous neutron spectra from various sources. The results obtained are discussed with detector configurations and the unfolding process.

Keywords: Neutron spectrometry, Unfolding, Dose estimation.

1. INTRODUCTION

Neutron spectrometry in particle accelerator is now a well established practice as a result of several important research programs. While this was evidently so for particle accelerator radiation physicist for past several decades, the subject has now found interest in medical physics, and even in conventional neutron dosimetry. Neutron spectrometry is practiced by active techniques such as time of flight [1] and proton recoil telescope or by passive techniques using Bonner sphere [2], activation methodology [3] etc. All these techniques require expertise and are thus difficult to be deployed on a routine basis. There are a couple of neutron spectrometers that are commercially available and in this work we study the performance of one such instrument (Microspec) [4] manufactured by M/s Bubble Technologies Inc. Canada.

The portable hand held readout system is capable of connecting to a wide range of detector probes. In the case of a neutron spectrometer, 2 probes are used to cover the energy region extending from thermal to 20 MeV. The region from thermal to 0.8 MeV is covered by a spherical ^{3}He proportional counter whose cross-section varies as 1/v above thermal energies. However, the spectroscopic information is poor in terms of energy resolution. Further the ^{3}He detector is shielded with a boron shell to achieve a flat response over the thermal and epithermal energy region as the dose equivalent curve is flat in this region. The manufacturer also make use of certain spectral matching techniques to extract neutron energy information that go beyond the traditional "peak resolution" techniques used in conventional spectroscopy. Above 0.8 MeV the neutron probe is a cylindrical organic liquid scintillator of 51 mm × 51 mm size, and this uses the pulse-shape discrimination technique to reject gamma signals. The pulse height data is unfolded using an algorithm and the final spectrum is obtained in predetermined bins. The energy region from thermal to 20 MeV is given in 18 bins. The neutron spectrum so obtained is folded with the fluence to dose conversion coefficients to obtain the dose equivalent.

2.   RESULTS AND DISCUSSIONS

The spectrometer was exposed to four different types of neutron sources (D-T neutron generator, Am-Be neutron source, CIRUS research reactor and FOTIA accelerator) and the results obtained from these experiments are discussed here. A real test for any spectrometer, especially those involving unfolding techniques, is to reproduce the emission from mono energetic sources. The results obtained from the instrument when exposed to 14 MeV neutrons from D-T reactions are shown in Fig. 1. The measurements were carried out at 0°, 45° and 90° with respect to the beam direction. The layout of the beam hall is such that the shield wall is closet in the forward direction. This results in a larger scattering component which reduces as the angle of measurement increases. From Fig. 1, it can be seen that the 14 MeV peak is reproduced reasonably well at all the angles. The area under the peak is less in the forward angles while the scattered component is seen to have significant contribution. This contribution however progressively reduces at larger angles. Another interesting component in the spectrum is the peak at about 2 MeV. This arises from the fact that the deuterium projectiles get embedded in the tritium target over a period of time and starts producing neutrons from D-D reactions which results in mono energetic emission at 2.2 MeV.

In Fig. 2, the neutron spectrum from the tomography beam line of the CIRUS research reactor is shown. The measurement was carried out in the heavily shielded enclosure and a few cm away from the direct beam line to avoid pile up effects. The result shows a peak at about 1-2 MeV and a large thermal energy neutron contribution. The highest energy extends up to about 4 MeV. The shape of the spectrum corresponds approximately to a Maxwellian and the peak is also reproduced reasonably well.

In Fig. 3, the spectrum measured from an Am-Be source is shown. Also shown in the Figure is the ISO standard spectrum whose bin width is adjusted to match the bin size of the measured spectrum. The measured and the ISO spectra [5] are normalized at the peak energy bin.

The results shown in Fig. 3 indicate that the shape of the spectrum above the peak is well reproduced but the low energy region is not reproduced well. This could possibly

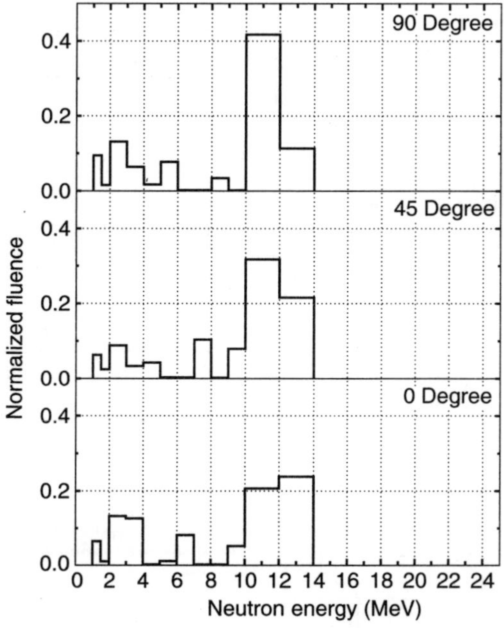

Fig. 1 *Normalized neutron spectra measured at 0°, 45° and 90° from a D-T neutron generator.*

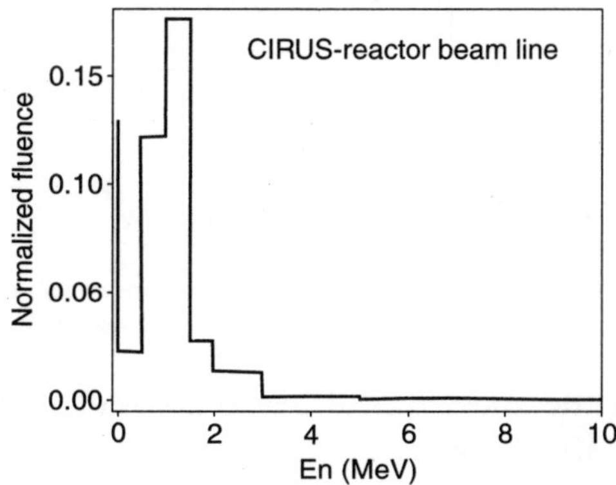

Fig. 2 *Normalized neutron spectrum at the CIRUS research reactor.*

be due to the bin structure of the instrument in the lower energy regions where the ^{3}He proportional counter is used to perform the measurement covering the energy range of thermal to 10 keV in one bin.

In Fig. 4, the neutron spectrum obtained form 5.6 MeV protons bombarding a thick Ta target measured at 30° with respect to the beam direction is shown. The measurement was carried out at the FOTIA accelerator facility in BARC. The overall spectrum appears similar to an evaporation neutron spectrum with the end point energy too appearing

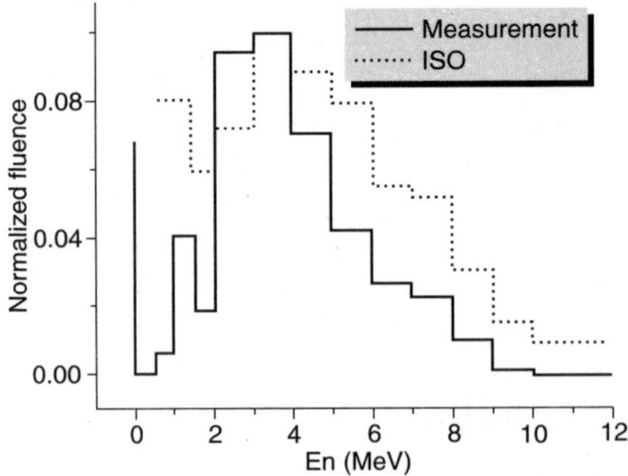

Fig. 3   *Normalized ^{241}Am-Be neutron spectrum: measured and compared with that from ISO report.*

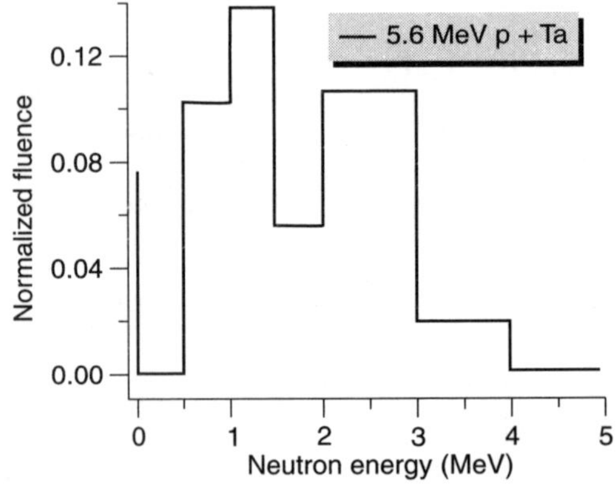

Fig. 4   *Normalized neutron spectrum from 5.6 MeV protons incident on a thick Ta target measured at 30 degree with respect to the beam direction.*

as expected. The drop in the spectrum at 1.5 to 2.0 MeV is however unphysical and could be due to the unfolding process.

3.   CONCLUSION

A commercially available (BTI Microspec) neutron spectrometer was used to measure the neutron spectrum from four different neutron sources namely a 14 MeV neutron generator, Am-Be neutron source, a particle accelerator and the experimental beam port of a research reactor. The instrument reproduces the shape of the spectrum satisfactorily in all the cases. The number of energy bins is too few in the thermal to 0.8 MeV range because of the boron shielding of the ^{3}He detector.

References

1. Sunil C, Saxena A, Choudhury R K, Pant L.M. Nucl. Instr. Meth A, Vol **534**, 518, (2004).

2. Tripathy, S.P., Bakshi, A.K., Sathian, V., Tripathi, S.M., Vega-carrillo, H.R., Nandy, M., Sarkar, P.K., Sharma, D.N, Nucl. Instr. Meth, 598 (2), pp. 556-560. (2009)

3. S.P. Tripathy, C. Sunil, M. Nandy, P.K. Sarkar, D.N. Sharma, B. Mukherjee, Nucl. Instr. Meth A Vol **583**, 421 (2007).

4. Ing, H. *et al.*, Radiat. Prot. Dosim. **126**, 238–24 (2007)

5. Technical reports series No. 403 Compendium of Neutron Spectra and Detector Responses for Radiation Protection Purposes. IAEA (2001).

Monoenergetic Neutron-induced Recoil Track Distribution in CR-39 Detectors: Construction of Response Matrix

S.P. Tripathy[1,*], G.S. Sahoo[1], Sunil C.[1], D.S. Joshi[1], F. Wissmann[2] and P.K. Sarkar[1]

[1]*Health Physics Division, Bhabha Atomic Research Centre, Mumbai, India*
[2]*Physikalisch-Technische Bundesanstalt, D-38116 Braunschweig, Germany*
E-mail: *sam.tripathy@gmail.com, tripathy@barc.gov.in*

ABSTRACT

In this work, an attempt is made to generate a response matrix from the area distribution of neutron-induced recoil tracks in CR-39 detectors exposed to different monoenergetic neutrons at PTB (Germany). The results are compared with those reported by other investigators. The response functions can be used for fast neutron spectrum unfolding. The dose calibration factors ($\mu Sv\ cm^2\ track^{-1}$) for these neutron energies are also presented which can be used for dose assessment directly from track density.

Keywords: Neutron dosimetry, CR-39 track detector, Response matrix, Spectrum unfoldingn.

Pacs No.: 87.53, 87.55.N, 29.40, 61.82.Pv

1.   INTRODUCTION

CR-39 detectors serve as passive integrated radiation detectors and have been routinely used for continuous measurements and total dose assessment in typical radiation environments where other active devices are difficult to be used [1]. Their insensitivity to RF-wave interference and other low LET radiations facilitates discrimination of neutrons from other components. This makes the detector unique for special application of neutron measurements in complex radiation fields such as high-energy particle accelerators [2], cosmic ray neutrons at high altitudes [3], etc.

CR-39 ($C_{12}H_{18}O_7$) detects neutrons primarily by the damaged trails from the recoil nuclei of its constituent atoms. These tracks can subsequently be revealed by suitable etching processes. A brief review on various etching techniques is reported elsewhere [4]. The track density can then be related to neutron dose equivalent subject to calibrations.

These detectors have been used reliably for neutron flux measurement, but to adequately calculate the equivalent dose in a mixed neutron field, one must characterize the energy spectrum of the neutrons, mainly due to energy dependence of the conversion coefficients [5]. Albeit, the neutron spectrometry using CR-39 involves tedious track counting procedures, some successful attempts are reported in literature [6-8]. An alternative method attempted by El-Sersy et al. [9] includes measurement of track length by counting from sides of CR-39 and then correlating the energy and range with the recoil track lengths. However, the counting needs to be done at an angle normal to the exposed surface, which is complicated and inconvenient. The track area measurement is much easier with automatic image analyzing systems [6, 10]. Proton recoil tracks are smaller compared to those due to ^{12}C and ^{16}O recoils, depending on their origin in etched layers of the detector. The larger tracks contribute to higher neutron energies [7].

In this work, an attempt is made to generate a response matrix from the area distribution of neutron-induced recoil tracks in CR-39 detectors exposed to different monoenergetic neutrons. The results are compared with those obtained by other investigators [6, 7]. Dose calibration factors in μSv cm^2 per unit track for all these neutron energies are reported, which can be applied to convert the track density in CR-39 detectors to dose equivalent. The response functions can be used to unfold the spectrum.

2. MATERIALS AND METHODS

CR-39 detectors (INTERCAST, 1.5 mm thick) were exposed to 5 different mono-energetic neutron sources (0.565, 1.2, 5.0, 8.0, 14.8 MeV) at PTB, Germany. The room scattered components were corrected by exposing another set of CR-39 detectors with a shadow cone between the source and the detectors. Source to detector distance in each case was 1 m.

The neutron-induced recoils tracks, as registered in CR-39 detectors, were developed by the optimized chemical etching procedure (6.25 N NaOH, 70°C, 9 h) [11]. The developed tracks were imaged at a magnification of 400 X using an automatic image analyzing system. A new profile compatible for the recoil track measurement was created for interactive based automatic batch processing of the images. The interactive measurements were required at critical steps of frame selection to have a better control on the selection, rejection and segregation of the tracks during the analysis process, which improves the signal-to-noise ratio. The parameters such as density, area, diameter, roundedness, major and minor axes, etc. were measured automatically for sufficiently large number of tracks in each detector surface. Net track density was obtained by subtracting the background tracks. Bulk-etch rate was obtained by gravimetric method.

The response functions can be used to unfold the spectrum.

3. RESULTS AND DISCUSSION

Fig. 1 shows the variation of dose equivalent per unit neutron fluence per track at different neutron energies. The reference values and shadow cone scaling factors (as provided by PTB for the mono-energetic neutron irradiation facility) were taken into account for the calculation of the calibration factors.

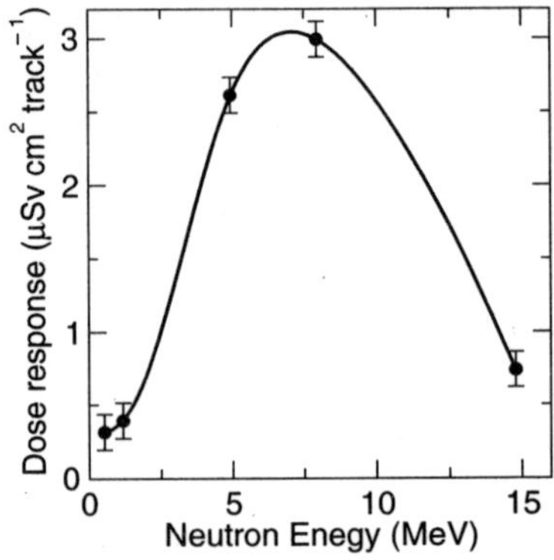

Fig. 1 *Dose equivalent response of CR-39 detectors exposed to mono-energetic neutrons.*

The sharp reduction in response at 14.8 MeV is because of the significant contribution from the carbon and oxygen recoil tracks. At higher neutron energies, while the elastic scattering cross section for hydrogen decreases, heavy recoils and charged reaction products (d, t, α and heavy ions) were increasingly detected. All the tracks having diameters between 2 μm to 30 μm were counted to balance between type-I and type-II errors.

Track area was considered to be more suitable parameter for distribution and analysis because it discards the ambiguity in automatic selection of opposite points in track perimeter for the measurement of diameter, major or minor axes by the software. Fig. 2 represents the response functions constructed from the distribution of recoil track area in CR-39 detectors for different neutron energies. Even if a mono-energetic source is used, a wide spectrum of different area was obtained due to the dependence on the neutron energy, the etching condition of the detector and the angle of scattering of the recorded proton recoil. Since the bin width characterizes the resolution of the distribution as well as the empty energy bins, so different bin widths were plotted and the optimized value obtained in this study is 25.

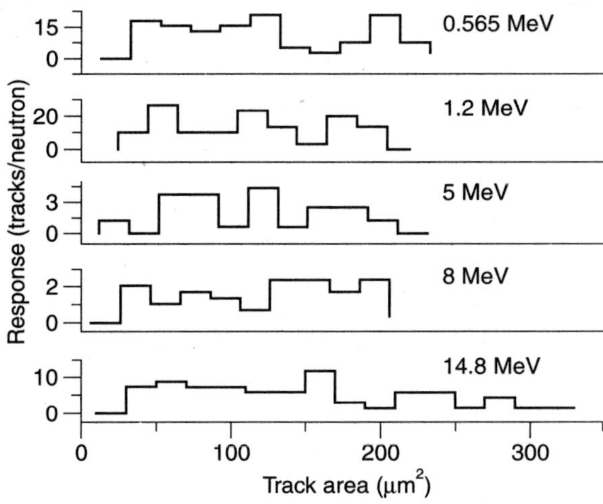

Fig. 2   *Response functions of the CR-39 detectors, as obtained in this work.*

The distributions as obtained in this study (Fig. 2) are compared with the results derived from the reported values by other investigators [6, 7] (Fig. 3). Difference in shape of the curves between Fig. 2 and Fig. 3 is due to larger bin width taken in this study, which reduces the frequency of empty bins.

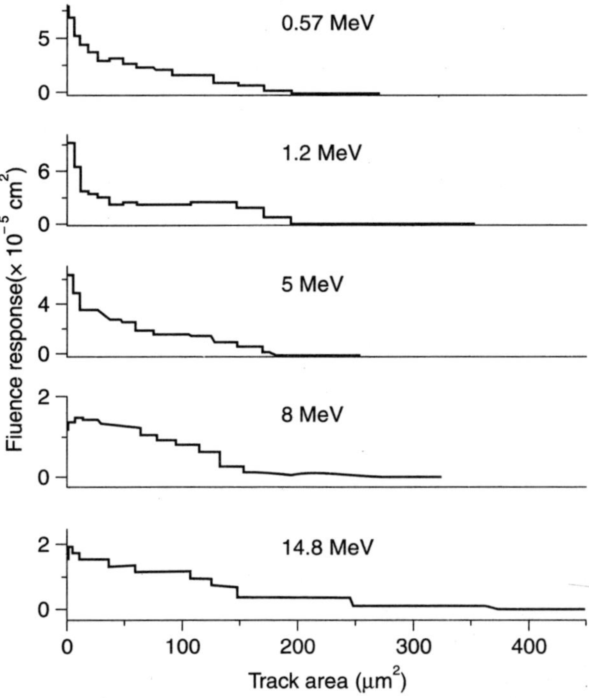

Fig. 3   *Response functions of the CR-39 detectors as taken from ref [6, 7], the track area is calculated from the plotted diameter values.*

This response matrix can be used to unfold the neutron spectrum from the track area distributions in CR-39 detectors exposed to different neutron sources. A systematic study on the unfolding procedure is under progress.

References

1. R L Fleischer, P B Price and R M Walker *Nuclear Tracks in Solids* (Berkely, USA: Univ. of California Press) (1975)

2. P K Sarkar *Radiat. Meas.* **45** 1476 (2010)

3. L Lindborg, D T Bartlett, P Beck, I R McAulay, K Schneur, H Schraube and F Spurny *EURADOS Working Group Report* (Luxembourg: European Commission) (2005)

4. S P Tripathy, R V Kolekar, C Sunil, P K Sarkar, K K Dwivedi and D N Sharma *Nucl. Instrum. Meth. in Phys. Res. A* **612** 421 (2010)

5. ICRP, 2007 *Recommendations of the International Commission on Radiological Protection ICRP Publication 103*, (Oxford: Ann. ICRP, Elesevier Science) (2007)

6. F. d'Errico et al. *Radiat. Meas.* **28** 823 (1997)

7. M. Luszik-bhadra et al. *Radiat. Meas.* **28** 473 (1997)

8. G W Phillips, J E Spann, J S Bogard, T VoDinh, D Emfietzoglou, R T Devine and M Moscovitch *Radiat. Prot. Dosim.* **120** 457 (2006)

9. A R El-Sersy, S A Eman and N E Khaled *Nucl. Instrum. Meth. B* **226** 345 (2004)

10. R Bedgoni *Radiat. Meas.* **36** 239 (2003)

11. I Lengar, J Skvarc and R Ilic *Nucl. Instrum. Meth. B* **192** 440 (2002)

Integrated Photon Dose Mapping in Indus-1 Experimental Hall using Ionization Chamber based Survey Meter

Vipin Dev[1*], S.K. Mishra[2], T.K. Sahu[2], Haridas G.[2], K.K. Thakkar[2], Gurnam Singh[1] and P.K. Sarkar[2]

[1]*Raja Ramanna Centre for Advanced Technology, India*
[2]*Health Physics Division, Bhabha Atomic Research Centre, Mumbai, India*
E-mail: *vipindev@rrcat.gov.in*

ABSTRACT

Integrated photon dose mapping of Indus-1 Synchrotron radiation Beam Lines (accessible area) was carried out using ion chamber based survey meters. The study shows that radiation levels in Indus-1 experimental hall is above permissible limits during injection mode of operation at some locations and hence entry prohibition to the area is confirmed during this mode of operation. However, during storage mode (user mode) radiation is of background level and no entry restriction is enforced.

Keywords: Ionization chambers, Synchrotron radiation beam lines, Dose mapping.

Pacs No.: 29.40.Cs, 41.60.Ap, 28.41.Te

1. INTRODUCTION

Indus-1 is a 450 MeV Synchrotron Radiation Source with critical wavelength of 61 A°. It comprises of pre-injector 20 MeV Microtron, 450 MeV Booster Synchrotron and storage ring (450 MeV). Indus-1 has three operating modes viz. Injection, storage and user mode. Indus-1 has five synchrotron radiation (SR) beam lines emanating from 4 bending magnets. A schematic diagram of Indus-1 along with different beam lines is shown in Fig. 1. Radiation dose measurement in the experimental hall was carried out in Injection mode of operation. Integrated dose measurement was performed as the radiation field during the injection process is of pulsed nature. In a pulsed field, dose rate measurement with a radiation detector show a pulsating field and hence correct information about the dose rate cannot be obtained. In this paper, bremsstrahlung x-ray photon dose measurement carried out is described and the results are discussed.

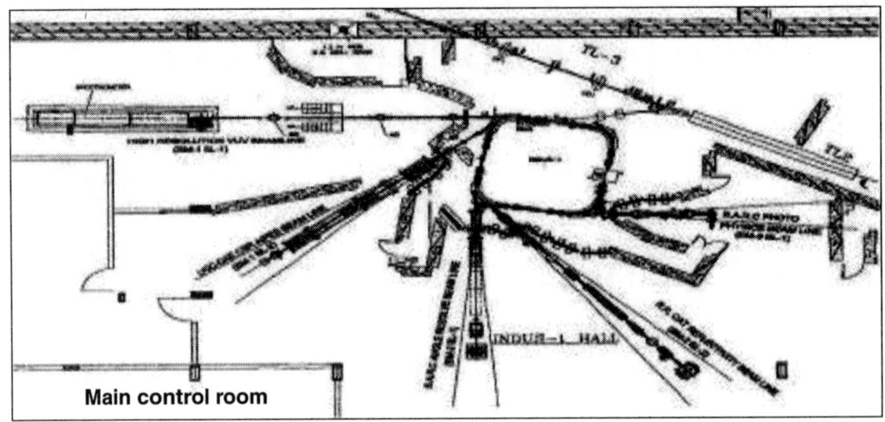

Fig. 1   *Schematic diagram of Indus-1 with SR beam Lines*

2.   DOSE MAPPING

Nine ion chamber survey meters used for experiment were calibrated first with Co-60 radioactive source (6.5 mCi). The response of all survey meters are found to be within ± 20%.

There after photon dose mapping at various locations of the SR beamline using the calibrated survey meters during the injection was carried out. Measurements were carried out in the integrated mode due to two reasons:

1 As the experimental hall area is inaccessible during injection mode.

2 Fluctuating dose rate due to pulsed radiation on account of injection.

Measurements were carried out at three different locations of five SR beam lines: top of hybrid shielding (HS) around the ring, central hole of the HS, experimental station and in the control room.

The measurement locations are shown in the schematic diagram given in Fig. 2. All the survey meters were positioned at the selected locations and injection process was started. At the end of injection process the radiation dose recorded by the survey meter was noted. It represent total dose received in Indus-1 experimental hall at the locations during the injection time. From the data, dose rate during injection was calculated.

3.   RESULTS AND DISCUSSION

Table 1 shows the experimental results at different locations of the beam line and the experimental conditions for the measurements are shown in the Table 1a.

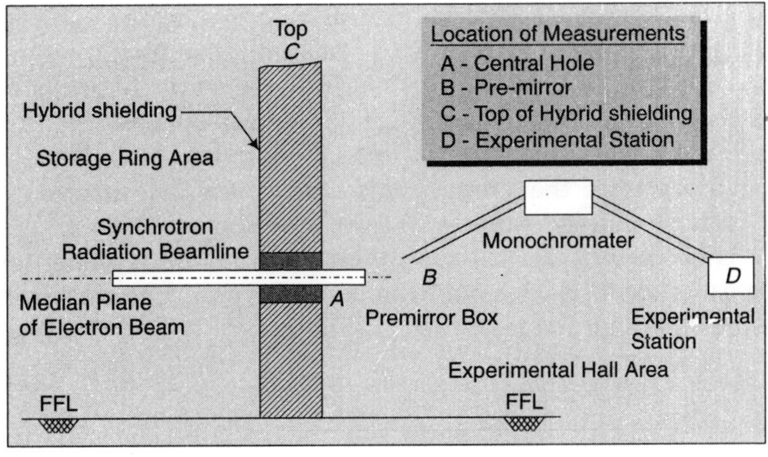

Fig. 2 *Locations of measurements*

Table 1 *Dose rates in Indus-1 experimental hall and control room during injection*

Location	Dose rate (mSv/h)		
	On top of hybrid shielding	Central hole of hybrid shielding	Experimental station
HRVUV beam line, BARC	0.23000	1.32000	0.00290
AIPES Beam Line UGC DAE	0.90000	0.41000	0.00120
ARPES Beam Line BARC	0.08000	0.00200	0.00170
Reflectivity Beam Line, RRCAT	0.08000	0.00700	0.00170
Photo Physics Beam Line BARC	0.06000	0.00300	0.00070
Control Room			0.00108

Note: HRVUV – High resolution vacuum ultraviolet, **AIPES:** Angle integrated Photo-electron spectroscopy, **ARPES:** Angle resolved photo-electron spectroscopy.

Table 1a *Experimental conditions during the measurement*

Parameters location	Parameter value		
	On top of hybrid shielding	Central hole of hybrid shielding	Experimental station
Booster DCCT (mA)	4.0	4.0	4.0
Avg Injection rate (uA/s)	208.0	500.0	500.3
Exposure Time (Sec)	275.0	600.0	215.0

Among the location in experimental hall maximum photon dose rate during injection was obtained near HRVUV Beam line area as this beam line comes in the direction of beam injection into Indus-1. On outer side of the hybrid shield, the central hole shows maximum of 1.32 mSv/h. As the central hole is shielded with lead shot filled bags, non-uniformity in shielding results in streaming of radiation. On top of hybrid shielding, a maximum of 0.90 mSv/h (unshielded dose as the hybrid shield is only up to 1.75 m height). At experimental station 2.9 μSv/h was obtained. Relatively less dose rate is observed at the experimental station, compared with other locations as it is typically, ~15 m away from the storage ring. During the measurements the injection current from Booster remained same. However injection rate was different for during measurement at the locations. For the central hole location the injection rate was higher whereas for the other locations the rate was poor. For poor injection rate the dose rate at central hole location may likely go up.

4. CONCLUSION

Following conclusions can be made from the study
1. It is observed that the radiation field at all locations of HRVUV BARC, AIPES UGC beam lines are comparatively higher than other beam lines as these beam lines are coming in the direction of injection line of Indus-1.
2. Among the locations studied, central hole in the hybrid shield gives the maximum dose rate.
3. The entry prohibition to Indus-1 experimental hall during injection is confirmed.

Acknowledgements

The authors wish to thank Dr P D Gupta, Director, RRCAT, Shri Narayan Kutty, Health Physics Division, BARC for providing guidance and support for work. We would also like to thank members of operation staff for their support.

Reference

1. "Experimental study of the radiation environment in Indus-1 storage ring using TLD badge" Nayak et al in proceedings of, National Conference on Indian Association of Radiation Protection, IARPNC (2008).

Neutron Spectrum Unfolding using Genetic Algorithm

V. Suman* and P.K. Sarkar

Health Physics Division, Bhabha Atomic Research Centre, Mumbai, India

E-mail: **vitisha@barc.gov.in*

ABSTRACT

To obtain Neutron spectrum from the measured data and using the response function of the measuring instrument many unfolding techniques are available. This paper discusses a genetic algorithm based unfolding technique that has been used to unfold the neutron spectra obtained from Li(p, n) Be and neutron response of an NE-213 liquid scintillation detector. The results so obtained were compared with those obtained using the established unfolding code FERDOR, matrix inversion calculations and Iterative method. It is shown that the genetic algorithm based solution has a very good match with the results obtained from iterative and matrix inverse methods and has a better result compared to the FERDOR output. The method also has potential advantages of speed, flexibility, computational simplicity and works well without need of any initial guess spectrum. This method appears to be a promising candidate for unfolding neutron spectrum for under determined problems where measured data are very less.

Keywords: Spectrum unfolding, Genetic algorithm, FERDOR, Fredholm integral.

1. INTRODUCTION

With the widespread use of nuclear reactors, accelerators and other neutron source facilities (in radiation therapy, material damage studies, production of radioisotopes for nuclear medicines etc.), neutron energy spectrum characterization is of great importance in such facilities. The precise knowledge of the neutron energy spectrum is important for various fundamental and experimental studies in different fields of nuclear physics and health physics for radiation protection calculations of the facility design, and dosimetry purposes.

Unfolding or deconvolution is a mathematical procedure for analyzing measured data, obtained through a detector system. Except for time of flight spectrometry any other neutron spectrum measurement technique either active or passive involves deconvolution of the measured data. Few of the active and passive detector systems to mention here are Bonner sphere spectrometer (BSS) also known as multisphere spectrometers, NE213 scintillation spectrometer, ^{3}He proportional counter, as active detectors and under passive detector comes neutron activation foil method. To unfold the neutron spectra, several mathematical approaches have been developed, among which the most popular are least-square, iterative and Monte Carlo methods [1]. There are few recently proposed methods such as methods using artificial neural networks (ANN) [2] and evolutionary processes based genetic algorithm (GA) [3] to unfold neutron spectra from detector count measurement and to perform neutron dosimetry.

Using the known information of fluence-response characteristics of the detector system, which varies with neutron energy, the neutron energy spectrum thus to an extent can be unfolded from the proton recoil scintillation pulse height distribution of the NE213 detector measurements. The detector's count rate $C(dN/dE)$, response matrix $R_\Phi(E', E)$ and the neutron spectrum $\Phi_E(E)$ are related through the Fredholm integral equation of the first kind

$$C = \int_{E_{min}}^{E_{max}} R_\Phi(E', E)\ \Phi_E(E)dE \tag{1}$$

This paper reports use of the genetic algorithm technique - an evolutionary method to search for a single globally optimized solution from a randomly generated population of prospective multi variate solution vectors. We compare the variate results obtained using the genetic algorithm with those obtained using the standard code FERDOR and an iterative technique and inverse matrix calculation.

2. BASIC PRINCIPLE OF GENETIC ALGORITHM

Genetic algorithm [4] is inspired from the natural selection and natural genetics; it mimics the Darwinian theory of Survival of the fittest. The basic principle of Genetic Algorithm deals with improving the result over a certain closed loop operations by successively incorporating the transitional results aiming to achieve a better solution. Genetic algorithm is a stochastic process and suitable for searching a single optimized solution – global optima or minima in a complex and multidimensional search space. The genetic algorithm search mechanism is probabilistic and is initiated through the creation of a large number of prospective solutions. The potential solutions are referred to as individuals and when encoded in binary strings or real values are called 'chromosomes'. All genetic algorithms contain four basic operations: initial population generation (reproduction), selection, crossover (mating) and mutation, where all are analogous to their namesakes in genetics. The worth of each individual within a population is numerically calculated and the parameter is referred to as 'fitness'.

To start with a large but fixed size of population is generated randomly. The fitness of each of the prospective solution is calculated which is represented by the parameter fitness. Under selection parent solutions with good fitness values are randomly selected, the parent with better fitness values stand the chance to be selected more than once therefore chances of duplicacy is there, the solutions with low fitness values are removed from the new population. For crossover pairs of the fittest parents are selected randomly and are crossed over (mated) at a randomly selected gene location to produce two offspring i.e. two new solutions. Individuals with higher fitness values are preferentially selected to mate so as to pass on the strong genetic material compared to rest in the population. The crossover results in exchange of high performing genetic info between two highly good solutions in hope to produce a pair of superior solutions. The solutions with lower fitness values also mate but they do so with a low frequency and with the passing generations due to evolutionary weightage they are not able to survive and are in the process killed off. In the meanwhile the solutions with better solutions and the information carried by them is carried forward to prospective generation to result in a solution with the best match with the requirement.

Genetic algorithm also mimics the process of genetic mutation by randomly changing the solution structure. Randomly a solution is selected from the population and a randomly selected gene is modified, introducing a new gene material into the population. The mutation helps the solution to reach the global minima and not to get stuck in any local minima. All the steps can be compiled in an algorithm

Simple Genetic Algorithm ( )

{

Initialize Population ( );

Evaluate Population ( );

While termination criteria is not reached

   {

   Select solution for next population;

   Perform crossover ( );

   Perform mutation ( );

   Evaluate population ( );

   }

}

All the processes through successive generations collectively lead to solution evaluation. As solution evolves better and better solutions with improved fitness values are produced. Genetic algorithm very efficiently and effectively explores through very large search spaces with very good timing. The representation of the methods is given in the methodology.

3. METHODOLOGY

The Fredholm equation (1) is discritised using the quadrature approximation

$$C_j = \sum_{g=1}^{n} R_{jg}\ \Phi g \quad \text{for } j = 1, 2, \ldots J \tag{2}$$

where, C_j is the recorded count in the jth channel, R_{jg} is the response matrix, coupling the jth pulse height interval with the gth energy interval, and ϕ_g is the total neutron fluence in the gth energy interval. The group of equations resulted from above equation generally does not have a unique solution, even if the experimental errors are neglected. Physically this is caused by the fact that part of the information contained in the response function is lost due the finite resolution of the measuring devices. To obtain a physically meaningful solution some kind of priori knowledge must be used.

In our case the individuals are the guess neutron spectra's and their goodness is judged how well they can match with the measured NE213 detector measurements. The prospective solutions were randomly generated between x_z and x_H and these constitute the population.

$$x_i = x_L + (x_g - x_L) \times \text{random number}$$

for
$$t = 1, 2, \ldots 6$$

The process of mating or crossover and mutation through successive generations leads to solution evolution. The process of crossover and mutation are shown below

(a) Crossover: Choosing two parents randomly from the population, say $x_1 x_2 x_3 x_4 x_5 x_6$ and $x'_1 x'_2 x'_3 x'_4 x'_5 x'_6$ and choosing a point randomly for crossover, say, 3rd position; then the daughters so generated will be $x_1 x_2 x_3 x'_4 x'_5 x'_6$ and $x'_1 x'_2 x'_3 x_4 x_5 x_6$

(b) Mutation: An individual is selected randomly from the population and a location in the individual is picked randomly to be modified by some value δ, say, we pick one of the above generated daughters $x'_1 x'_2 x'_3 x_4 x_5 x_6$ (only a representation) and choose the 5th value for modification

$$x_5^{new} = x_5 \pm \delta$$

after mutation the new individual will be

$$x'_1 x'_2 x'_3 x_4 x_5^{new} x_6$$

All the daughters so generated are included in the population, they are checked for their fitness, sorted and then the unfit ones are removed from the population keeping the population size constant. At the end of first generation the obtained population is passed on to the next generation and this is repeated till the goal is satisfied. As solution evolves, better solutions with higher fitness values are produced. In our case genetic algorithm searches for better fitness individuals i.e. solutions that reproduce good agreement with the detector measurements.

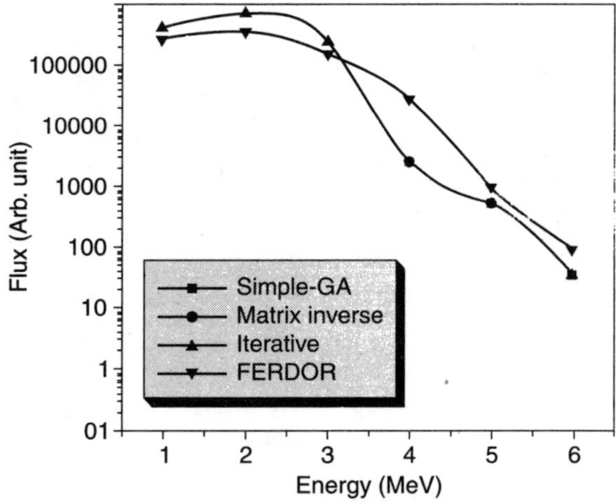

Fig. 1 *Spectrum obtained by NE213 liquid scintillator unfolded with genetic algorithm, along with matrix inversion method, Iterative method and FERDOR code.*

4. RESULTS AND DISCUSSION

We explored a completely determined problem i.e. we used a set of six measurements and a response matrix was constructed for six energy intervals. A population was constructed with 100 randomly formed individuals. The fitness was evaluated for the individuals and were sorted accordingly. The individuals were selected based on fitness maintaining the population size constant at 100, with a probability of a better solution to be selected more than once resulting in duplicacy. Crossover was carried out with a crossover probability of 0.7 which was done randomly selecting two parents and randomly picking a position to interchange. Mutation was carried out randomly with a mutation probability of 0.1, the goal was set as 1% and the maximum number of generations was kept to be 500. The spectrum unfolded using GA-based spectra unfolding method is shown in Fig.1 in comparison with the neutron field parameters unfolded using FERDOR code, iterative method and matrix inversion calculation.

The spectrum unfolded using genetic algorithm shows a good match with the one unfolded using matrix inversion and iterative method and it resembles closely with the FERDOR unfolded spectra.

An important observation was that when the flux so calculated by various methods is folded back to get the activity for the respective methods and the genetic approach taken by us gives a very good match with the experimental measurements but in case of FERDOR approach the folded spectra results in huge differences in folded back values and the measurements as shown in the Fig. 2, the lower region of the curve is plotted on log scale and is shown in the Fig. 3.

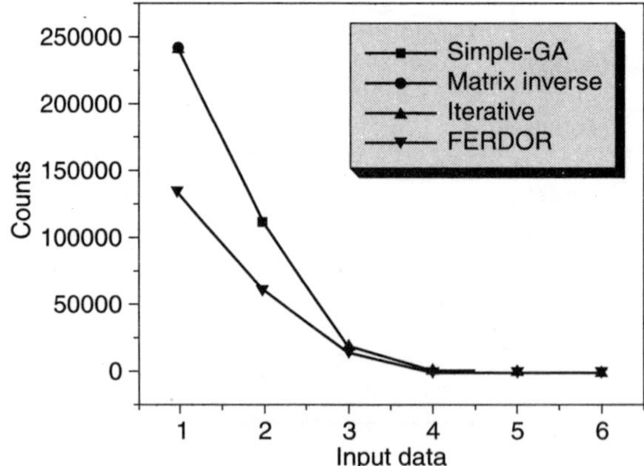

Fig. 2 *Comparison between the folded back spectra in case of GA, matrix inversion, Iterative and FERDOR outputs.*

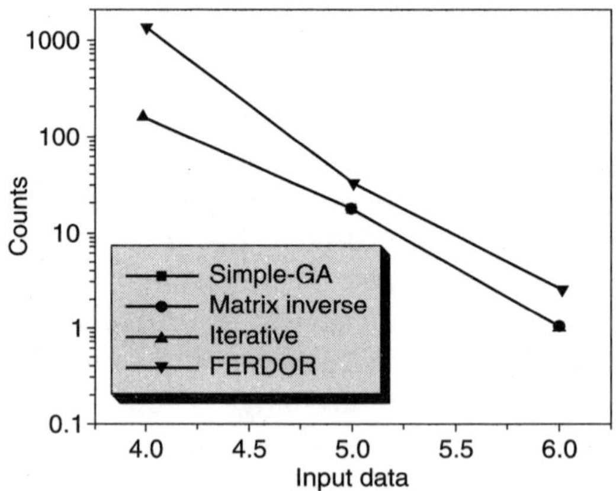

Fig. 3 *Comparison between the lower part of Fig. 2 on log scale.*

The GA approach overcomes the problem of inconclusive outputs generated using matrix inversion methods in case of under-determined problem, as few of the calculated flux values come out to be negative, resulting in an unphysical solution. The plot of the fitness with successive generation is shown in Fig. 4 it can be seen that the population we start with is completely random with very high percentage error value of order 10^5 which finally over the generations reduces to an error of just 1%.

The run time for the code was of order few seconds which further reduces if population size is reduced but then the no of generations to get the same result will increase. Genetic algorithm also lifts the constraint for supply of some initial guess spectra, as it starts with a completely randomly generated solution set and at the end converging to the desired result.

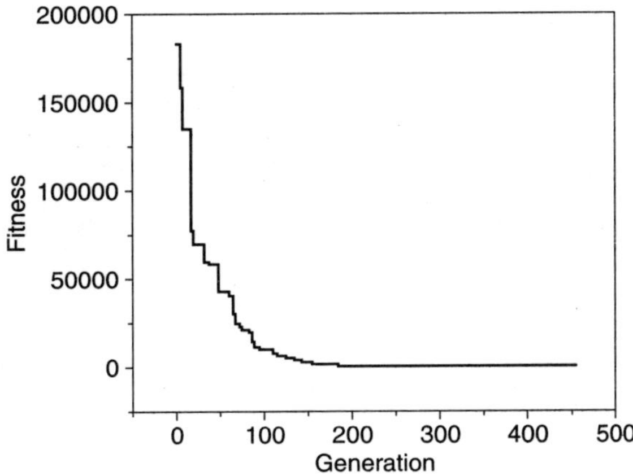

Fig. 4 *Curve for fitness versus generations.*

Acknowledgments

The authors are thankful for the help provided by Dr. C. Sunil and encouragement from Dr. A. K. Ghosh, Director, Health, Safety and Environment group is thankfully acknowledged.

References

1. M. Matzke, Report PTB-N-19, Physikalisch-Technissche Bundesanatalt, Braunschweig, 1994.
2. A. Sharghi Ido, et al., Applied Radiation and Isotopes 67 (2009) 1912.
3. B. Mukherjee, Nuclear Instruments and Methods in Physics Research A 432 (1999) 305.
4. J.H. Holland, Adaptation in Natural and Artificial Systems, University of Michigan Press, Ann Arbor, 1975.

Study of Dose Deposition in CR39, Makrofol and PMMA and its Applications in Hadrontherapy

Summit Jalota, Ashavani Kumar* and Asim Mahakul

National Institute of Technology, Kurukshetra, India

E-mail: *ashavani@yahoo.com*

ABSTRACT

The estimation of energy deposition of O^{8+}, C^{6+} and He^{2+} ion beams in CR39 $(C_{12}H_{18}O_7)n$, PMMA $(C_5H_2O_8)_n$ and Makrofol $(C_{16}H_{14}O_3)_n$ polycarbonates media were made and other characteristics are studied by using simulation toolkit: $Geant_4$. The energies of the ion beams are so adjusted that the range of ions are limited to about 5 cm. Peak-to-entrance ratios lie in between 7.2 to 9.2. Higher peak-to-entrance ratio leads to higher deposition and less secondary fragmentation reactions.

Keywords: Bragg peak, $Geant_4$, Hadrontherapy, Linear energy deposition, Makrofol, CR39 and PMMA.

Pacs No.: 87.50

1. INTRODUCTION

The interaction and propagation of energetic heavy ions in matter is of great interest in the fields of nuclear, medicine and particle physics. Recently, the interaction and transport of light ions in tissue-like matter became of particular interest in hadrontherapy [1]. Water as tissue like media is assumed by many researchers [2-4]. There is no doubt about that the water is the major contents in human body, but we cannot ignore the presence of carbon and calcium like materials in the body. Particularly, in hadrontherapy, we study the heavy ions energy loss in human tissues like media and its relative biological effects (RBE). Heavy charged particles passing through shielding material will result the energy loss of the particles mainly due to interaction and fragmentation. Therefore, the study of irradiation of heavy ions on plastic materials, consisting carbon, hydrogen and oxygen present in various proportions, will be supportive study of the energy loss and its RBE in various types of tissues present in human body. The study of depth-dose profile for various ions at intermediate energies in extended tissue like media CR39 $(C_{12}H_{18}O_7)_n$, PMMA $(C_5H_2O_8)_n$ and Makrofol $(C_{16}H_{14}O_3)_n$ polycarbonates provides the better platform for hadrontherapy. To estimate the energy deposition in these media and other characteristics are studied by using simulation toolkit: Geant4. In the present study, CR39, Makrofol and PMMA polycarbonates are used as target material in simulation study to plot linear energy deposition per mm per particle versus depth. Moreover, the knowledge about depth-dose profile of energetic charged particles in these plastic materials is of great importance for the study of treatment planning. As ion therapy requires best knowledge of both physical and biological dose distribution, the present study of effective depth-dose deposition profile under the Bragg's peak, considering polymers as media, will provide a good platform for the treatment planning and future scope in hadrontherapy.

2. METHODOLOGY

The simulation of depth-dose distributions were made by Monte Carlo simulation using GEANT4-toolkit [5]. It is based on object-oriented technology to achieve transparency of physics implementation and provide the possibility of validating the physics results. For precise calculations and good estimations, highly mono-energetic pencil beam is used to focus on desired position. This may prevent the damage of healthy tissues in unwanted portions. The geometry of targets/media assumed is very simple. These are three dimensional boxes whose two dimensions are equal and third axis parallel to ion-beam is longer to get dose deposition beyond range. Target material chosen is of uniform density for better results.

3. CODE DESCRIPTION

The major categories of processes provided by Geant4 are electromagnetic, hadronic and transportation. All these processes are included in the present simulation work. To study recoils and reaction products, elastic scattering interaction and inelastic reactions are included as well as the Binary Cascade model [5] for one-on-one collision.

4. RESULTS AND DISCUSSIONS

We obtained the response of O^{8+}, C^{6+} and He^{2+} ion beams in PMMA, CR39 and Makrofol polycarbonates medium with densities $1.18g/cm^3$, $1.32g/cm^3$ and $1.23g/cm^3$ respectively. In order to get almost the same depth position (~5 cm), we adjusted the beam energy to achieve the Bragg Peaks at the same location, yet their entrance dose and peak height are different depending on their RBE. The depth-dose profiles for various ions at different energies are shown in Fig. 1-3. The position of the Bragg's peaks is selected at a distance of about 5 cm by keeping in mind the average locations for the sensitive body organs from the surface of the skin. The entrance dose, Bragg's peak position, peak to entrance ratio at various energies in PMMA, CR39 and Makrofol polycarbonates are presented in the Tables. 1-3.

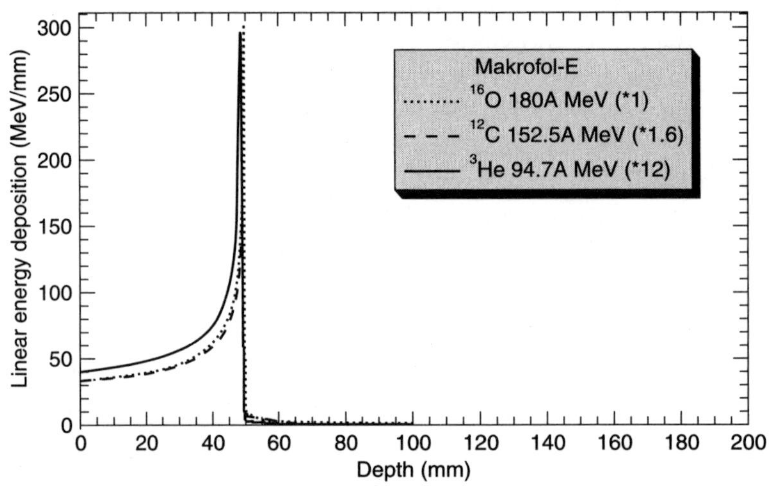

Fig. 1   *Calculated depth-dose distribution for (a) beam of ^{16}O (180A MeV), ^{12}C (152.5A MeV) and ^{3}He (94.7A MeV) in Makrofol polycarbonate.*

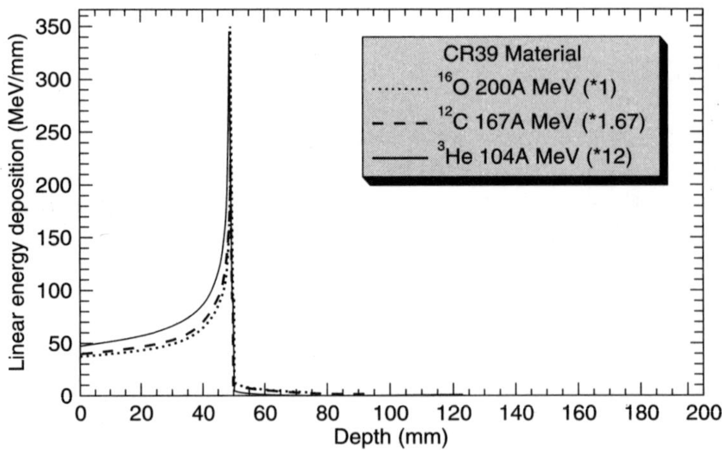

Fig. 2   *Calculated depth-dose distribution for (a) beam of ^{16}O (200A MeV), ^{12}C (167A MeV) and ^{3}He (104A MeV) in CR39 polycarbonate.*

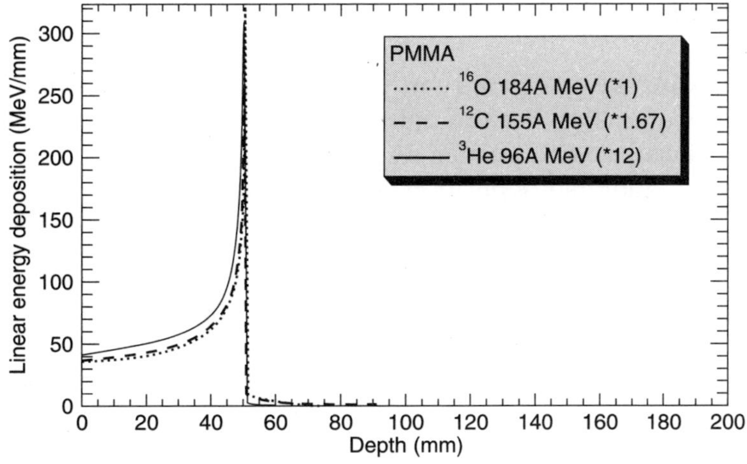

Fig. 3 *Calculated depth-dose distribution for (a) beam of ^{16}O (200A MeV), ^{12}C (167A MeV) and ^{3}He (104A MeV) in PMMA polycarbonate.*

Table 1

PMMA			
Ion-beam	Energy (A MeV)	Entrance dose (MeV/mm)	Bragg's peak position (mm)	Peak-to-entrance ratio
^{16}O	184	34.676	50	9.0
^{12}C	155	21.91	50	8.9
^{3}He	96	3.489	50	7.7

Table 2

Makrofol polycarbonate			
Ion-beam	Energy (A MeV)	Entrance dose (MeV/mm)	Bragg's peak position (mm)	Peak-to-entrance ratio
^{16}O	180	32.84	50	9.2
^{12}C	152.5	20.76	50	8.9
^{3}He	180	3.35	50	7.2

Table 3

CR39			
Ion-beam	Energy (A MeV)	Entrance dose (MeV/mm)	Bragg's peak position (mm)	Peak-to-entrance ratio
^{16}O	200	36.25	50	8.8
^{12}C	167	22.86	50	8.6
^{3}He	104	3.59	50	7.4

5. CONCLUSION

Peak-to-entrance ratios are calculated for ^{16}O, ^{12}C and ^{3}He ion beams in PMMA, CR39 and Makrofol polycarbonates. Peak-to-entrance ratios lie in between 7.2 to 9.2. Higher peak-to-entrance ratio leads to higher deposition and less secondary fragmentation reactions and hence ignorable deposition beyond range.

References

1. S. Agostinelli, *et al.*, *Nucl. Instrum. Meth. A*, **506**, 250 (2003).
2. J.W. Wilson, *et al.*, *Health Phys.*, **68**, 50(1995)
3. M. Kramer and M. Scholz, *Phys. Med. Biol.* **51**, 1959 (2006).
4. U. Amaldi, *Nucl. Phys. A*, **751**, pp. 409c-428c, Apr. 2005.
5. C. Z. Jarlskog and H. Paganetti, "*IEEE Trans. Nucl. Sci.*, **55**, 1018 (2008).

A Non-destructive Scattering Technique for Measurement of Pulmonary Edema and Mandibular Bone Density

Amandeep Sharma, B. Singh and B.S. Sandhu*

Physics Department, Punjabi University Patiala, India

E-mail: *balvir@pbi.ac.in*

ABSTRACT

The objective of present work is to explore the use of scattered gamma photons for medical diagnosis, especially for pulmonary edema and osteoporosis. In the present study, for the density measurement of lung phantom and mandibular bone, scattering of 59.54 keV gamma photons are studied using an HPGe detector. Phantoms simulating lung and mandibular density are prepared by mixing appropriate amount of saw dust and K_2HPO_4 with distilled water respectively. The regression lines, obtained from experimental data for intensity ratio of Rayleigh to Compton scattered gamma rays, provide the density measurements for lung and mandibular bone. The present non-destructive technique has the potential for a measure of pathological state like pulmonary edema and density for mandibular bone. A portable non-invasive system described presently may be used for various industrial applications also.

Keywords: Phantom, Rayleigh to compton scatterings, Intensity ratio, Pulmonary edema, Mandibular bone density.

Pacs no.: 13.60.-Fz, 78.70.-g, 32.80.-t

1. INTRODUCTION

Pulmonary edema, is swelling and/or is abnormal extravascuiar water storage in the lungs, leads to impaired gas exchange and may cause respiratory failure, and if unchecked, to profound disability and even death. Metabolic bone diseases like osteoporosis lead to loss of bone mineral from both compact and trabecular bone which eventually leaves the affected bone vulnerable to traumatic fracture. Billy and Fred [1] developed a Compton densitometer for measuring pulmonary edema by using ^{57}Co radioactive source and an HPGe detector. The bone mineral density of the mandible had been investigated by Morgan et al [2] to study bone resorption following tooth loss and to determine the relationship between mandibular and skeletal bone mineral density. Singh et al [3-4] had already applied this intensity ratio technique successfully to measure the effective atomic number of samples of scientific interest, and to measure stable iodine content of tissue. In the present study, intensity ratio method is used to measure pulmonary edema and mandibular bone density. The major advantage of intensity ratio technique is that by taking intensity ratio of Rayleigh to Compton scattered photons, a number of parameters such as absolute source strength of radioactive source and solid angles subtended by the source and detector at the sample are eliminated in the expression of ratio technique, otherwise these parameters introduce large amount of error in the final measured results.

2. THEORY

In scattering experiments, for a gamma ray flux impinging on a target (phantom in present study), there is significant probability for Rayleigh (coherent) scattering to occur in addition to well-known Compton (incoherent) scattering. Rayleigh (or elastic) scattering is the process in which the scattered gamma ray has same energy as that of incident gamma ray, and is predominant at low incident gamma ray energies, small scattering angles and high atomic number of the target.

The Compton peak has energy less than incident gamma ray energy, and its FWHM (Full width at half the maximum height) is larger than a photo-peak at same energy mainly because of angular aperture of the spectrometer, multiple scattering in the target and the so-called Compton profile.

The Rayleigh to Compton scattering cross-section ratio [4] becomes

$$ R = \frac{d\sigma_R}{d\sigma_{comp}} \propto \frac{|F(q,\ Z)|^2}{S(q,\ Z)} \tag{1} $$

where $F(q,\ Z)$ and $S(q,\ Z)$ are the atomic form factor and incoherent scattering function respectively, with q representing the momentum transferred to the electron and Z is the atomic number of the target. Rayleigh to Compton scattering ratio has a power relation to Z in the region of elemental interest and this power dependence is based upon the ratio F^2/S. Hubbell et al. [5] have provided theoretical data for calculation of Rayleigh to Compton cross-section ratio from the parameters for $F(q,\ Z)$ and $S(q,\ Z)$.

3. EXPERIMENTAL SET-UP AND MEASUREMENTS

Fig. 1 shows the principle of the method used in present work for measurement of Rayleigh and Compton scattered photons in narrow beam geometry. The principle of measurements is to observe the intensities of Rayleigh and Compton scattered gamma photons at a particular scattering angle using a high resolution semiconductor gamma detector.

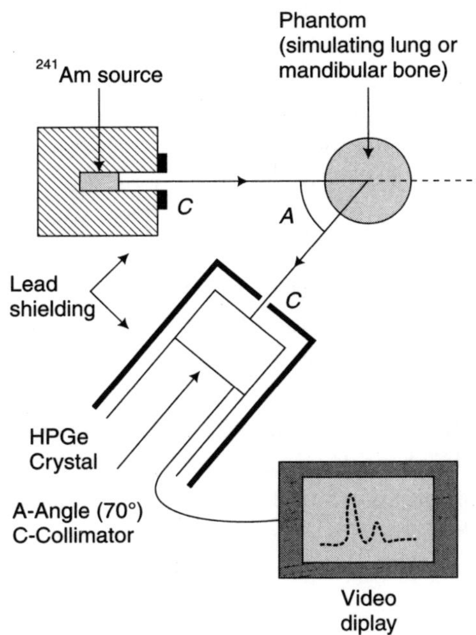

Fig. 1 *Experimental set up*

A high purity germanium detector (HPGe) is housed in a lead castle similar to that used for the source and collimated in an identical manner. Lung and bone phantoms are prepared mixing appropriate amount of saw dust and K_2HPO_4, respectively, in distilled water. The solutions are placed in plastic containers positioned at the centre of rotation of the source and detector assemblies. A well collimated beam of 59.54 keV gamma-rays from of ^{241}Am source (strength 7.4 GBq) irradiates the phantom. The distances of centre of phantom under study from the source collimator (radius 0.4 cm) and detector collimator (radius 0.4 cm) are kept 120 mm and 100 mm respectively. The scattered gamma rays are detected by a high resolution HPGe semiconductor detector (75 cc in volume) placed at scattering angles of 110°, to the primary incident beam. The coherent to incoherent intensity ratios are measured for each phantom having different density values. The phantom-in scattered spectra are recorded for a period of 1 ks by placing different density phantoms. The data are accumulated on a PC based ORTEC Mastreo-32 Multi channel analyzer (MCA).

4. RESULTS AND CONCLUSION

The intensity ratio of Rayleigh to Compton scattered peaks are corrected for photo-peak efficiency of the HPGe gamma detector, and absorption of photons in air column present between phantom and detector, and self absorption in the phantom. In Fig. 2, Rayleigh to Compton intensity ratio, for phantoms simulating lung, is found to be decreasing linearly with increase in phantom density. This is because an increase in phantom density results in more scattering centres for gamma interactions through Compton scattering process while owing to low effective atomic number (7.18-7.31 for various lung phantoms used in present work) of the phantom there is not much appreciable change in intensity of Rayleigh peak. This results in decrease in intensity ratio with increasing phantom density. The departure of some of the data points from linear nature is due to non uniformity of packing of mixture within thin plastic container. The best-fitted regression line of Fig. 2 serves as calibration curve in our experiment and provides the lung density of unknown sample of interest. The intensity ratio originating from K_2HPO_4 solutions (simulating mandibular bone density) increases linearly (Fig. 3) with increase in concentration of K_2HPO_4 in distilled water within experimental errors as Rayleigh scattering contributes significantly in this case owing to enhanced Z_{eff} because of presence of Potassium (Z = 19) and Phosphorus (Z = 15). The best-fitted regression line of Fig. 3 serves as calibration curve in our experiment and provides the mandibular bone density of unknown sample.

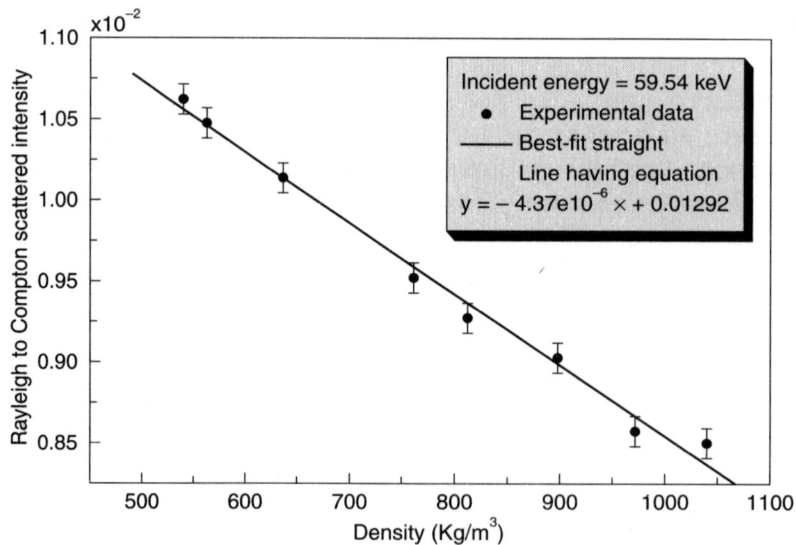

Fig. 2 *Experimental variation of Rayleigh to Compton intensity as function of lung phantom density*

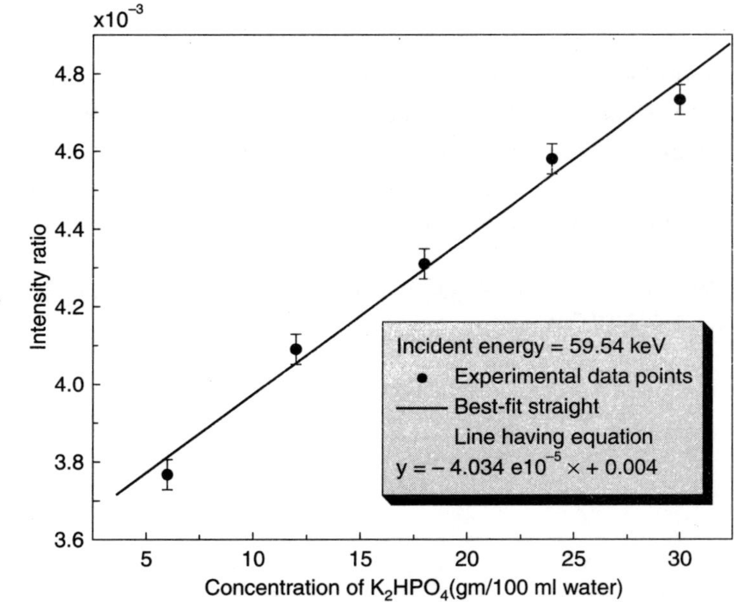

Fig. 3 *Experimental variation of Rayleigh to Compton intensity as function mandibular bone phantom density*

The results of present experiment indicate that the technique has potential for a measure of excess water storage in lungs (pulmonary edema) and is also a measure of mandibular bone density for either bone resorption studies or as a predictor of osteoporosis. The present measurements provide a low dose inexpensive alternative to the use of dental radiographs in studies of mandibular bone density. Medical diagnostic techniques based on gamma rays scattering are relatively new and many are still under development. The use of a gamma emitting radioactive source to provide the radiation beam offers the advantages over an X-ray system of constant beam intensity, mono-energetic radiation, lower patient dose and apparatus that can be made small and portable.

As assessment of the clinical utility of the techniques requires further investigation, so we believe that our experimental findings with regard to the pulmonary edema will be useful to other investigators in improving their experimental design for clinical purposes. Moreover, by making use of a strong radioactive source and multi-element (Circular or planar) gamma detector array (scattered intensity will be collected simultaneously from different regions of lung) the measuring time can be reduced to an acceptable limit (less than 50 sec) for practical use in clinical situations. There is also need to measure the dose at the phantom surface for comparison with dose received (100 mR) over the entire chest during chest X-ray and in other medical diagnostic techniques used for this purpose.

References

1. Loo Billy W. and Goulding Fred S. *IEEE Trans. Nucl. Science* **33** 531 (1986)

2. H.M. Morgan, J.T. Shakeshaft and S.C. Lillicrap *Br J Radiol* **72** 1069 (1999)

3. M.P. Singh, Amandeep Sharma, Bhajan Singh and B.S. Sandhu *Radiation Measurements* **45** 960 (2010)

4. M.P. Singh, Amandeep Sharma, Bhajan Singh and B.S. Sandhu, *Nuclr. Instr. & Meth. A* **619** 63 (2010)

5. J H Hubbell, W J Veigele, E H Briggs, R T Brown, D T Cromer and R Howertan *J. Phys. Chem. Ref. Data* **4** 471 (1975)

Studies on Effects of 8 MeV Electrons on P-channel MOSFETs

Godwin D' Souza[1*] and K.M. Balakrishna[2]

[1]*St. Joseph's College, Lalbagh Road, Bangalore, India*
[2]*Department of Studies in Physics, Mangalore University, India*
E-mail: *go4godwin@yahoo.com*

ABSTRACT

P-channel MOSFETs have important applications in analog or mixed signal circuits because of their constant current drive capability, linearity and very small drain resistance. Radiation evaluation on P-channel MOSFET was carried out by exposing them to 8 MeV electron beam under unbiased conditions. After each exposure, samples were electrically characterized. A systematic variation in interface traped charges (ΔV_{Nit}) and oxide traped charges (ΔV_{Not}) of MOSFETs due to threshold voltage shift (ΔV_{Th}) with different electron doses is presented in this paper.

Keywords: Irradiation, P-Channel MOSFETs, Trapped charges, Mobility carriers.

1. INTRODUCTION

Semiconductor devices are the backbone of high tech systems employed in commercial, industrial, space and defense systems. They are often subjected to various types of radiations. Ionizing radiations can alter the properties of the semiconductor devices for desirable or undesirable changes depending upon the dose imparted. The devices can fail due to large cumulative dose or due to large transient pulse of radiation. Hence the study of radiation on semiconductor devices like MOSFETs and bipolar devices, which are used in defense and space applications, is very important. Particle radiations such as electrons, protons and neutrons interact with devices and change their characteristics by forming electron-hole pairs and by exciting lattice atoms. The effect depends upon the particle species and energy.

Ionizing radiation often causes detrimental effects on the characteristics of MOS devices and circuits. The threshold voltages, current drives and leakage currents of MOS transistors change as function of number of factors such as the total dose of radiation received and its energy, the bias applied during the radiation. The geometry, type and method of fabrication of the transistor, the dose rate at which radiation is delivered, the temperature during irradiation, bias time and temperature after the irradiation is completed. Changes in the properties of transistors can lead to significant change in the characteristics of the integrated circuits of which they are the primary elements. Circuit properties that are typically affected include functionality, leakage currents, timing, input and output switching levels (TTL compatibility), output drives, operating voltage and frequency. The manner and degree to which these properties are altered depends on the same factor listed above for individual MOS transistors

The P-channel MOSFETs are used in commercial/industrial applications at lower power dissipation levels to accomplish analog functions. MOS devices are important components used in electronic equipment, in defence and space systems. When these MOSFETs are exposed to radiation, interface trap density will increase and thereby disturbing the basic properties of the devices. The effects are felt more severely in the form of shifts in the threshold voltage, mobility, lifetime degradation, increase in the leakage current and transconductance etc.

2. EXPERIMENTAL METHODS

The irradiation studies on P-channel MOSFETs (IRF 9540N and IRF 9530N) were carried at Microtron Centre Mangalore University with 8 MeV electrons [1]. Graded electron doses were delivered to the devices at 50 Hz pulse repetition rate at a distance of 30 cm from the beam exit port. The characterization of all the P-channel MOSFETs was carried out at ISAC, Bangalore using TESEC 881-TT/A meter [2].

3. RESULTS AND DISCUSSIONS

When MOS devices are exposed to radiation, the high energy electrons easily pass through the device. These high energy electrons deposit energy into the device through electronic excitations, which will produce ionization or breaking bonds and displacement of atoms along its path during irradiation process. In MOSFETs, the role of the $321Si/Sio_2$ interface is very important in determining the device performance. Some of the radiation-induced electron-hole pairs quickly undergo recombination and are not available for any further radiation effect, and some of the positively charged holes make slow dispersive transport towards the Si/Sio_2 interface where they are trapped in deep hole traps. Thus some of the holes are captured by trapping sites typically located within 5 mm of the Si/Sio_2 interface, causing a long-term negative voltage shift. Thus deeply trapped but are discharged over time by electron tunneling in from the silicon substrate and either recombining with or compensating the trapped holes. The electron trapping in Sio_2 is negligible because the capture cross- sections of electron traps are small by a factor of 10^4 the mechanism of generation of interface trapped charge is as follows.

The sub threshold behaviour of 8 MeV electron-irradiated MOSFETs as a function of radiation dose for $V_{GS} = 0$ V during irradiation was studied. In the MOSFETs the drain to source current (I_D) flows even at $V_{GS} = 0$ V because of the in-built channel between source and drain, so it is necessary to apply positive voltage to the gate to completely deplete the channel.

The threshold voltage shift (ΔV_{Th}) due to interface trapped charge (ΔV_{Nit}) was separated from that due to oxide trapped charge (ΔV_{Not}) by sub threshold measurements using the technique proposed by Mc Whorter and Winokur [3]. Using this method it is possible to split the threshold voltage shift (ΔV_{Th}) into a contribution due to interface trapped charge (ΔV_{Nit}) and a contribution due to trapped oxide charge (ΔV_{Not}), where $\Delta V_{Th} = \Delta V_{Nit} + \Delta V_{Not}$ from sub threshold current measurements. It is then possible to determine the increase in the number of interface states, ΔN_{it} (cm^{-2}) and trapped oxide charge ΔN_{ot} (cm^{-2}). ΔN_{it} represents the increase in the total number of interface traps between mid gap and threshold. The shift between sub threshold curves at the mid gap voltage represents ΔN_{ot}.

$$\Delta N_{ot} = \Delta V_{ot}\, C_{ox}/q \tag{1}$$

and $$\Delta Nit = \Delta V_{it}\, C_{ox}/q \tag{2}$$

where $q = 1.6 \times 10^{-19}$ c, ΔN_{ot} is the effective oxide trapped charge density in cm^{-2} as projected to the interface and C_{ox} is the oxide capacitance per unit area. The variation of ΔN_{ot} and ΔN_{it} as a function of dose for the devices irradiated with 8 MeV electrons is given in Fig. 1 and Fig. 2 respectively. It is observed that ΔN_{ot} and ΔN_{it} have been increased significantly after irradiation for all the devices. Where ΔN_{ot} is the irradiation-induced oxide trapped charge.

The mobility of carriers (μ) in the *P*-channel was determined from the maximum or peak Transconductance (g_m). The Transconductance of the MOSFET is defined as the rate of increase in Id per unit increase in V_{GS} at fixed V_{DS}.

$$g_m = \Delta I_D / \Delta V_{GS} \text{ where } V_{DS} = \text{constant } (0.1) \tag{3}$$

There are two sources of additional charged defects following irradiation, trapped holes and interface traps, which act to scatter the carriers moving in the channel of MOS device and thereby degrade g_m. The variation g_m peak up to the total dose of 15 kGy irradiated at $V_{GS} = 0$ V, from which we observe around 75% decrease in g_m peak.

The mobility of electrons (μ) in the *P*-channel was estimated from g_m and the relation between g_m and μ is given by

$$g_m = (Z\, \mu C_{ox}/L)\, V_{DS} \tag{4}$$

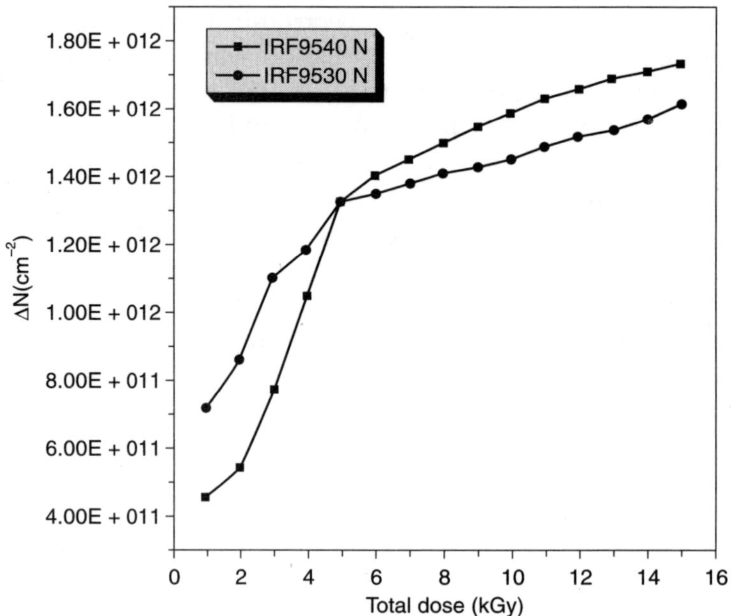

Fig. 1 ΔN_{ot} as a function of total dose.

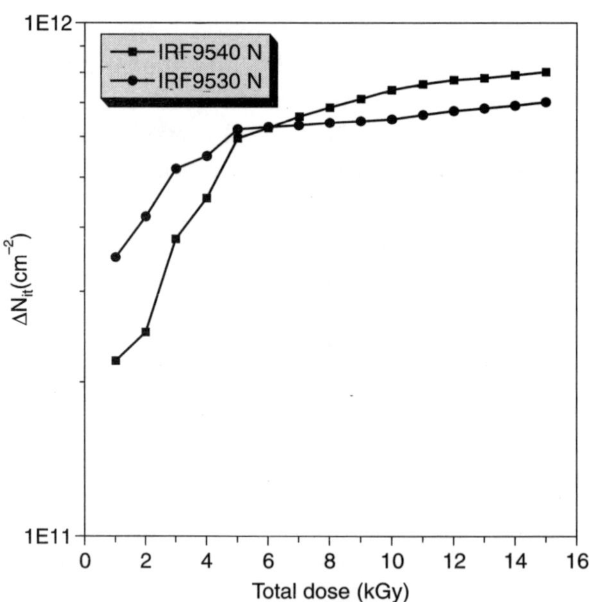

Fig. 2 ΔN_{it} as a function of total dose.

where Z is the width of the device, L is the length of the device, μ is the mobility and C_{ox} is the oxide capacitance per unit area. Rearranging the above expression for μ, it is known as the field mobility (μ_{FE}) and is given by

$$\mu_{FE} = L\, g_m/(Z\, C_{ox}\, V_{DS}) \tag{5}$$

The mobility of electrons in the P-channel reduced after irradiation because of radiation-induced charges at the interface produces coulomb scattering centers, which affect the motion of carriers in the channel. The mobility degradation was attributed mainly due to the interface trapped charge (N_{it}). The effect of oxide trapped charges (N_{ot}) which lie further away from the inversion layer was considered to be negligible.

Following the mobility model of Sun and Plummer [4], the effective mobility, μ, was related to the build-up of interface trapped charge, ΔN_{it}, through the equation

$$\mu = \mu_o/(1 + \alpha\, \Delta N_{it}) \tag{6}$$

Where μ_o is the pre-irradiation value of mobility and α is a constant. This model was fairly successful in explaining the observed data of decrease in μ. It has been observed that ΔN_{ot} can also play an important role in degrading μ. Several authors have suggested the use of a modified version of equation (6)

$$\mu = \mu_o/(1 + \alpha_{it}\, \Delta N_{it} + \alpha_{ot}\, \Delta N_{ot}) \tag{7}$$

The experimental results of Zupac et al [5] on irradiated and annealed MOSFETs convincingly demonstrate that only taking ΔN_{it} into account will not adequately fit the data, therefore ΔN_{ot} also has to be considered. It is observed that α_{ot} is not consistent and similar result was also observed by Dimitrijev et al [6]. Hence as inferred by earlier investigations, the effects of ΔN_{ot} on mobility is negligible and an increase ΔN_{ot} should lead to an increase in inversion layer hole mobility. The values of ΔN_{it} obtained in the present work are consistent for all, the devices and agree with the results.

4. CONCLUSION

The threshold voltage (V_{Th}) of the irradiated MOSFETs decreased significantly during irradiation caused more degradation. The interface trapped charge (ΔN_{it}) and oxide trapped charge (ΔN_{ot}) were calculated from the sub threshold measurements and ΔN_{ot} was found to be higher compared to ΔN_{it} after exposure to a total dose of 15 kGy. The peak transconductance (g_m peak) was extracted from the I_D versus V_{GS} and the mobility of carriers in the channel was estimated from g_m peak. More than 75% degradation was found in transconductance (g_m) after receiving a total dose of 15 kGy.

References

1. Ganesh, Prashanth K.C, Nagesha Y.N, Gnanprakash A.P, Umakanth D, Manjunath Pattabi and Siddappa K Indian Journal Physics 73S 177 (1999)

2. Bhat B. R and Sahu R.P Journal of Space Tech., vol.3 doc. 36 (1993).

3. Mc Whorter P. J. and Winokur P. S. , "Simple technique for separating the effects of inter-face traps and trapped-oxide charge in MOS transistors", Applied Physics Letters, Vol 48, No 2, P-133 13 January (1986).

4. Sun C. and Plummer J. D, "Electron Mobility in inversion and accumulation layers on thermally oxidized silicon surfaces', IEEE Tra. Electron Devices, Vol ED-27, NO 8, P-1497, August (1980).

5. Zupac D, Galloway K. F. Schrimpf R. D. and Augier P., "Effects of radiation Induced oxide-trapped charge on inversion-layer hole mobility " Journal of Applied Physics Letters, Vol 60, No 25, P- 1356, (1992).

6. Dimitrijev S. and Stojadinovic N., "Analysis of CMOS transistor instabilities", Solid –State Electronics, Vol.30, No 10, P- 991, (1987).

The Effect of 8 MeV Electrons on 2 N2907A PNP Transistor

Godwin D' Souza[1*] and K.M. Balakrishna[2]
[1]St. Joseph's College, Lalbagh Road, Bangalore, India
[2]Department of Studies in Physics, Mangalore University, India
E-mail: *go4godwin@yahoo.com*

ABSTRACT

The PNP transistor 2N 2907A was irradiated with 8 MeV electron beam for different doses, to understand the displacement damage. It was observed that, the collector current and collector gain decrease for higher dose rates due to oxide trapping at Si-SiO$_2$ interface. The gain degradation was also studied for different doses of electrons and it was observed that gain of a transistor was reduced significantly. This degradation in the gain can be attributed to the fact that the increased base current due to the recombination centers introduced by the high energy electrons. The displacement damage factor was calculated using the gain. Similar study was carried out using Co-60 gamma source compared with 8 MeV and the details of these studies are presented in this paper.

Keywords: Displacement damage, Oxide trapping, Degradation, Recombination centers.

1. INTRODUCTION

Ever since the discovery of semiconductor devices, they have become integral part in the development of electronic components. Semiconductor electronic components are widely used in circuitry of gadgets, instruments etc. having innumerable applications. Thus application of semiconductor devices has reached enormous potential.

It is in this background the question arises sturdy/stable or effective are these semiconductor components when the standard parameters vary. It is to be precise, what happens to semiconductor electronic components when they are subjected to various radiations (e.g. X-rays, gamma rays, heavy ions, electrons etc). It becomes utmost important whether the given parameter remains unchanged or undergo changes when the electronic components under the influence of different types of radiations. The degradation due to radiation depends on various parameters like radiation dose, dose rate energy content, exposed time, type of component, radiation species etc. [1].

This brings us to the taste of assessing the damage caused by these radiations by analyzing the performance of the devices under strict conditions and careful monitoring during the process of radiation and after when we consider bipolar junction transistor (BJT), the gain (h_{FE}) and subsequently its collector characteristics effect becomes important after bombarding it with radiations. In fact in some cases to heavy dosage components can fail completely leading to breakdown of the instrument in use. The improvement of semiconductor components to withstand radiations without change in its characteristics thus becomes important. Electrons being the main component of study here it when excited by radiations shift the orbital electron to higher energy orbit or dislodge completely from the orbit. This leads to change in carrier concentration and shift in BJT characteristics.

2.   EXPERIMENTAL METHOD

The PNP transistor 2N 2907A was irradiated with 8 MeV electron beam for different doses, to understand the displacement damage. It was observed that, the collector current and collector gain decrease for higher dose rates due to oxide trapping at $Si-SiO_2$ interface. The gain degradation was also studied for different doses of electrons and it was observed that gain of a transistor was reduced significantly. This degradation in the gain can be attributed to the fact that the increased base current due to the recombination centers introduced by the high energy electrons. The displacement damage factor was calculated using the gain. Similar study was carried out using Co-60 gamma source compared with 8 MeV and the details of these studies are presented in this paper. The various parameters of 2N 2907A, PNP transistor, were measured after exposing to 8 MeV radiation source at Microtron Centre, Mangalore University. The same parameters were also studied for the device after exposing to Co-60 gamma source at ISAC Bangalore using Semiconductor Parameter Analyzer-4145B [2].

3.   RESULTS AND DISCUSSION

The PNP transistor 2N2907A was characterized as a function of dose. The measurement of collector characteristics of the device at a = constant current shows significant changes due to electron irradiation. A similar characteristic is observed for Co-60 also. In the characteristic curve the collector current in the saturation region decrease considerably as the accumulated dose increase, but collector emitter saturation voltage V_{CE} (sat) remains almost constant Fig. 1 shows the variation of forward current gain h_{FE}, estimated at a constant base current of 50 μA as a function of accumulated dose. The current gain decreases considerably due to radiation. The most common effect of radiation a semiconductor devices is the gain degradation. This degradation in discrete BJTs can basically occur in two ways i.e. Bulk degradation and degradation by ionization. The bulk degradation occurs due to the atomic displacement in the bulk of the semiconductor when incoming energetic particle transfers momentum to atoms of the target silicon when transferred energy is sufficient the atoms of silicon can be ejected from its location leaving a vacancy or defect. The displacement damage is a bulk effect deep inside the semiconductor and produces an increase in the number of recombination centers. Recombination centers in the base region of the transistor reduce the minority carrier's life-time and so increases the base current and decrease the gain. In addition to recombination centers, displacement damage produces generation centers, trapping centers and scattering centers. Generation centers increases the reverse leakage current across the pn junction. This displacement damage is a bulk effect deep inside the semiconductor and produces an increase in the number of recombination centers. Recombination centres in the base region of the transistor reduces the minority carrier life-time and so increases the base current and decrease the gain. In addition to recombination centres, displacement damage produces generation centres, trapping centres and scattering centres. Generation centres increases the reverse leakage current across the pn junction. Trapping centres remove charge carriers. Scattering centres decrease the mobility of charge carriers. Both contribute to increase in resistance. Reverse breakdown voltage of the pn junction get increased slightly after irradiation.

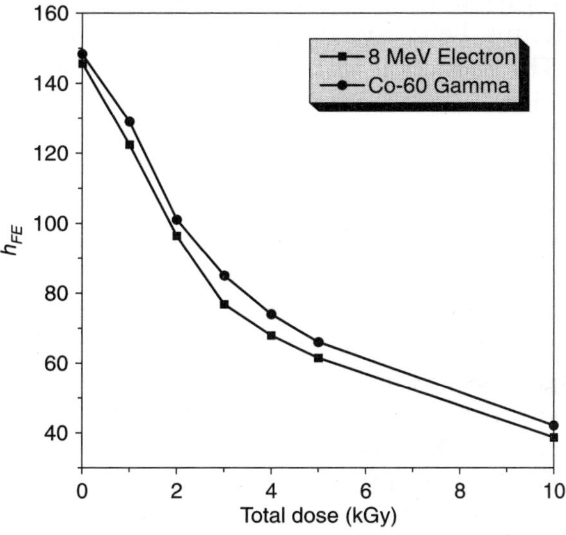

Fig. 1 h_{FE} *as a function of total dose.*

There are several factors, which combine to determine transistor current gain(β), emitter efficiency, surface recombination velocity, recombination in either field, recombination in the base region and conductivity modulation[3]. Of these the recombination in the emitter field region and in the base region are two dominant factors, which influence the transistor current gain.

When BJTs are exposed to radiation, the current gain of the transistor decreases as the accumulated dose or fluence increases. The main cause for gain degradation is the displacement of atoms in the bulk of the semiconductor. This damage produces an increase in the number of recombination centers and therefore reduces minority carrier lifetime. Another cause for current gain degradation is the ionization in the oxide passivation layer, particularly that part of the oxide covering the emitter-base junction. The gain degradation can be represented by the equation

$$\Delta\left[\frac{1}{\beta}\right] = \frac{1}{\beta} - \frac{1}{\beta_0} \tag{1}$$

Where β_0 and β are the gain values before and after irradiation.

Gain degradation is often analyzed by plotting the change in reciprocal gain $\Delta(1/\beta)$, versus radiation fluence. The term $\Delta(1/\beta)$ is known as the gain degradation figure. The effects of bulk and surface damage can be separated as follows

$$\Delta\left[\frac{1}{\beta}\right] = \Delta\left[\frac{1}{\beta}\right] b + \Delta\left[\frac{1}{\beta}\right] s \tag{2}$$

where the suffixes, b and s refers to the bulk and surface contributions respectively. However, while the bulk contributions may be reasonably predicted from the analysis of minority carrier lifetime behavior, the surface contributions is highly dependent upon process factors. The reduction in h_{FE} with incident particle fluence is given by Messenger Spratt equation.

$$h_{FE} = h_{FEO}/(1 + h_{FEO} \, K\varphi) \tag{-3}$$

where h_{FEO} and h_{FE} are the gain values before and after irradiation φ is the fluence and K is the displacement damage constant. In order to calculate a displacement damage factor, it is necessary to convert from dose (rad(s)) to effective electron fluence. The rad equivalent 1 Mev electron fluence is about 4.08×10^7 p/cm^2 [4]. Fig. 2 shows variation of displacement damage K as a function of accumulated dose. The average value of K, calculated from the above equation agrees with the value reported in the literature for similar family of the device. The agreement between theoretical fit and experimental data is fairly good. Thus, the results observed indicate that the gain degradation in the investigated device is primarily due to displacement damage produced in the bulk of the semiconductor [5]. The displacement damage due to radiation also causes other important effects in bipolar transistors for e.g. the increase in collector-base leakage current (I_{CBO}), breakdown voltage. These parameters have been measured before and after irradiation. The I_{CBO} increases due to radiation. This increase in leakage current is due to the ionization in the surface oxide, particularly the region over the collector-base junction. At the same time the breakdown voltage of the transistors are exhibiting only a small increase due to radiation. This small increase in breakdown voltage is attributed to a decrease in the free charge carrier concentration. The increase in the I_{CBO} and breakdown has been observed on bipolar devices.

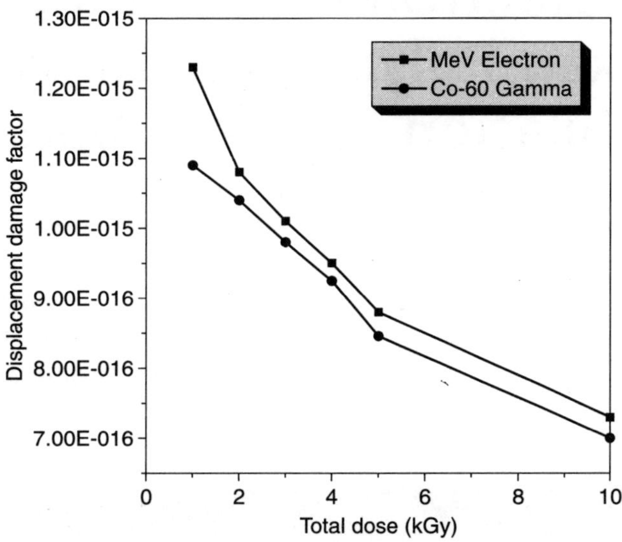

Fig. 2 *Displacement damage as a function of total dose.*

4. CONCLUSION

The commercial bipolar transistor 2N2907A studied appears to be sensitive to radiation. The forward current gain of the transistor decreases more than 80% as the electron and gamma accumulated dose. The gain degradation observed in primarily due to displacement damage produced in the bulk of semiconductor and similar to that of the other transistor series.

References

1. Schrimpf R.D., "Recent Advances in Understanding Total-Dose Effects in Bipolar Transistors", proceeding of Third European Conference on Radiation and its Effects on Components and Systems, pp.September, 1995, Arcahon, France.
2. Bhat B.R., Radiation Hardness Specifications for Electronics Components,ISRO-ISAC-ST-0087, 2003.
3. Dale C, Marshal P.W and Wolicki E.A "High energy Electron Induced Displacement Damage in Silicon" IEEE Trans. Nicl. Sci. Vol 35, (1988).
4. Messenger G C and Ash M.S. The Effects of Radiation on Electronic Systems. (1986)
5. "Radiation Design Handbook" published by European Space Agency-PSS 01-669 section 7, May (1999)

8 MeV Electron Induced Modification in Lexan Polycarbonate by Positron Annihilation Lifetime Spectroscopy

K. Hareesh[1], P. Ramya[2], C. Ranganathaiah[2] and Ganesh Sanjeev[1,*]

[1]*Microtron Centre, Department of Studies in Physics, Mangalore University, Mangalagangotri, India*

[2]*Department of Studies in Physics, Mysore University, Manasagangotri, India*

E-mail: *ganeshsanjeev@rediffmail.com*

ABSTRACT

The microstructural and thermal properties modifications of the Lexan polycarbonate irradiated to 8 MeV electron doses of 100 and 225kGy have been studied using Positron annihilation lifetime spectroscopy (PALS), Fourier transform infrared spectroscopy (FTIR) and Thermogravimetric analysis (TGA). PALS result shows o-Ps lifetime and free volume size decreases initially and then increases for further increase in electron dose. The o-Ps intensity was found to increase initially and then decreases for further electron dose. The chemical changes in electron irradiated polymers due to chain-scission and reconstruction as observed from FTIR spectroscopy ascertains the PALS results. The increase in thermal decomposition temperature after irradiation was observed from TGA due to bond cleavage and reconstruction.

Keywords: Lexan polycarbonate, Electron irradiation, PALS

Pacs No: 61.80.jh, 61.82.Pv, 82.50.Bc, 66.30 hk,

1. INTRODUCTION

Polymers have their own importance in the field of Solid State Nuclear Track Detector (SSNTD) and in other allied science field. Among the several types of these detectors, polycarbonate detector such as Lexan detector is used for recording fission fragment tracks. Lexan polycarbonate (Lexan) is a promising polymer having lot of desirable properties for different application. Its chemical composition and 3D network is as shown below.

Electron irradiation of polymers results in modification of physical and chemical properties of the polymer. The physical changes include microstructure, glass transition temperature, optical band gap, electrical conductivity etc. and the chemical changes include gas evolution, main chain scission, radical-radical combination etc. [1]. In amorphous polymer characterization, the free volume determination plays an important role. The better technique to get information about free volume is positron annihilation lifetime spectroscopy (PALS) [2]. In the present work, the microstructural and thermal modifications in Lexan induced by 8MeV electrons have been studied using PALS, FTIR and TGA technique.

2. EXPERIMENTAL

Commercially available Lexan films having thickness 200 μm were irradiated at doses 100 kGy and 225 kGy in air sealed in polyethylene bags. The irradiation was carried out at room temperature at Microtron Centre, Mangalore University, India using 8MeV electrons having beam current 20 mA; pulse repetition rate 50Hz and pulse width 2.3 m*s. The dose delivered to the sample was measured using current integrator calibrated against appropriate radiation dosimeters. The free-volume size in the polymers was measured from positron annihilation lifetime spectroscopy using a conventional fast-fast coincidence system with a time resolution of 220 psec. A 17 Ci ^{22}Na source, which was deposited on 0.127 mm thick Kapton foil and sandwiched between two discs of the sample (~1 mm thick) under study, was placed between two BaF_2 scintillation detectors. Each spectrum contained more than 10^6 counts that were accumulated in approximately 1.5 to 2 h. The measured lifetime spectra were analyzed using the computer program PATFIT88. FTIR spectra were recorded in the transmission mode using spectrophotometer NICOLET 5700, FTIR, USA in the wave number range $400-4000$ cm^{-1} having a resolution of 4cm^{-1}. TGA was carried out using SDT Q600 TA Instruments, USA, heated at a rate of 10°C/min under Nitrogen atmosphere.

3. RESULTS AND DISCUSSION

The measured positron lifetime spectra for pristine and electron irradiated Lexan films were analysed interms of three lifetime component τ_1, τ_2, τ_3 and corresponding intensities I_1, I_2, I_3 and are listed in Table 1. Here τ_3 is the longest lived component with intensity I_3 and is attributed to the ortho-positronium (o-Ps) atoms in the free volume sites of amorphous regions of polymer via pick-off annihilation.

Table 1 *Parameters of PALS*

Dose (kGy)	$(\tau_1 \pm 0.005)$ ns	$(\tau_2 \pm 0.015)$ ns	$(\tau_3 \pm 0.013)$ ns	$(I_1 \pm 1.9)$ %	$(I_2 \pm 1.2)$ %	$(I_3 \pm 0.26)$ %	$(V_f \pm 1.1)$ Å^3
Pristine	0.171	0.46	2.00	44.60	31.20	24.22	97.60
100	0.168	0.43	1.98	37.95	34.85	27.20	95.74
225	0.192	0.49	1.99	49.11	27.58	23.32	96.66

The radius R of free volume hole can be obtained by o-Ps lifetime (τ_3) using semi empirical relation given by

$$\frac{1}{\tau_3} = 2\left[1 - \frac{R}{R_0} + \frac{1}{2\pi}\,\text{Sin}\left(\frac{2\pi R}{R_0}\right)\right]$$

where $R_0 = R + \Delta R$ and ΔR is the fitting parameter and a value $\Delta R = 0.166$ nm was determined by fitting above equation with experimental τ_3 values to data from molecular materials like zeolites with known hole sizes. Here o-Ps lifetime in the electron layer of thickness ΔR is the spin-averaged Ps lifetime of 0.5ns. The free volume size (V_f) can be calculated as $V_f = (4/3)\,\pi R^3$.

Fig. 1 shows the variation of o-Ps lifetime and free volume size V_f against electron dose. From this figure, it can be seen that both o-Ps lifetime τ_3 and free volume size V_f decreases up to 100 kGy and then increase slightly for 225 kGy. This o-Ps lifetime variation is sensitive to local molecular environment and is directly correlated to the size of free volume region in which it is localized. From Fig. 1 it is observed that o-Ps lifetime τ_3 decreases which is related to the decrease in the free volume as a result of the formation of new bonds or cross-linking. This decrease in τ_3 may be due to sudden rupture of single bond and formation of double bond. It can be seen in FTIR results, the irradiation of Lexan by a dose of 100 kGy results i±n the formation of conjugated bonds and free radicals. After 100 kGy, it seems that the chain scission process starts dominating over the cross-linking process and thereby increasing free volume size.

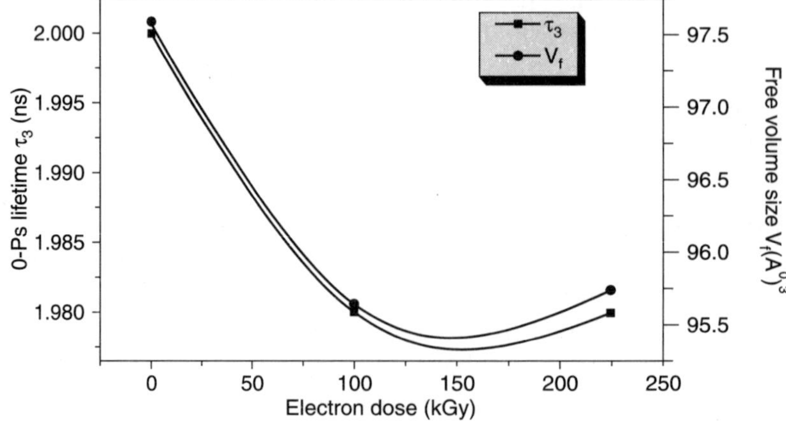

Fig. 1 *Plot of o-Ps lifetime and free volume size vs. electron doses.*

It is observed from Fig. 2 that o-Ps intensity increases for an electron beam of dose 100 kGy, which suggests the redistribution of the molecules. This would effectively reduce the free volume size by splitting the larger size free volume in to smaller ones and increases the number of such holes as the cross-linking hinders the molecular chain mobility. Therefore, decrease in τ_3 and V_f have been observed [3].

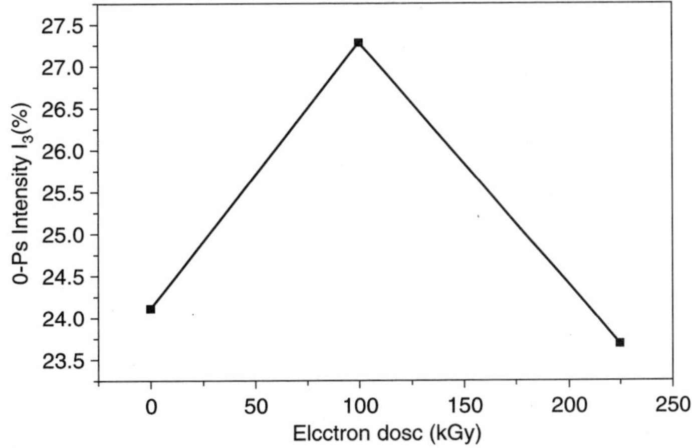

Fig. 2 *Correlation of o-Ps Intensity with electron doses.*

From FTIR spectra as shown in Fig. 3, it was observed that for pristine Lexan, the bands for methyl ($-CH_3$) group appears at $2976 cm^{-1}(v_{as})$, $2876 cm^{-1}(v_s)$, 1469 cm$^{-1}(\delta_{as})$ and 1366, 1380 cm$^{-1}(\delta_s)$. The carbonyl ($-C=O$) stretching band absorbs at $1760 cm^{-1}$. The asymmetric stretching of the aromatic ether ($C-O-C$) absorb at $1240 cm^{-1}$ and the symmetric stretching absorbs at 887 cm^{-1}[4]. Apart from these various shifts in the peak positions, on irradiation new C=C band appears at 1599, $1496 cm^{-1}(v_{C==C})$, 1012, 942 cm$^{-1}(\beta_{C=CH})$ and $824 cm^{-1}(\gamma_{C=CH})$. On increasing the electron dose, the rate of chain cleavage increases as a result the formation of C=C bond also increases, and these small carbon enriched units may agglomerate into carbonaceous clusters. From the overall discussion of FTIR results, it can be concluded that the chemical changes arise due to the rearrangements of ions to form a stable product after chain cleavage, which is dominant process at lower dose, and crosslinking is the dominant process at higher dose.

The decomposition behavior of the polymer was studied by thermogravimetric analysis as shown in Fig. 4. In all the cases (pristine and irradiated) a stable zone (no weight loss), slow decomposition zone, fast decomposition zone and residual decomposition zone were observed. Thermal decomposition temperature (T_d) for pristine sample was found to be 495°C. On irradiation, T_d decreases to 502°C for 100 kGy and 510°C for 225 kGy electron irradiated sample due to breakage of few bonds in polymer structure [5].

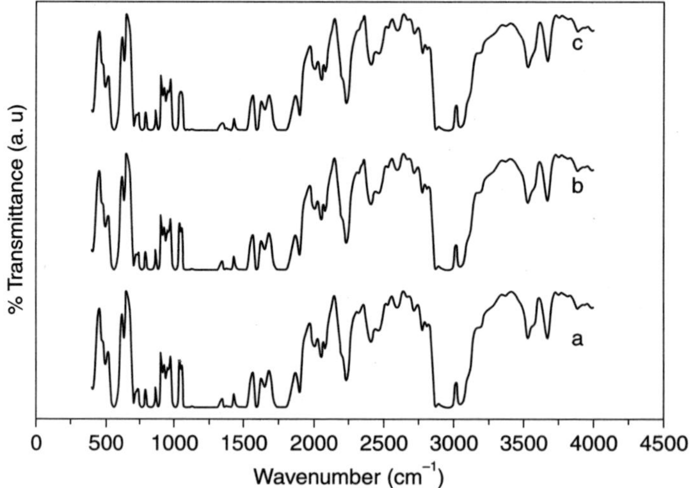

Fig. 3   *FTIR spectra for (a) pristine and 8 MeV electron irradiated Lexan with doses (b) 100 kGy, (c) 225 kGy.*

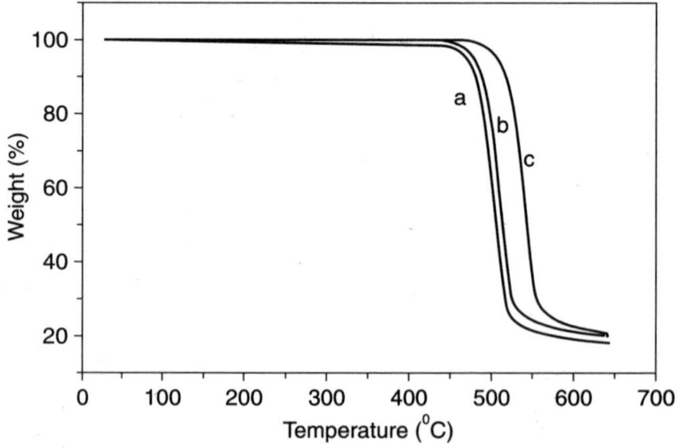

Fig. 4   *TGA thermogram of (a) pristine and electron irradiated Lexan with doses (b) 100 and (c) 225 kGy.*

4.   CONCLUSION

PALS result shows that the changes in o-Ps lifetime, free volume size and o-Ps intensity with electron dose is due to breakage of bonds and formation of carbon cluster by irradiation. The FTIR spectroscopy show the bond breaking processes after irradiation, and it also shows that the damage rate is low for this polymeric structure. TGA thermogram indicates that the thermal decomposition temperature increases after irradiation due to bond cleavage and also due to cross-linking of some of the degraded molecules.

Acknowledgements

One of the authors (KH) would like to thank BRNS-DAE, Govt. of India for financial assistance. The authors wish to thank research group at Microtron Centre for their help and support during the course of this work.

References

1. A Chapiro *Radiation chemistry of polymeric systems in: high polymers* (Newyork: Interscience publishers) **15** (1962)
2. Y C Jean *Microchem. J.* **42** 72 (1990)
3. E H Lee, G R Rao and L K Mansur *Radiat. Phys. Chem.* **55** 293 (1999)
4. G Aruldhas *Molecular structure and spectroscopy* (New Delhi: Prentice-Hall of India) (2004)
5. B N Jang, A Charles and Wilkie *Polym. Degrad. Stab.* **86** 419 (2004)O

Testing of *in Situ* Emergency Field Instruments with a Subject Who Underwent Medical Diagnostic Test

M. Manohari*, R. Mathiyarasu, V. Rajagopal and B. Venkataraman

Radiation Safety Section, Radiological Safety and Environmental Group
Indira Gandhi Centre for Atomic Research, Kalpakkam, India
E-mail: **manohari@igcar.gov.in*

ABSTRACT

During a rare event of nuclear accident estimation of radioactive iodine in human thyroid is needed for deciding counter measures. Many radiation monitoring field instruments like scintillometer, thyroid monitor are calibrated for this *in situ* monitoring purpose. Usually these equipments once calibrated are tested periodically with standard sources to ensure their performance. For testing such post accidental *in situ* monitoring equipments, measurement on a person who has undergone medical diagnostic procedure with radioactive iodine can also be gainfully used. Radioactive iodine activity present in thyroid of one such subject was estimated with pre-calibrated instruments meant for such *in situ* post accidental monitoring. In order to simulate the condition of thyroid with varying levels of radioactivity, the measurements with two types of instruments are carried out at different time period after the medical diagnostic testing. The measured activity by these instruments agreed within ±5%. The activity in the subject is further confirmed by shielded chair wholebody counting measurement and also theoretically by MCNP simulation. This paper details the measurement procedure and the salient results obtained.

Keywords: Thyroid measurement, in *situ post* accidental monitoring, Radioiodine.

Pacs No: 28.41.Te

1. INTRODUCTION

In case of hypothetical nuclear accident, estimation of radioactive iodine in human thyroid is also needed for the evaluation of dose to the individuals. Portable radiation monitoring instruments like scintillometer, thyroid monitor are calibrated for this rapid *in situ* monitoring purpose [1]. Usually these instruments once calibrated are tested periodically with standard sources to ensure their performance. Persons who have undergone medical diagnostic procedure with radionuclide can also be gainfully utilized for testing such post accidental *in situ* monitoring instruments [1]. These people make excellent test subjects for testing the emergency response instruments that would be used in the event of an accidental or intentional release of radioactivity that may result in intake of radioactive iodine by members of public. The level of activity present in their body over a period of time would simulate the prevailing field conditions vis-à-vis the varying levels of intake of radioactive iodine by the members of the public.

2. MATERIALS AND METHODS

2.1 Instruments

Two portable radiation monitoring instruments are used for this study. A scintillometer housing a 50mm x 50 mm NaI(Tl) detector [ECIL make, model 141D] is used. A thyroid monitoring system [ECIL make of model MDS 33C] based gamma ray spectrometer (GRS) with single channel analyzer (SCA) is also used. These instruments were calibrated using standard Thyroid Phantom of Radiology Support Devices, Inc. [2]. Scintillometer was calibrated in terms of (mR/h)/MBq and the thyroid monitoring system in terms of cps/kBq [3].

2.2 Measurement With Subject

A female worker had undergone a medical diagnostic procedure involving radioiodine volunteered for this study. The measurements started 2 days after the medical diagnostic testing and the measurement was continued till the thyroid activity reached below detection levels of the instruments. The subject was positioned in an armed chair. The measurements were made with scintillometer & thyroid monitoring system.

2.3 Observed Data And Estimated Activity

The measured activity data is presented in Table 1. The measurements are made at different time intervals to simulate the various levels of activity in the thyroid. The activities estimated by the instruments are matching within ±15%. This also ensures the response linearity of the instruments to the activity levels tested presently.

Table 1 *Estimated activity in subject's thyroid*

Days after testing	Estimated Activity (kBq)		
	Scintillometer	*Thyroid monitor*	*SC**
6	300	303	....
8	275	250	....

Contd...

Contd...

13	165	170	188
18	108	110	115
20	90	86	90
22	75	80	80
25	53	60	63
26	47	46	53
28	40	36	45
32	29	32	30

* Shielded Chair whole body counter

2.4 Measure of Actual Activity in The Subject

2.4.1 Shielded Chair Wholebody Counter

The actual activity present in the subject's thyroid is measured using shielded chair (SC) whole body counting system [4] to confirm the measurement of in situ monitoring systems.

2.4.2 Instrument Response Using MCNP

To verify the activity present in the thyroid of the subject Monte Carlo simulations have also been done using MCNP code, which takes into account detailed characteristics of the source, detector and the scatter, in calculating the pulse height spectrum [5].

The desired result in the simulation component of this study is a pulse height spectrum since it produces the distribution of the energy deposited in a "cell", i.e. the γ-ray energy spectrum in a physical model of a detector. Pulse height spectra simulations are implemented in MCNP in the so-called "f8 tally". The detector geometry is modeled with the MCNP code, which simulates the detection process to obtain the spectral shape [6,7].

The activity estimated from measurement using thyroid monitoring was 300 kBq and for the same counts MCNP estimated 314.5 kBq. MCNP estimation is more by 4.8%.

2.4.3 Spectrum Response Comparison

The MCNP predicted spectrum & the observed spectrum are compared in Fig. 1. In the photo peak energy region the prediction and measurement match very well. The matched response gives the confidence that the activity measured in the thyroid is correct.

3. CONCLUSION

The actual activity present in thyroid was estimated by shielded chair wholebody counting system and further confirmed by MCNP response spectrum simulation. The

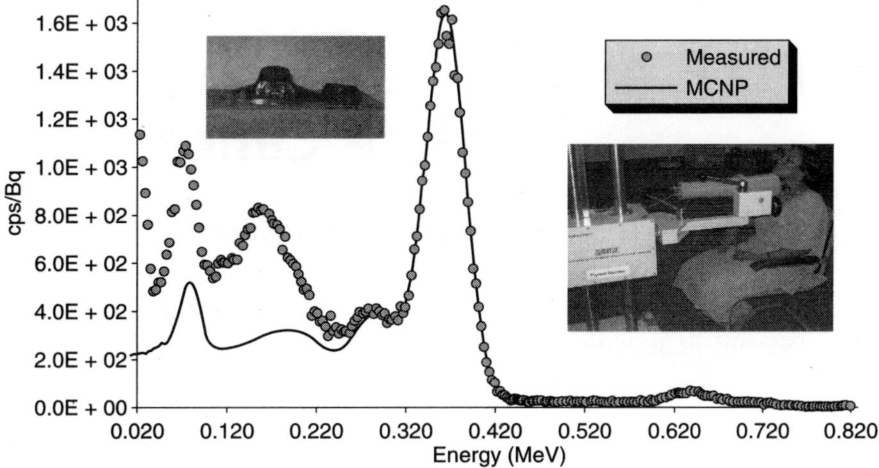

Fig. 1   *Special response comparison*

activity estimated by in situ monitoring instruments matched within 15% of actual activity from tens of kBq to hundreds of kBq.. This study confirmed that the subjects, who are willing to volunteer, can be used for testing *in situ* emergency radiation monitoring instruments.

References

1. Using Radioactive People to Calibrate / Test Emergency Response Equipments, *Gary H. Kramer and Barry M. Hauck,* European Conference on Individual Monitoring of Ionizing Radiation, Athens, 8-12 March, 2010.

2. http://www.rsdphantoms.com/hp_fission.htm

3. Technical write up on thyroid phantom measurements, IGC Highlights, 2004.

4. Wholebody Counting Manual, RSD-WBC-01 Rev-0, 2005.

5. Briesmeister, J.F. (Ed.), MCNP™ - A General Monte Carlo N-Particle Transport Code Version 4C. Los Alamos National Laboratory, Los Alamos, New Mexico, LA-13709-M Manual, 2000.

6. A study of Thyroid Radioiodine Monitoring by Monte Carlo Simulations: Implications for Equipment Design, *Gary H Kramer, Michael J Chamberlain and Suzanne Yiu,* Phys. Med. Biol. 42 (1997), 2175 – 2182.

7. Variability of Radioiodine Measurements in the Thyroid, *J. Damet, F.O. Bichud, C. Bailat, J.P. Laedermann, S. Baechler,* European Conference on Individual Monitoring of Ionizing Radiation, Athens, 8-12 March 2010.

PIXE Analysis of Blood Serum of Palate Cancer Patients

P. Sarita[1*], G.J. Naga Raju[1], M. Ravi Kumar[1], A.S. Pradeep[1] and S. Bhuloka Reddy[2]

[1]Department of Physics, GIT, GITAM University, Visakhapatnam, India
[2]Swami Jnanananda Laboratories for Nuclear Research, Andhra University, Visakhapatnam, India
E-mail: sarita0309@yahoo.co.in

ABSTRACT

Particle induced X-ray emission technique was used to study the change in concentrations of trace elements in the blood serum of palate cancer patients when compared to that of control subjects and to correlate the excess and deficiency states of certain trace elements with the etiology of palate cancer. The PIXE measurements were carried out using a 2.5 MeV collimated proton beam from the 3 MV Tandem Pelletron Accelerator at Institute of Physics, Bhubaneswar, India. The elements Ti, V, Cr, Mn, Fe, Co, Ni, Cu, Zn, As, Se, and Br were identified and their concentrations were determined. The serum of the cancerous group displayed significantly increased concentrations of V ($p < 0.005$) and Cu ($p < 0.05$) and significantly lowered concentrations of Fe ($p < 0.0005$), Zn ($p < 0.01$), and Se ($p < 0.01$). The various findings presented in this paper give guidelines for future study into the possible roles and interactions of essential trace elements in the carcinogenic process.

Keywords: PIXE, Trace elements, Palate cancer.

Pacs No: 87.19.xj, 87.56.bd, 87.64.K-.

1. INTRODUCTION

Trace elements have been extensively studied in recent years to assess whether they have any modifying effects in the etiology of cancer. The importance of trace elements in cancer was reported by *Schwartz* [1] which opened the door for new diagnostic and therapeutic endeavours in many areas of medicine, especially in the areas of oncology. In this work, Particle induced X-ray emission (PIXE) technique was used for trace elemental analysis of blood serum of patients suffering from cancer of the palate. Palate cancer refers to the cancerous growth that affects the roof of a person's mouth. Cancer of the palate is highly prevalent in the North-Coastal Andhra region of India. Reverse smoking is a specific etiologic factor for cancer of the palate. In reverse smoking, the lit end of the cigarette is placed in the mouth so that an intense heat is generated during smoking. The aim of this work was to correlate the excess and deficiency states of certain trace elements with the etiology of palate cancer.

Respected for its practical accuracy and detection range of parts per million, PIXE has enjoyed a secure place in the analytical arsenal of the nuclear physics laboratory. It serves as an excellent tool for trace elemental analysis of biological samples, which are available in very small amounts. Owing to these advantages PIXE technique was chosen for elemental analysis in this work.

2. MATERIALS AND METHODS

2.1 Study Subjects

In the present work, PIXE technique was used for trace elemental analysis of blood serum of palate cancer patients and healthy controls. Twenty four patients of both sex (males 14 and females 10) affected by palate cancer and who had not received any treatment comprised the palate cancer group. The ages of these patients ranged from 44 to 65 years; mean age being 54.1 ± 8.1 years. Twenty of these patients were habituated to reverse smoking. Fifty apparently healthy volunteers (30 females and 20 males with ages ranging from 25 to 62 years; mean age being 44.6 ± 9.9 years) who had not had any medication served as controls.

2.2 Sample Collection and Preparation

The blood samples of palate cancer patients were collected from Lion's Cancer Treatment and Research Centre, Visakhapatnam. The whole blood samples obtained from all the patients and controls were collected in separate vacutainer tubes and then centrifuged at 3000 rpm for 10-15 minutes. For each sample, the supernatant containing the serum was aspirated with an air displacement pipette and stored at -20°C until further biochemical determinations. Further details of sample preparation are described in one of our earlier works [2].

2.3 Experimental Details and Data Analysis

The PIXE measurements were carried out using a 2.5 MeV collimated proton beam from the 3 MV Tandem Pelletron Accelerator at Institute of Physics, Bhubaneswar, India. The characteristic X-rays emitted by the elements present in the target were recorded by a high resolution Si(Li) detector (energy resolution 180 eV FWHM at 5.9 keV) positioned perpendicular to the beam axis. Fig. 1 depicts the typical PIXE spectra recorded for palate cancer patients and control subjects. The obtained PIXE spectra were analyzed using Guelph PIXE (GUPIX) software package [3]. Using this software package, 12 trace elements were identified and their concentrations were estimated in each sample. The trace elemental concentrations were averaged separately for all the serum samples of the two studied groups and are furnished in Table 1.

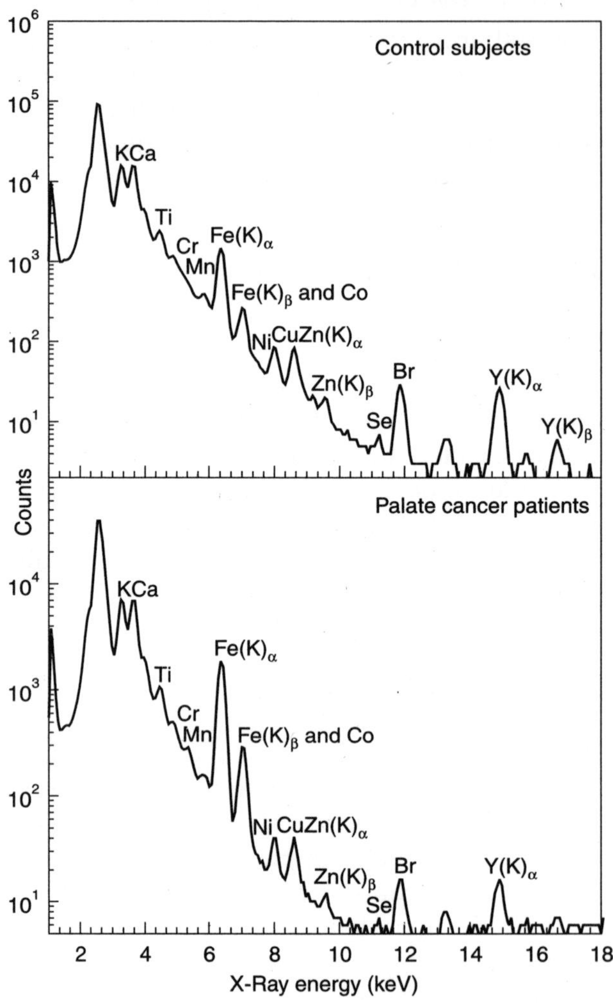

Fig. 1 *PIXE spectra for control subjects and palate cancer patients.*

Table 1 also contains the corresponding standard deviations and significant values for each element. The relative concentrations of different elements in normal and cancer subjects are shown in Fig. 2.

Table 1 *Comparison of elemental concentrations in blood serum of healthy controls and palate cancer patients.*

Elements	Concentration± standard deviation (ppm)		p value
	Control subjects(CN)	*Palate cancer patients(CP)*	
Ti	420.9 ± 17.10	420.6 ± 26.51	>0.05
V	33.94 ± 12.23	56.35 ± 17.97	<0.005
Cr	18.57 ± 7.24	18.22 ± 8.23	>0.05
Mn	39.91 ± 5.13	24.68 ± 8.03	>0.05
Fe	344.02 ± 7.11	228.04 ± 9.13	<0.0005
Co	2.25 ± 1.13	1.12 ± 1.39	>0.05
Ni	7.29 ± 3.18	8.11 ± 2.81	>0.05
Cu	24.55 ± 3.31	31.00 ± 4.26	<0.05
Zn	31.81 ± 3.64	18.74 ± 5.27	<0.01
As	0.89 ± 0.62	2.35 ± 1.74	>0.05
Se	2.15 ± 2.05	1.41 ± 0.63	<0.01
Br	42.43 ± 6.81	55.54 ± 12.22	>0.05

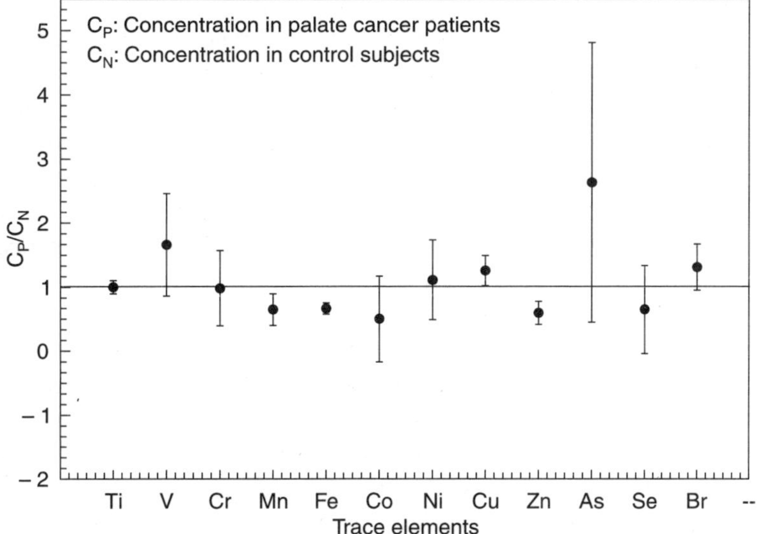

Fig. 2 *Relative concentrations of trace elements in the sera of control subjects and palate cancer patients.*

The accuracy of this technique was checked by analyzing the International Atomic Energy Agency (IAEA) certified reference material, freeze-dried animal blood A-13, and National Institute of Standards and Techniques (NIST) certified reference material, bovine liver (1577b) in the same experimental conditions as that of the serum samples. The results obtained for animal blood and bovine liver are presented in our earlier papers [2,4]. From the obtained results, it is inferred that the measured values are in good agreement with the certified values. This shows the accuracy and reliability of the present experimental set-up and use of GUPIX software package in the data analysis. The Mann–Whitney test, a nonparametric statistical unpaired-sample test, was used to examine the data for significant differences between control subjects and palate cancer patients. Statistical significance was set at $p < 0.05$. This test reveals the significant differences in serum trace elemental content between the two studied groups.

3.   RESULTS AND DISCUSSION

On comparing the average concentrations of trace elements in the serum samples of the palate cancer patients with control subjects, it was observed that the serum of the cancerous group displayed increased concentrations of V, Ni, Cu, As, and Br but lowered concentrations of Mn, Fe, Co, Zn, and Se (Table 1). A statistically significant difference was found for serum $V (p < 0.005)$, $Fe (p < 0.0005)$, $Cu (p < 0.05)$, $Zn (p < 0.01)$, and $Se (p < 0.01)$ between the two groups studied. The copper to zinc ratio for the palate cancer group was 1.65 ± 0.52, which was more than two times the value for normal subjects (0.77 ± 0.14).

The present findings of elevated copper levels in the serum of palate cancer patients suggest that excess copper might have led to the initiation and promotion of cancer in these patients by causing oxidative DNA damage [5, 6]. High levels of copper in the serum of cancer patients might have also possibly led to tumour progression through angiogenesis [7, 8]. Serum zinc was found to be significantly depressed in the cancer patients compared to healthy controls. These observed low levels of zinc lend support to the hypothesis that a low zinc status enhances the risk of developing cancer through the increase in oxidative stress and DNA damage along with an inability to adequately signal DNA repair mechanisms [9]. Other possible mechanisms of carcinogenesis associated with zinc deficiency include the decrease in antioxidant defense capacity and decline in immunological competence [10, 11]. The higher serum copper to zinc ratios observed in this work might be used as a valuable predictor for the presence of cancer. The observed deficiency of selenium in the serum of palate cancer patients is consistent with selenium's role as a potential chemopreventive agent for most of the cancers. Selenium's ablility to provide considerable protection against cancer can primarily be attributed to its antioxidant effects *via* glutathione peroxidase (GSH-Px) [12].

4.   CONCLUSION

The various findings presented here give guidelines for future study into the possible roles and interactions of essential trace elements in the carcinogenic process. Anyhow, studies at the cellular and molecular level are required in order to substantiate the observed elevated or deficient levels of trace elements in initiating, promoting, and inhibiting palate cancer. It is anticipated that similar studies from all corners of the world would ultimately lead to the development of novel therapeutic agents to treat palate cancer.

Acknowledgements

One of the authors Dr. P. Sarita acknowledges the financial support provided by University Grants Commission – Department of Atomic Energy Consortium for Scientific Research, Kolkata Centre to carry out this work during the course of her Ph.D degree programme in the Dept. of Nuclear Physics, Andhra University. The authors also thank the authorities and staff of Ion Beam Laboratory, Institute of Physics, Bhubaneswar, for providing the Pelletron accelerator facility and for rendering technical assistance. The authors acknowledge the management and staff of Lion's Cancer Treatment and Research Centre, Visakhapatnam for providing the blood serum samples of palate cancer patients.

References

1. M K Schwartz *Cancer Res.* **35** 3481-3487 (1975)
2. P Sarita, G J Naga Raju, A S Pradeep, T R Rautray, B Seetharami Reddy, S Bhuloka Reddy and V Vijayan *J. Radioanal. Nucl. Chem.* In press (2011).
3. J A Maxwell, W J Teesdale and J L Campbell *Nucl. Instr. and Meth. in Phys. Res. B* **95** 407-421 (1995)
4. G J Naga Raju, P Sarita, G A V R Murty, M R Kumar, B S Reddy, S Lakshminarayana, K P Chand, A D Prasad, S B Reddy, V Vijayan, P V B R Lakshmi and G Satyanarayana *Eur. J. Cancer Prev.*, **6** 108-115 (2007)
5. D T Sawyer, In: A E Martell and D T Sawyer (eds.) New York: *Plenum Press.* pp. 131–148 (1987).
6. T Theophanides and J Anastassopoulou *Crit. Rev. Oncol. Hematol.* **42** 57–64 (2002)
7. GF Hu, *J. Cell Biochem.* **69** 326–335 (1998)
8. A Parke, P Bhattacherjee, R M Palmer and N R Lazarus *Am. J. Pathol.* **130** 173-178 (1988)
9. E Ho, C Courtemanche and B N Ames *J. Nutr.* **133** 2543-2548 (2003)
10. E Ho *J. Nutr. Biochem.* **15** 572– 578 (2004)
11. D Galaris and A Evangelou, *Crit. Rev. Oncol. Hematol.* **42** 93–103 (2002)
12. J Neve, *Experentia*, **47** 187–193 (1991)

Assessment of DNA Damage Induced on Tumor Cells by Proton Beam from Folded Tandem Ion Accelerator

Praveen Joseph[1], N.N. Bhat[2], S. Santra[3], R.G. Thomas[3], S.K. Gupta[3] and Y. Narayana[1*]

[1]Department of Studies in Physics, Mangalore University, Mangalagangotri
[2]RPAD, Bhabha Atomic Research Centre, Mumbai
[3]NPD, Bhabha Atomic Research Centre, Mumbai
E-mail: narayanay@yahoo.com

ABSTRACT

Potential use of protons and other charged particles for cancer treatment have several advantages over low LET radiation. Single cell gel electrophoresis (Comet assay) is a powerful tool to detect the radiation induced DNA damages in single cell level and can be used to measure the DNA damage induced by protons at the single cell level. In the present study the effect of different doses of proton beams from the Folded Tandem Ion Accelerator (FOTIA) on induction of DNA damage on human tumor adeno carcinoma (HT 29) cells has been analysed using alkaline comet assay. Dosimetry was carried out using two surface barrier semiconductor detectors and the flux was adjusted to be 1.6×10^6 protons cm^{-2} s^{-1} and the fluence was varied from 5×10^7 to 7.5×10^8 protons cm-2 to deliver different doses from $0 - 15$ Gy. The energy of particles at the sample surface was also calculated theoretically using the Monte Carlo code SRIM 2008.04. The average energy of the proton beam used was 3.2 MeV and the corresponding LET was estimated to be 12.5 keV μm^{-1} within the cell layer. The mean dose per nucleus per particle and the mean number of impacts per nucleus for different fluence were calculated. Distributions of olive tail moment (OTM) as a function of fluence were studied. The measured radiation dose due to proton beam was compared with theoretical values.

Keywords: Proton beam, Folded Tandem Ion accelerator, Comet assay, DNA damage

Pacs No.: 87.53 Bn

1. INTRODUCTION

Radiation plays an important role in life science and medicine. Much of radiobiology is done with photons because they are readily available and they can penetrate a fair amount of material before loosing their effectiveness, and thus are easily encapsulated. But the behavior of charged particles (protons, light and heavy ions) are different and they can be confined by magnetic and electric field and thus be accelerated and directed as needed [1]. Potential use of protons and other charged particles in the improvement of cancer treatment has been known for some time to offer several advantages over low LET radiation cancer therapy [2].

Several techniques with different sensitivity to single-strand breaks and/or double strand breaks were applied to detect DNA breaks generated by ionizing radiation. Tests that assess DNA damage in single cells might be the appropriate tool to estimate damage induced by particles, facilitating the assessment of heterogeneity of damage in a cell population. The single cell gel electrophoresis (comet assay) is a sensitive method to evaluate DNA damage at single cell level and is ideal for human investigations. It provides a very sensitive method for detecting strand breaks and measuring DNA damages in single cells and provides a unique opportunity to investigate intercellular differences in DNA damage. Among the various versions of the assay the alkaline version (pH of the unwinding and electrophoresis buffer > 13) enables detection of the broadest spectrum of DNA damage, and it can detect double and SSBs, alkali-labile sites that are expressed as single-strand breaks and single-strand breaks associated with incomplete excision repair [3]. In the present study the alkaline comet assay has been used to assess the effect of different doses of proton beam from the folded tandem ion accelerator (FOTIA) on human tumor adeno carcinoma (HT 29) cells.

2. METHODS AND MATERIALS

2.1 Radiation Source and Dosimetry

Proton beam irradiation was carried out using 3.2 MeV Folded Tandem Ion Accelerator (FOTIA). The primary proton beam from the FOTIA was collimated using an adjustable slit to reduce the fluence and then diffused using a gold foil. The diffused beam was channeled to the exit window made of 20 μm titanium foil of 3-cm diameter to get uniformly distributed irradiation area. The target to be irradiated was positioned at a distance of 11 mm from the exit window. Before the irradiation, a silicon surface barrier (SSB) detector was positioned at the sample position and the beam energy as well as the uniformity was measured. Another SSB detector placed inside the scattering chamber at a forward angle of 80° to the primary beam, after the gold foil, served as a monitor detector. The ratio of the monitor detector counts to that of the flux measured using SSB detector at the sample position was measured by multiple trials and the calibration factor was obtained. Monitor detector counts and the measured ratio were used for delivering the required fluence to the samples.

2.2 Sample Preparation and Radiation Treatment

HT 29 cells were grown in 25 ml cell culture dishes (Falcon) and maintained in Dulbecco's modified essential medium (DMEM) supplemented with 10% fetal calf serum (FCS; Himedia) at 37°C in a humidified atmosphere containing 5% CO_2. Experiments were performed on the day of 70-80 % confluence. Cultured cells were harvested with 0.025 % trypsin EDTA and seeded in frosted slides after embedding in 0.75 % low melting point (LMP) agarose (Sigma). Cells in slides were mounted vertically at the beam exit window for the irradiation and samples were irradiated for different time duration to get the required fluence. The cells were exposed to different doses with the help of monitor detector counts under normal atmospheric pressure at 24°C.

2.3 Comet Assay

To quantify the DNA damage the comet assay was carried out under alkaline conditions, by the methods outlined by Singh et al.[4]. After irradiation the slides were covered with 0.75% LMP agarose. Slides were immersed for 12 h in ice-cold freshly prepared lysis solution [2.5 M NaCl, 100 mM Na_2EDTA, 10 mM Tris–HCl, (Sigma), pH 10] with 1% Triton X-100 (Sigma) and 10 % dimethyl sulfoxide. The slides were then placed on a horizontal gel electrophoresis tank facing the anode. The unit was filled with fresh electrophoresis buffer at pH 13 (1mM Na_2EDTA/300 mM NaOH). To allow denaturation of DNA the slides were left for 30 min in the buffer and the electrophoresis was performed at a field strength of 0.74 V cm^{-1}, 180 mA for 22 min at 4 °C under dim light. After electrophoresis, the slides were rinsed gently three times with a neutralization buffer (0.4 M Tris–HCl, pH 7.5) to remove excess alkali and detergents. Each slide was stained with SYBR Green II and covered with a cover slip. A total of 100 randomly captured comets per sample (50 cells on each of two replicate slides) were examined at 40 X magnification using a fluorescent microscope. Images of the comets were acquired and the quantification of the DNA strand breaks in each cells were performed using the Comet assay software project (CASP). The ability of the comet assay to detect differences in DNA damage between control and irradiated groups and between two irradiated groups was analyzed using Student's t-test (Origin 8.0 software)

3. RESULTS AND DISCUSSIONS

3.1 Dosimetry

A 6 MV FOTIA was used for proton irradiation studies. The beam profile of the FOTIA and energy of the beam was measured using a collimated surface barrier semiconductor detector calibrated with a standard alpha source. The proton fluence at various points in the irradiation area is plotted in Fig. 1. The graph clearly indicates the uniformity of the ion fluence in the irradiation area.

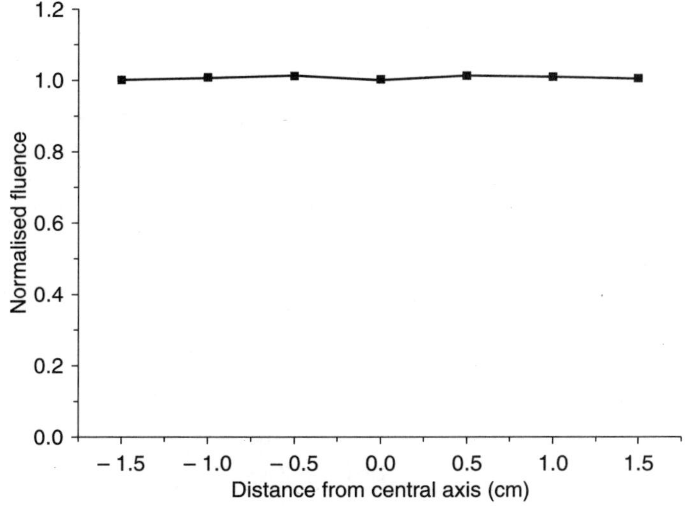

Fig. 1 *Distribution profile of proton beam in horizontal direction*

The stopping power and range of particles at the sample surface was also calculated theoretically using the Monte Carlo code SRIM [5], which is in agreement with the measured values within experimental uncertainties. The average LET corresponding to the beam energy and hence the flux of beam required to deliver a particular dose in tissue was also calculated using the same code. Flux was calculated to be 1.6×10^6 protons cm^{-2} s^{-1} and the fluence was varied such that the cells received the required dose. Fluence was then used to calculate dose with the help of specific ionization curve constructed for the given energy distribution and composition of the cell line. The average energy of the proton beam used in this study was measured to be 3.2 MeV and the corresponding estimated LET was 12.5 keV μm^{-1} within the cell layer. Initial portion of Bragg's curve was used for dose estimation since the cell layer used for irradiation was in monolayer geometry with thickness less than 20 μm, which helped to avoid steep increase in LET within the sample. The proton beam of intermediate LET (12.5 keV μm^{-1}) were used for the study which has got a penetration of 240 μm in tissue.

3.2   DNA Damage On Cells

The samples were irradiated with different doses ranging from 0 to 15 Gy of proton particles. Alkaline comet assay was applied to quantify the induced DNA damage and the comets were quantified by the Olive tail moment (OTM). Distributions of OTM as a function of fluence and experimental dose for DNA damage induced by protons were analyzed. The variation in OTM with dose is shown in Fig. 2 and the control value is normalized to one. A statistically significant increase in OTM in comparison with control is observed. A clear shift and dispersion of comet distributions towards high OTMs was observed with increasing fluence and mean dose.

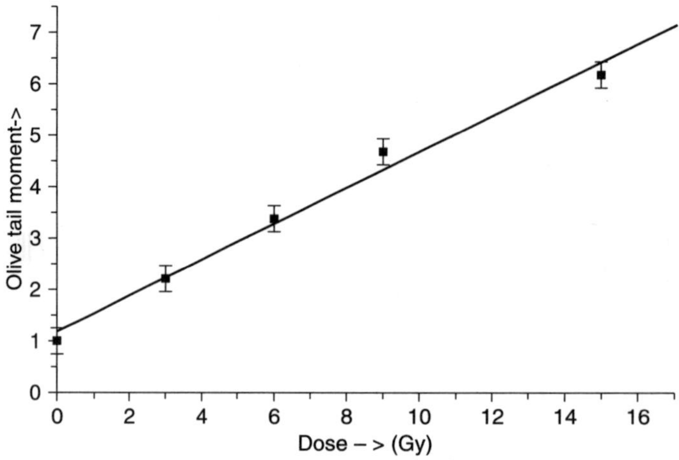

Fig. 2 *The normalized Olive Tail Moment (OTM) with dose in HT 29 cells after proton irradiation.*

4. CONCLUSION

The comet assay can be used to get information on DNA damage induced by protons in single human HT 29 cells. The LET of the beam was calculated and the fluence required for 1 Gy of dose was found to be 5×10^7 particles cm^{-2}. The mean dose per nucleus per ion is found to be 11.3 mGy. A clear shift and dispersion of comet distributions towards high OTMs was observed with increasing fluence and mean dose. Cell proliferation varies linearly with proton fluence and a linear increase in OTM shows that DNA damage increases linearly with dose.

References

1. Hartmut F and W Sadrozinski *Nuclear instruments and methods in physics research section A: Accelerators, spectrometers, detectors and associated equipment,* **514**, 1-3, 224-229 (2003)

2. W Wieszczycka and H Scharf Physical and radiobiological properties of hadrons, in: *Proton Radiotherapy Accelerators,* London: World Scientific ed **318** 24–47 (2001)

3. A.Hartmann, E.Agurell, C.Beevers, S.Brendler-Schwaab, B.Burlinson, P.Clay, A.Collins, A.Smith, G.Speit, V.Thybaud and R.R.Tice. Recommendations for conducting the *in vivo* alkaline Comet assay *Mutagenesi,* .**18** no.1, 45–51 (2003)

4. N P Singh, M T McCoy, R R Tice and E L Schneider *Exp. Cell Res.* **175** 184–91 (1988)

5. J F Ziegler, J P Bierserk and U Littmark Stopping power and ranges of ions in matter, Pergamon Press, New York. (1985)

Indoor Radiation Dose Enhanced by Underground Radon Diffusion

R.P. Chauhan

Department of Physics, National Institute of Technology, Kurukshetra, India
E-mail: *chauhanrpc@gmail.com*

ABSTRACT

The indoor radiation dose received by humans is mainly due to radioactive radon gas and its progeny. There are different factors of radon entrance in to a building environment; emission of radon from building materials used in construction, convection through cracks and openings in the building and diffusion from soil through pore space of construction materials etc. The transport phenomenon of radon through diffusion is a significant contributor to indoor radon entry. The diffusion of radon in dwellings is a process mainly determined by the radon concentration gradient across the building material structure between the radon source and the surrounding air. Keeping this in mind the radon diffusion studies have been made through some building materials like soil, brick powder, stone powder and cement. Simultaneously the indoor radon levels and annual effective dose received by the residents residing in dwellings with soil, bricks, stone and cemented flooring have also been measured. The results indicate that indoor radiation dose is higher in the dwellings with flooring material having higher radon diffusion through them.

Keywords: Radon, progeny, Annual dose, Diffusion, Building materials.

Pacs No.: 87.55.N

1. INTRODUCTION

The radon and its progeny constitutes more than 50% of the dose equivalent received by general population from all sources of radiation, both naturally occurring and man-made [1]. When radon decays it forms its progeny ^{218}Po and ^{214}Po, which are electrically charged and can attach themselves to tiny dust particles, water vapours, oxygen, trace gases in indoor air and other solid surfaces. These daughter products remain air-borne for a long time and can easily be inhaled into the lung and can adhere to the epithelial lining of the lung, thereby irradiating the tissue. Bronchial stem cells and secretion cells in airways are considered to be the main target cells for the induction of lung cancer resulting from radon exposure. The exposure of population to high concentrations of radon and its daughters for a long period lead to pathological effects like the respiratory functional changes and the occurrence of lung cancer [2].

The radon entry into a building is mainly due to the following factors [3]; a) emanation from building materials b) convection via cracks and openings and c) diffusion from soil via the pore space of building materials. The emanation from building materials is of less concern when room ventilation is few air exchanges per hour but when it is less the radon concentration may exceed the action level. The radon entry via cracks and openings may be suppressed by sealing the radon entry points. The diffusion from soil via the pore space of building materials is reduced when thickness of the basement slab is about three times the radon diffusion lengths [4]. The transport phenomenon of radon through diffusion is a significant contributor to indoor radon entry [5].

Radon diffusion through material media obeys the equation:

$$N = N_0 \exp. \left(-\sqrt{\lambda/D}\right) X \tag{1}$$

where N is the concentration of radon at any time t at a distance X from source, N_0 is the concentration of radon at source and λ is the decay constant of radon.

If N_1 and N_2 are the radon concentrations at distances X_1 and X_2 from source respectively.

Using eq^n (1) the diffusion coefficient D is given by:

$$D = \lambda[(X_2 - X_1)/\ln (N_1/N_2)]^2 \tag{2}$$

Eq^n (2) can be used to calculate radon diffusion coefficient through material medium.

The diffusion length can be calculated using the eq^n:

$$L = \sqrt{D/\lambda}, \tag{3}$$

where D is radon diffusion coefficient and λ is decay constant of radon.

In the present study, radon diffusion coefficients and diffusion lengths through building materials viz.; soil, brick powder, stone powder and cement have been calculated using eq^n (2) & (3).

2.   EXPERIMENTAL DETAILS

The diffusion chamber designed for the study of radon diffusion through different building materials consisted of a hollow plastic cylinder of inner diameter 25 cm and length 50 cm deployed vertically. The uranium ore was used as radon source covered with latex membrane fixed at the bottom of the cylinder in the cavity. Open-ended cylindrical diffusion tubes of diameter 1.5 cm and of length 15 and 25 cm were installed in hollow plastic cylinder fixed with radon source. Different materials were filled in pulverized form in different open-ended diffusion tubes up to the height of 10 cm and 20 cm. A piece of LR-115, type-II plastic track detector was fixed at the top of each diffusion tube such that sensitive side of the detector always faced the source. The system was left undisturbed for a period of 30 days. The packing density of each sample was also calculated by taking mass over volume ratio. All samples were subjected to similar process of exposure as described above.

For the measurement of radon and its progeny concentration radon-thoron twin dosimeter cups were used. The radon-thoron dosimeter cup has three different modes namely bare mode, filter mode and membrane mode. Three pieces (1cm × 1cm) of LR-115 solid-state Nuclear Track detectors were fixed in the dosimeters and were suspended in the dwellings for three months. The bare mode detector registers tracks due to radon, thoron and their progeny, the filter mode detector registers tracks due to radon and thoron while the membrane mode detector registers tracks only due to radon. The dosimeters were suspended at a height a height of about 1.5 m in order to evaluate the annual average indoor radon levels.

At the end of the exposure time, the detectors were removed and subjected to a chemical etching process in 2.5N NaOH solution at 60^0C for 90 minutes. The detectors were washed and dried and the tracks produced by the alpha particles were observed and counted under an optical Olympus microscope at 600X. A large number of graticular fields of the detectors were scanned to reduce statistical errors. The measured track density (Track/cm^2/day) was converted into radon and thoron concentration using calibration factors [6]. Radon and thoron progeny levels in mWL have also been calculated using indoor equilibrium factor as 0.4 for radon and 0.1 for thoron from UNSCEAR [7]. Annual dose received by the inhabitants in the dwellings under study in mSv was estimated using the relation [8-9]:

$$D = [(0.17 + 9\ F_R)\ C_R + (0.11 + 32\ F_T)\ C_T] \times 7000 \times 10^{-6}$$

where, F_R = equilibrium factor for radon; C_R = radon concentration; F_T = equilibrium factor for thoron and C_T = thoron concentration

3. RESULTS AND DISCUSSION

The values of diffusion coefficient and diffusion length calculated for different materials are shown in Table 1. The values are found to be least for cement, which shows that cement is the least, permeable to radon flow as compared with the other building materials studied. Similar results have been reported [10-11] for cement, soil etc. The minor differences may be due to the difference in the nature, grain size and porosity of the materials. The annual effective doses received by residents due to radon, thoron and their progeny are shown in Table 2. The results indicate that the levels are higher in dwellings with soil flooring compared with cement, stone and brick flooring. It may be due to the more underground radon diffusion through soil as compare to compare to other building materials. A positive correlation has been observed between underground radon diffusion and annual effective dose shown in Fig. 1.

Table 1 *Radon Diffusion coefficient and diffusion length for different building materials*

Diffusing medium	No. of samples	Packing Density $\times 10^{-3}$ (kg/m^3)	Diffusing Coefficient $\times 10^{-6}$ (m^2/s)	Diffusion Length (m)
Soil	3	1.4	3.3 ± 0.3	1.2 ± 0.1
Brick Powder	4	1.3	2.3 ± 0.4	1.1 ± 0.1
Stone Powder	3	1.7	2.1 ± 0.5	1.0 ± 0.1
Cement	4	1.5	1.2 ± 0.1	0.7 ± 0.1

Table 2 *Radon, thoron and their progeny levels in some dwellings of Haryana.*

Type of dwelling	No. of dwellings	Radon conc. (Bq m^{-3})	Thoron conc. (Bq m^{-3})	Progeny levels (mWL) Radon	Thoron	Annual Dose (mSv)
Soil	6	104 ± 12	69 ± 10	11.2	1.8	4.3
Brick	6	97 ± 11	65 ± 6	10.5	1.8	4.1
Stone	8	81 ± 12	48 ± 9	8.7	1.3	3.3
Cemented	12	66 ± 6	27 ± 5	7.1	0.7	2.2

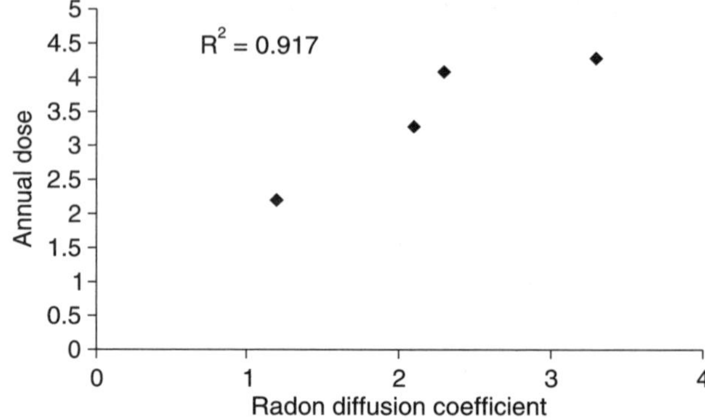

Fig. 1 *Correlation between radon diffusion and annual effective dose*

4. CONCLUSION

Based on the experimental investigations, the following conclusions are drawn:

- For the building materials studied during present investigations, cement is found to be the least permeable to radon flow.

- The levels are higher in dwellings with soil flooring compared with cement, stone and brick flooring due to the more underground radon diffusion through soil as compare to other materials.
- On the basis of these investigations it is suggested that the cemented houses are safer than mud houses from the health hazard point of view from radon and its progeny.
- Positive correlation has been observed between radon diffusion and annual indoor dose ($R^2 = 0.917$)

References

1. BEIR V (Report of the Committee on the Biological effects of Ionizing Radiation) *Natl. Acad. of Sciences.* Natl. Acad. Press, Washington, DC (1990)
2. BEIR VI (Report of the Committee on the Biological effects of Ionizing Radiation) *Natl. Res. Council.* Natl. Acad. Press, Washington, DC (1999)
3. K Kovler, A Perevalov, V Steiner and E Rabkin *Health Phys.* **86 (5)** 505 (2004)
4. G Keller, B Hoffman and T Feigenspan *Sci.Total Env.* **37** 85 (2001)
5. K J Renken and T Rosenberg *Health Phys.* **68 (6)** 800 (1995)
6. K P Eappen, T V Ramachandran, A N Saikh and Y S Mayya *Rad. Prot. Env.* **24 (1&2)** 410 (2001)
7. UNSCEAR (United Nations Scientific Committee on the Effects of Atomic Radiation), *Exposures from Natural Sources of Radiation, A/Ac, 82/R* p 511 (1992)
8. J Sannappa, M S Chandra Shekara, L A Sathish, L Paramesh and P Venkataramaiah *Radiation Measurement* **37** 55 (2003)
9. Y S Mayya, K P Eappen and K S V Nambi *Radiat. Prot. Dosim.* **77(3)** 177 (1998)
10. S Singh, J Kumar, B Singh, and J Singh *Radiation Measurement* **30** 461 (1999)
11. M Tuffail, S M Mirza, M K Chugtai, N Ahamad, H A Khan *Nucl. Track. Rad. Meas* **19**: 427 (1991)

Study of Radon Concentration and its Influencing Factors in the Lower Atmosphere, Bangalore, South India

G.V. Ashok[1], N. Nagaiah[2*] and N.G. Shiva Prasad[3]

[1]*Government College (Autonomous), Mandya, India*
[2]*Department of Physics, Banaglore University, Bangalore, India*
[3]*Government First Grade College, Srirangapatna, India*
E-mail: *nagaiahn@rediffmail.com

ABSTRACT

The measurements of radon in the outdoor air were carried out with a time interval of 2 hours for a period of 3 days in every month for one year (oct 2007-sept 2008) near the department of Physics, Jnanabharathi, Bangalore. The Low Level Radon Detection System (LLRDS) was used to measure the concentrations of radon. The influence of exhalation rate(X_1) and meteorological parameters such as humidity(X_2), and temperature(X_3) on the variation of radon concentration(Y) was studied using multiple linear regression technique with the model given by $Y = \beta_1 X_1 + \beta_2 X_2 + \beta_3 X_3 + \varepsilon$, ε being the random error. The R-squared value (co-efficient of determination) is found to be 0.936. Average daily pattern of radon and its exhalation rate in the lower atmosphere featured a minimum in the late afternoon and a maximum in the early morning hours. While seasonally the concentrations attain a maximum average in winter and a minimum average in rainy. The annual mean value of radon concentration and inhalation dose were found to be 7.8 Bq m^{-3} and 0.076 mSv/y respectively.

Keywords: Radon, Exhalation rate, Diurnal, Seasonal, Dose.

1. INTRODUCTION

Over the past few decades, there has been a large scientific interest in the study of environmental radon. One of the main reasons is its associated health hazard because it is found to be the second leading cause of lung cancer after cigarette smoking [1]; another is its widespread use as a good natural tracer of vertical dispersion. The outdoor radon concentrations found during surveys of France were as high as 400 Bq m^{-3}, being equal to action level recommended by the ICRP-1993 [2]. Therefore it is important to know the levels of radon in our living environment.

Some researchers have studied the outdoor radon levels and their variation at different regions of the world for health risk assessment and to understand the vertical dispersion in the atmosphere [3-5]. But no such study was carried out so far in this part of the country and hence it appears to be the first of its kind in this area.

The concentration of radon in the lower atmosphere is governed by several geophysical and meteorological parameters. In the present study the measurement of diurnal and seasonal variations of radon & its exhalation rate were carried out for a period of 1 year. During the measurement period the meteorological parameters such as temperature and humidity were noted carefully. The observed diurnal variations were studied with respect to these meteorological parameters and the exhalation rates prevailing during the time of measurements of radon concentration using the multiple linear regression technique. From the measured concentrations, the inhalation dose to the general public was also estimated.

2. MATERIALS AND METHODS

In the present work, LLRDS (Low Level Radon Detection System) was used for the estimation of radon concentration. The description of the method and procedure for the measurement of radon concentration using LLRDS is as explained elsewhere [6].

The exhalation rate of ^{222}Rn was estimated by employing the accumulation chamber method. The open end of the chamber was buried to a depth of 150 mm at the selected location. The ^{222}Rn exhaled from the soil was allowed to accumulate in the chamber. Then the radon gas collected in the chamber was transferred to the LLRDS and ^{222}Rn concentration was then estimated from which the exhalation rate was determined [7].

3. RESULTS AND DISCUSSION

The radon exhalation rate was found to be high during night & early morning hours, followed by the decrease after sunrise and reaches a minimum during afternoon hours (Fig. 1). This may be due to the fact that, the rate of exhalation of radon from the soil depends on the emanation factor which is known to increase with the moisture content of the soil. The temperature drop at night may causes condensation in the soil. The interstitial water due to condensation will absorb the radon atoms recoiling from the soil particle and due to the low solubility of radon in water, the radon emanation rate will increase, thereby giving rise to higher radon concentration in the morning. The exhalation rate ranges from 5.49 to 14.6 mBq m^{-2} S^{-1} with a mean value of 10.1 ± 2.79 mBq m^{-2} S^{-1}. The mean value is slightly lower than the estimated mean worldwide flux of ^{222}Rn of 16 mBq m^{-2} s^{-1}[8]. It can be seen from Fig. 1 and 2 that, there exist a positive correlation between radon concentration & relative humidity and the negative correlation with ambient temperature. The higher radon concentrations during early morning hours and lower concentrations in the afternoon (Fig. 1) can be attributed to temperature-induced turbulence and changes in the exhalation rates. The radon concentration in air ranges from 3.7 to 31.5 Bq m^{-3} with a mean value of 11.83 ± 9.59 Bq m^{-3} showing the diurnal variation of the order 8. To study the influence of regression variables exhalation rate(X_1), humidity(X_2), and temperature(X_3) on radon concentration(response variable-Y) the statistical analysis of the data was carried out using multiple linear regression technique (through the Origin) and it is given by the model $Y = \beta_1 X_1 + \beta_2 X_2 + \beta_3 X_3 + \varepsilon$, where ε is a random error and β's are the regression coefficients respectively. The assumptions of the above model are: (a) The relationship between each of the predictor variables and the dependent variable (Rn concentration) is linear. (b) $\varepsilon \sim NID$ $(0, \sigma^2)$, i.e., the error term is Normally and Independently Distributed with mean 0 and variance σ^2. Further the errors are uncorrelated with the predictors.

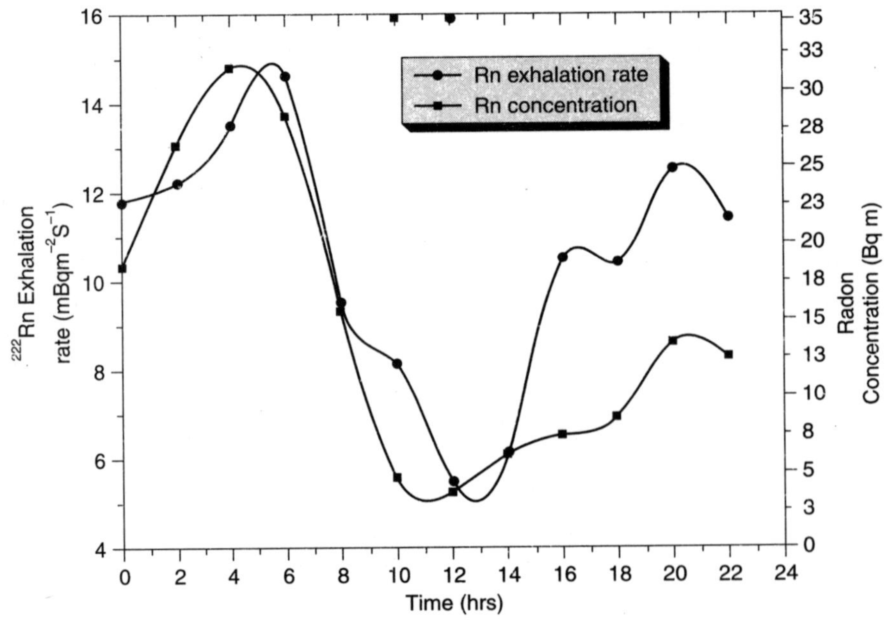

Fig. 1 *Diurnal variation of Rn concentration and its exhalation rate*

The R-squared value (co-efficient of determination) which can be interpreted as the the variance in y(Rn concentration) that can be obtained from the predictor variables is computed. In the present analysis R–squared value is found to be 0.936 which reveals that about 93.6% of the variation in radon concentration can be explained by the exhalation rate, relative humidity and temperature. Also, the QQ plot of residuals

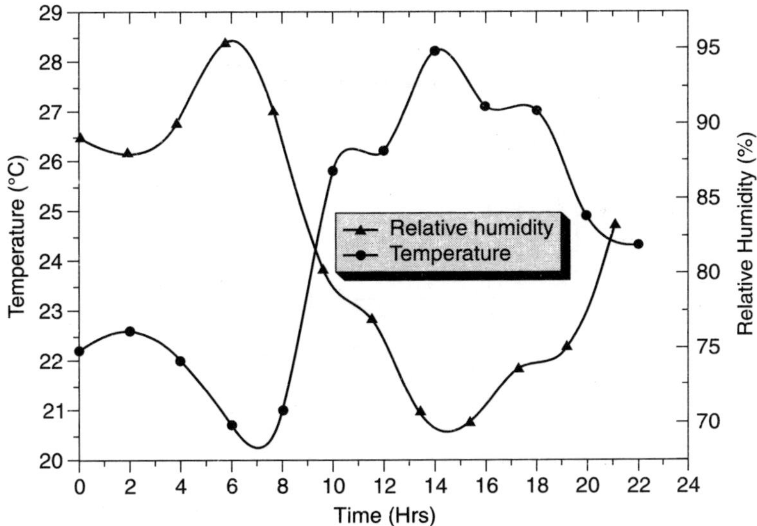

Fig. 2 *Diurnal variation of temperature and relative humidity*

showed a straight line which satisfies normality assumption. Statistical test for the significance of the regression coefficients was carried out and the results are shown in Table 1. It can be observed that *P*-value corresponding to exhalation rate and temperature is less than 5% level of significance. Hence, it clearly indicates that both the regression co-efficients have significant effect on radon concentration. But for humidity the P-value is > 0.05, it may be due to the fact that predictor variables exhalation rate and humidity are inter correlated.

Table 1 *Statistical test for the significance of the regression coefficients*

Predictor variables	β	Std. Error	t-test statistics	P-value
Exhalation rate (mBq m^{-2} S^{-1})	1.689	.593	2.849	.019
Relative humidity (%)	.268	.126	2.129	.062
Temperature (°C)	-1.024	.271	-3.775	.004

Fig. 3 shows the average concentration of ^{222}Rn in all the months of a measurement period. It was observed that, the concentration was high in the month of January (12.29 Bq m^{-3}) and low (5.02 Bq m^{-3}) in the month of June.

3.1 Seasonal Variation

The seasonal variations of radon concentrations are presented in Fig. 4. It can be seen from the figure that, the concentrations of radon during winter were much higher than these during summer and rainy. The radon concentrations during winter,

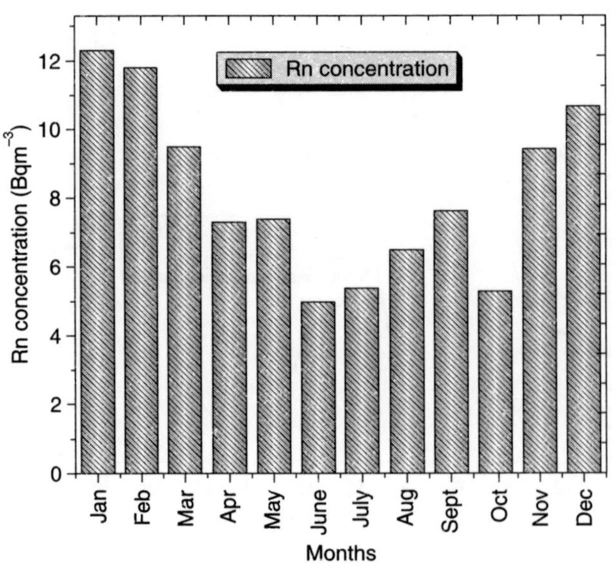

Fig. 3 *Monthly variation of radon concentration.*

summer and rainy vary between 4.06–24.2 Bq m^{-3}, 4.2–12.8 Bq m^{-3} and 3.5–12.4 Bq m^{-3}, respectively.

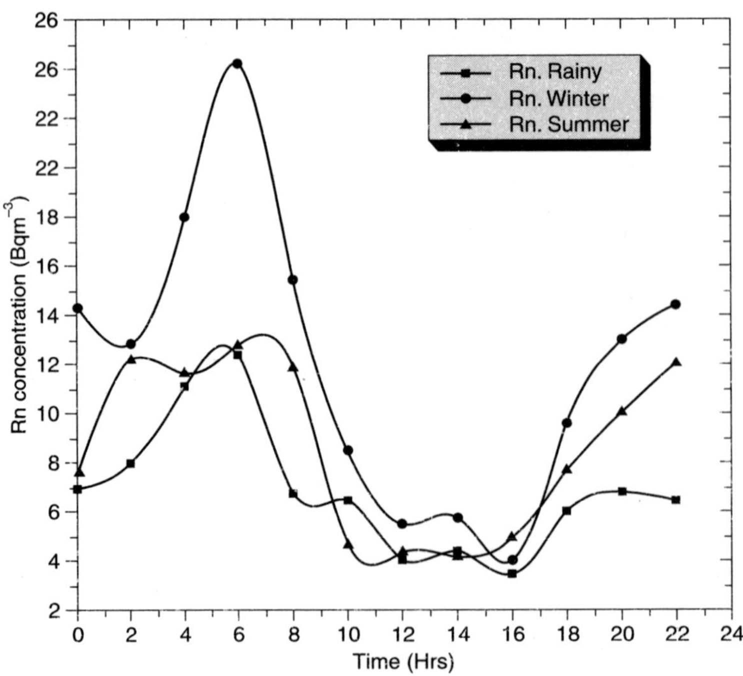

Fig. 4 *Seasonal variation of Rn concentration*

3.2　Radiation Dose

A mean value of outdoor radon concentration was found to be 7.8 ± 2.53 Bq m^{-3} and is well within the worldwide, population-averaged radon concentration of 10 Bq m^{-3} for outdoors [9]. With this an attempt was made to establish the inhalation dose to the population of the region and is found to be 0.076 mSv/y.

4.　CONCLUSION

The *R*–squared value obtained from the multiple regression analysis is found to be 0 .936. The P-value corresponding to the predictor variables such as exhalation rate and temperature is less than 5% level of significance. Hence both the regression co-efficients have significant effect on radon concentration. Nocturnal accumulation of radon and high concentrations during early morning hours confirms the strong influence of temperature-induced turbulence on outdoor radon concentration. The concentration of radon during winter was found to be much higher than that during summer and rainy. The annual mean value of radon concentration and inhalation dose were found to be 7.8 Bq m^{-3} and 0.076 mSv/y respectively.

References

1. National Academy of Sciences, Health Effects of Exposure to Radon: BEIR VI, National Academy Press, Washington, 1998.
2. ICRP 65: Ann. ICRP., **23(2)** (1993).
3. Grasty Health Phys. **66 (2)** 185-193 (1994).
4. S Oikawa, N Kanno, T Sanada, N Ohashi, M Uesugi, K. Sato, J Abukawa and H.Higuchi J Environ. Radioact. **65** 203-213 (2003).
5. L Sesana, E Caprioli, GM Marcazzan J. Environ Radioact. **65** 147-160 (2003).
6. Srivastava, G.K., Raghavaiah, M., Khan, A.H. and Kotrappa, P.A. A low level radon detection system. Health Phys. 46(1), 225-228 (1984).
7. M S Chandrashekara, J Sannappa, L Paramesh Atmospheric Environ. **40**, 87-95 (2006).
8. UNSCEAR, Report to the general assembly with scientific annexes. New York (2000)
9. UNSCEAR, Report to the general assembly with scientific annexes. New York (1993).

Statistical Analysis and Effect of Physico-chemical Parameters on Radionuclides in Soil and Sediments of Sharavathi River

K.M. Rajashekara[1*], V. Prakash[2] and Y. Narayana[3]
[1]Department of Physics, SJC Institute of Technology, Chickballapur, India
[2]Department of Physics, SDM Institute of Technology, Ujire, D.K, India
[3]Department of Studies in Physics, Mangalore University, Mangalagangothri, India
E-mail: *km_rajashekar@yahoo.co.in

ABSTRACT

Detailed studies on radionuclides concentration in riverine environs of Sharavathi River, a major river of coastal Karnataka, to study the distribution and seasonal variation in this region. The riverbank soil and sediments are collected in pre-monsoon, post-monsoon and monsoon seasons, were analyzed for natural radionuclides by gamma spectrometry. The range of measured activities differed widely as their presence in riverine environment depends on their physical and geo-chemical properties. Therefore, an attempt is made in the present investigation to studies on the statistical analysis and physico-chemical parameters that govern the distribution of radionuclides in the different seasons of these environs.

Keywords: Natural radioactivity, Radiation monitoring, Seasonal variation, Geochemical parameters, River.

Pacs No.: 28.41.Te, 92.40.Lg, 02.50.-r

1. INTRODUCTION

The distribution of naturally occurring radionuclides mainly uranium (^{238}U), Thorium (^{232}Th), potassium (^{40}K) and other radioactive elements, depends on the distribution of rocks from which they originate and the rocks which concentrate them. Exposure to ionizing radiation from natural sources is a continuous and unavoidable feature of life on the earth. The major sources responsible for this exposure are due to the presence of naturally occurring radionuclides in the earth's crust [1]. Published data on natural radionuclide series for the riverine environs are sparse and no reference citing data for coastal Karnataka. The Sharavathi river originates in western Ghats, flow due east or northeast initially, then take a sudden turn to the west, flow down the steep western slopes of the Ghats and after meandering through the short plateaus join the sea passing through the coastal plain. The length of the Sharavathi River is 128 km and catchment area of 3,592 km². Several tributaries join the Sharavathi and four hydroelectric power plants with a total installed power of 1469 MW are in operation along the river (Figure 1). The aim of this programme is to measure the gamma radiation level, know the distribution of source-rock materials containing elevated levels of radionuclides and understand the physical and geochemical process that effects on the natural radionuclides.

2. MATERIALS AND METHODS

2.1 Sample Collection

Presently the sampling stations were identified along river Sharavathi (Fig. 1). The soil samples from banks of the river and sediment samples from the river are collected during pre-monsoon, post-monsoon and monsoon seasons following standard procedure [2].

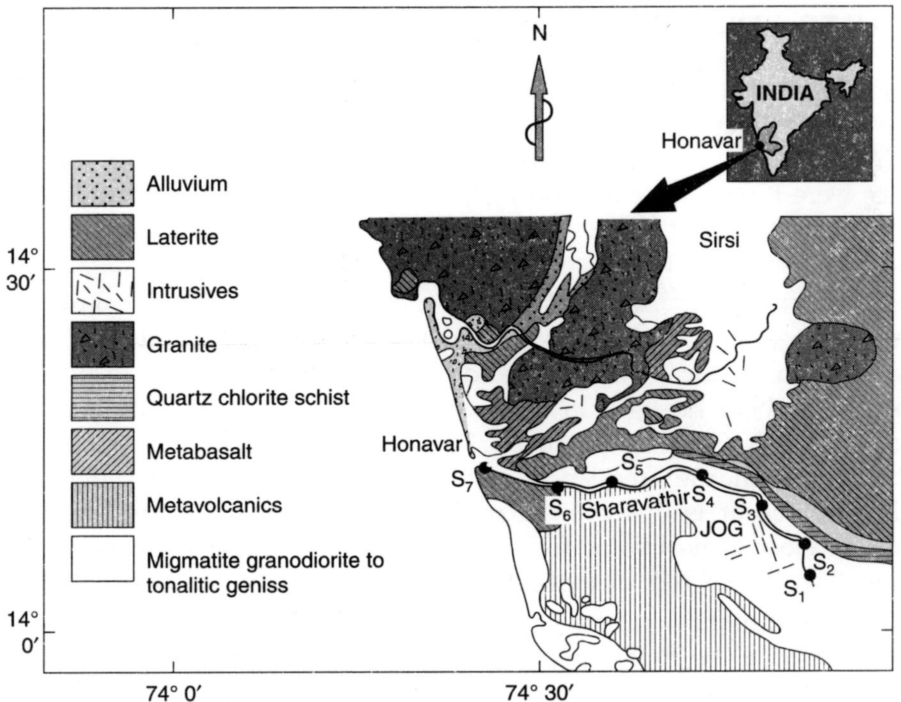

Fig. 1 *Map of the sampling stations*

The soil and sediment samples collected were brought to the laboratory, dried, ground and sieved to get below 250 microns and filled in air tight plastic containers and stored to ensure equilibrium between radium and its short lived daughters.

2.2 Physico-chemical Parameters of the Samples

In the present study organic matter content of the soil and sediment was determined by the weight loss-on-ignition method [3] at an ignition temperature of 550°C for 24 h. The soil and sediment pH were determined by standard procedure. Measurement of pH was carried out using a EUTECH pH Scan 35 pH meter. For separation of sand, silt and clay fractions, about 20 g of sediment were taken from the samples. The residual fraction was used for pipette analysis to obtain the percentage of silt and clay based on Stokes law [4].

2.3 Activity Determination

The concentrations of natural radionuclides in the soil and sediment samples returned to the laboratory were determined employing a high efficiency NaIs(Tl) gamma ray spectrometer [5]. The gamma ray spectrometer consists of a 5″ × 5″ NaI (Tl) detector coupled to a 4 K MCA.

3. RESULTS AND DISCUSSION

The seasonal variation measurements of natural radionuclides like ^{226}Ra, ^{232}Th and ^{40}K activity in soil samples of pre-monsoon season were found to be 80.4 Bq kg^{-1}, 15.9 Bq kg^{-1} and 596.9 Bq kg^{-1}. For post monsoon season the mean values of ^{226}Ra, ^{232}Th and ^{40}K activity were found to be 64.0 Bq kg^{-1}, 14.8 Bq kg^{-1} and 573.8 Bq kg^{-1}. In monsoon season the mean values of ^{226}Ra, ^{232}Th and ^{40}K activity were found to be 54.5 Bq kg^{-1}, 18.3 Bq kg^{-1} and 454.1 Bq kg^{-1} respectively.

The mean values of ^{226}Ra, ^{232}Th and ^{40}K activity in sediment samples for pre-monsoon season were found to be 99.3 Bq kg^{-1}, 22.8 Bq kg^{-1} and 633.5 Bq kg^{-1}. For post monsoon season the ^{226}Ra, ^{232}Th and ^{40}K activity were found to be 72.0 Bq kg^{-1}, 11.6 Bq kg^{-1} and 493.4 Bq kg^{-1}. In monsoon season the ^{226}Ra, ^{232}Th and ^{40}K activity were found to be 53.9 Bq kg^{-1}, 14.8 Bq kg^{-1} and 465.7 Bq kg^{-1} respectively. Figs 2-7 shows the corresponding frequency distributions of the activities detected for the cited radionuclides. It is observed that the skewness and kurtosis coefficients are approximately close to null value for some samples indicating the existence of normal distribution (Fig. 2) and rest of that are log-normal distribution (Figs. 3–7).

The activities of natural radionuclides in soil and sediment samples in pre-monsoon season were found to be high compared to post-monsoon and monsoon seasons. The concentration of radium activity in post-monsoon, pre-monsoon and monsoon shows

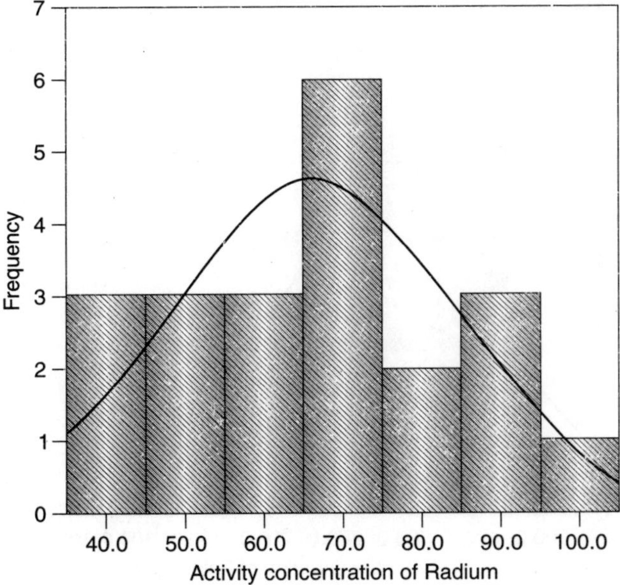

Fig. 2

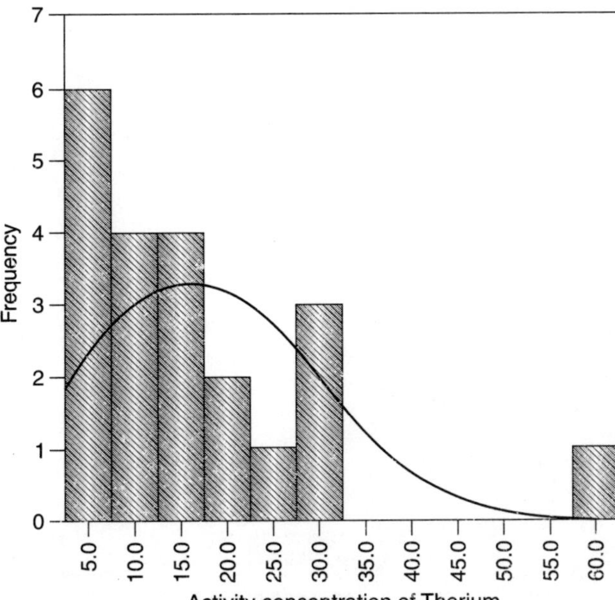

Fig. 3

8
6
4
2
0
Frequency
300.0 400.0 500.0 600.0 700.0 800.0 900.0 1000.0
Activity concentration of Potassium

Fig. 4

12
10
8
6
4
2
0
Frequency
40.0 60.0 80.0 100.0 120.0 140.0 160.0 180.0 200.0
Activity concentration of Radium

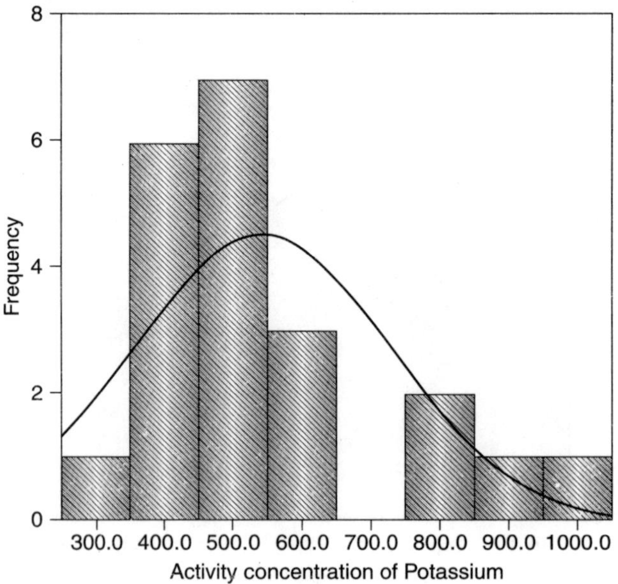

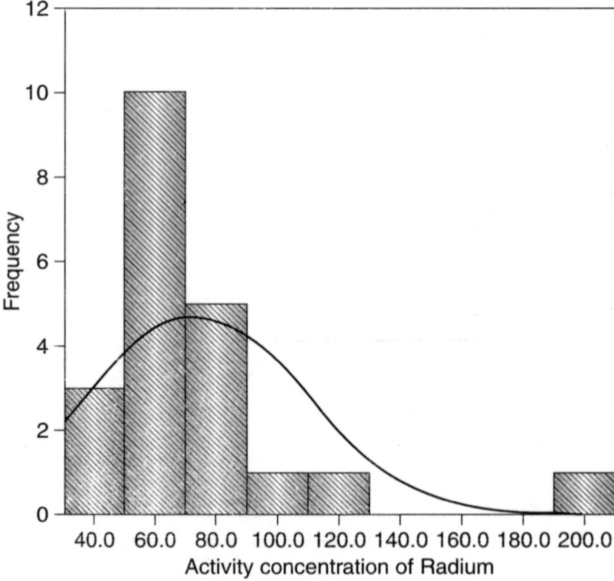

Fig. 5

8
6
4
2
0
Frequency
0.0
5.0
10.0
15.0
20.0
25.0
30.0
35.0
40.0
45.0
50.0
55.0
Activity concentration of Thorium

Fig. 6

7
6
5
4
3
2
1
0
Frequency
300.0
400.0
500.0
600.0
700.0
800.0
900.0
1000.0
Activity concentration of Potassium

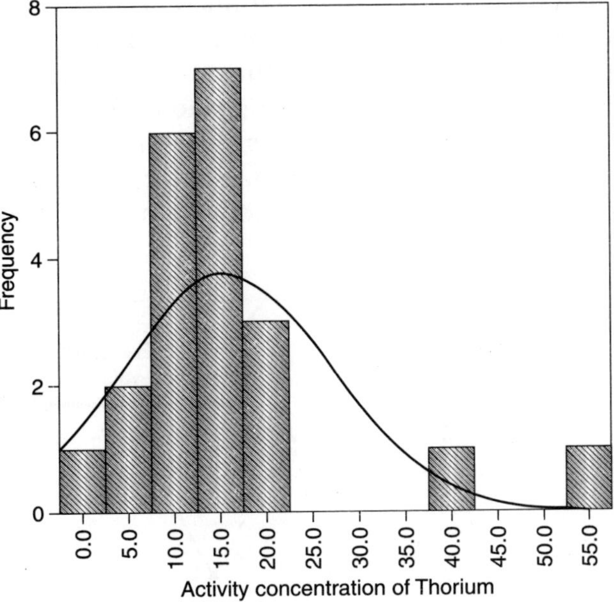

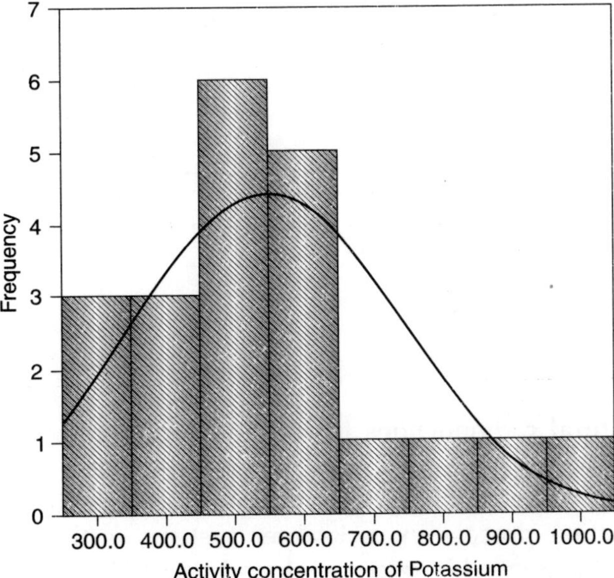

Fig. 7

high values. In the river basin are formed by alkaline rocks of volcanic origin. The alkaline rocks of volcanic origin aquifers, have deep sources, are highly mineralized and relatively the pH values in soil and sediment ranges from 6.8 to 7.6; 6.8 to 8.0. The variation of ^{226}Ra activity and pH from origin to estuary in sediment samples are shown in Figs. 8–10. The correlation coefficients between the activities of radionuclides and silt/clay and organic matter contents shows no significant correlation, however, a good correlation was observed between activity and pH only in the sediment samples.

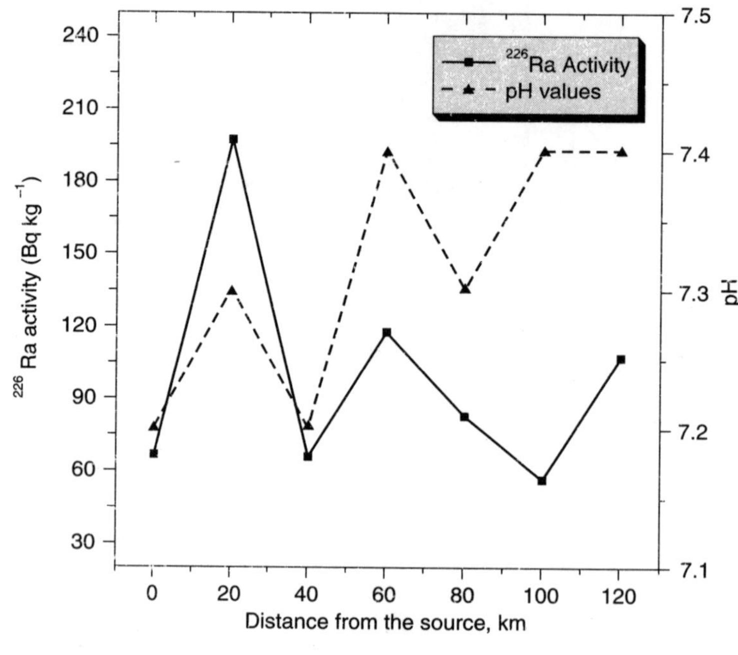

Fig. 8

4. CONCLUSION

The activities of natural radionuclides in soil and sediment samples in pre-monsoon season were found to be high compared to post-monsoon and monsoon seasons. The concentration of radium activity shows high values in all the seasons. The ^{226}Ra prevailing solubilization process in rock-water systems is the chemical reaction and relatively higher contact period between rock and water. As *U* series radionuclides are largely more soluble than Thorium, it is often found in deficient with more respect to Thorium in the solid surface environment. The concentrations of these radionuclides are not in good order due to differential erosion, highly dissected landscape, river capture and existence of waterfalls of great magnitude. This river is a classic example

^{226}Ra activity (Bq kg^{-1})
pH
Distance from the source, km
^{226}Ra Activity
pH values

Fig. 9

^{226}Ra activity (Bq kg^{-1})
pH
Distance from the source, km
^{226}Ra Activity
pH values

Fig. 10

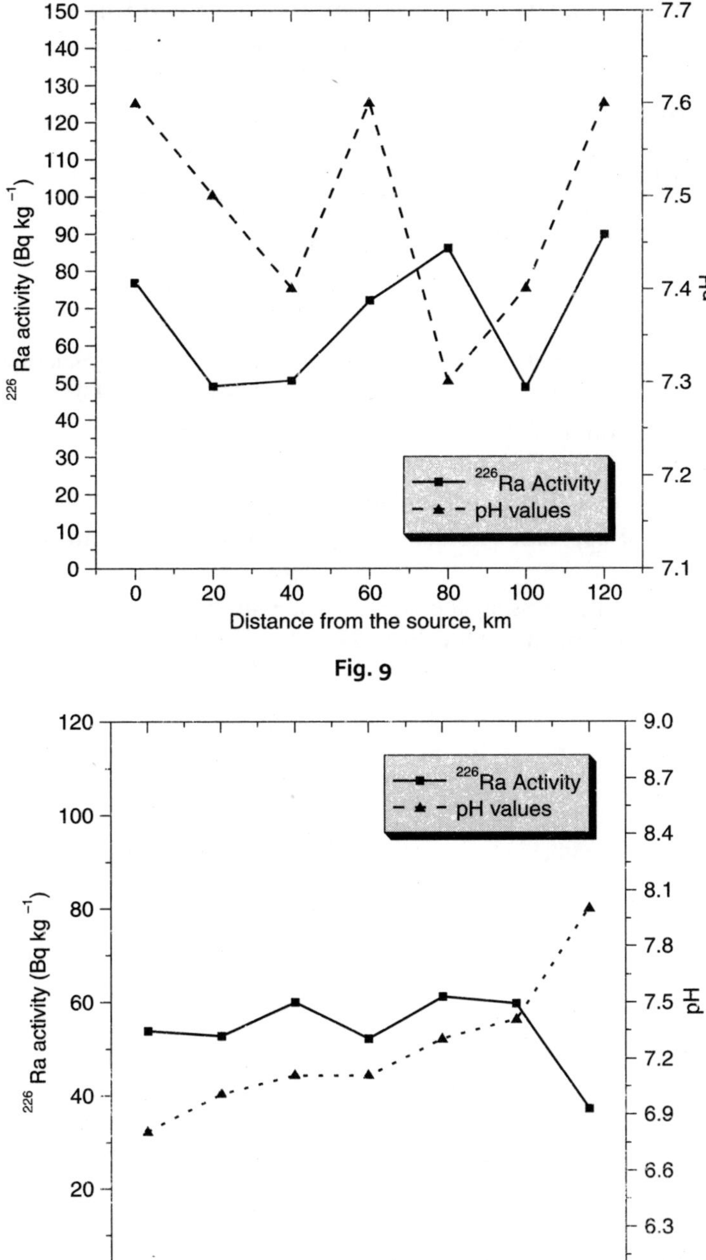

of piracy of an east flowing river over the gentle old plateau top, by headword erosion of a swift west-flowing river over the western slopes of the Ghats.

Acknowledgments

The authors are grateful to Prof. K Siddappa, Former Vice-Chancellor, Bangalore University and Dr. T Munikenche Gowda, Principal, SJC Institute of Technology, Chickballapur for their encouragement and support.

References

1. UNSCEAR *Sources and effects of ionizing radiation* United Nations Scientific Committee on the Effects of Atomic Radiation, New York. United Nations (1993)
2. EML procedure manual (Edited by Herbert, L.V. and Gail de Planque. 26th Edn.) *Environmental Measurement Laboratory* (1983)
3. M H Lee, C W Lee and B H Boo *J. Environ. Radioact.* **37** 1 (1997)
4. C L Roy *A practical approach to sedimentology* (Published by Alan and Unwin, London; Boston) (1987)
5. M C Abani *Methods for processing of complex gamma ray spectra using computers* (In: Refresher course in gamma ray spectrometry, June 27- July 1, BARC, India) (1994)

Identification of Bio-indicators for Impact Assessment in the Environment of Mangalore, Coastal Karnataka

V. Prakash[1*], K.M. Rajashekara[2] and Y. Narayana[3]

[1]Department of Physics, SDM Institute of Technology, Ujire, D.K, India
[2]Department of Physics, SJC Institute of Technology, Chickballapur, India
[3]Department of Studies in Physics, Mangalore University, Mangalagangothri, India
E-mail: *prakashamv@rediffmail.com

ABSTRACT

The activity of important radionuclides ^{232}Th, ^{226}Ra and ^{40}K were analysed in twelve medicinal plants collected from Mangalore and surrounding region by gamma spectrometry. The activity in the representative soil sample is also measured in order to study the transfer of radionuclides from soil to plant. In plant the average values of ^{232}Th, ^{226}Ra and ^{40}K activity were found to be 0.8 Bqkg^{-1}, 5.0 Bqkg^{-1} and 37.1 Bqkg^{-1} respectively. In soil the average values of ^{232}Th, ^{226}Ra and ^{40}K activity were found to be 54.7 Bqkg^{-1}, 64.2 Bqkg^{-1} and 384.3 Bqkg^{-1} respectively. The average values of transfer coefficient for ^{232}Th, ^{226}Ra and ^{40}K were found to be 0.02, 0.08 and 0.10 respectively. The ^{232}Th activity was below detection level for most of the plant samples, though the activity was significant in soils associated to these plants. The significant activity of ^{226}Ra in both plant and associated soil shows the higher root uptake of this radionuclide from soil. All the plants and associated soils showed significant ^{40}K activity. The plant *Mamia suregia* showed higher transfer coefficient for all the three radionuclides. The plant can be used as bio-indicator for the future monitoring of these radionuclides. The absorbed gamma dose rates prevailing in the study area were also measured using portable scintillometer. The results of these systematic investigations are presented and discussed in this paper.

Keywords: Gamma-ray detectors, Gamma-ray, Radiation effects.

Pacs No.: 07.85.Fv, 29.40.-n, 07.85, 61.82.Rx, 87.50.-a, 87.53.-j

1. INTRODUCTION

Mangalore city and the nearby areas of the coastal Karnataka are heading to become a region of major industrial activity centre. In view of this detailed studies on the radioactivity and trace element concentration in terrestrial, aquatic and atmospheric environs of the region has been carried out. As part of the program the activities of important radionuclides ^{232}Th, ^{226}Ra and ^{40}K in twelve medicinal plants, which are commonly found in the region, were analysed by gamma spectrometry. Medicinal plants are important for pharmacological research and drug development, not only when plant constituents are used directly as therapeutic agents, but also when they are used as basic materials for the synthesis of drugs.

To study the transfer of radionuclides from soil to plant, the activity in the representative soil sample was measured. The mobility and availability of radionuclides depend on several factors such as geochemical, biological and climatic conditions. The transfer factor, defined as the ratio of the concentration of the radionuclide in plant to the concentration of the radionuclide in soil, was calculated. From the concentration of radionuclides in soil the radium equivalent activity Ra_{eq} (single quantity which represents the specific activities of ^{232}Th, ^{226}Ra and ^{40}K) was estimated. Detailed gamma radiation survey was also been carried out in the environment using portable plastic scintillometer. The results of these systematic investigations are presented and discussed in this paper.

2. MATERIALS AND METHODS

2.1 Sample Collection

In the present study, 12 ayurvedic medicinal plants and associated soils collected from Moodabidri, near Mangalore, were analyzed for the concentration of natural radionuclides ^{232}Th, ^{226}Ra and ^{40}K. About 1 kg of soil sample was collected from the plant sampling site and brought to the laboratory. All the samples were carefully processed following standard procedure [1]. The soil samples were oven dried, ground, sieved and filled in air tight plastic containers and stored for one month to ensure secular equilibrium between ^{226}Ra and its short lived daughters. The containers were sealed carefully to avoid the escape of gaseous ^{222}Rn and ^{220}Rn [2]. These samples were subjected to gamma spectrometry analysis.

Approximately 5 kg of each plant was collected in a big polythene bag, brought to the laboratory, washed with running water to free from pollutants and cut into pieces. The samples were then ashed at 500°C in a muffle furnace. The ashed vegetation samples were filled in air tight plastic vials and stored for one month to ensure secular equilibrium between ^{226}Ra and its short lived daughters. These samples were subjected to gamma spectrometric analysis.

2.2 Experimental Set-up

Gamma spectra from the samples are recorded using a high efficiency 5″ × 5″ NaI (Tl) detector coupled to a 4K Multi-channel analyzer. The gamma ray spectrometer was calibrated using different standards and the efficiency of the detector was determined experimentally [3]. In order to compare the specific activities of materials containing different concentrations of radium, thorium and potassium, the radium equivalent activity concentration index Ra_{eq} was calculated [4].

3. RESULTS AND DISCUSSION

The results of activity of ^{232}Th, ^{226}Ra and ^{40}K in plant and associated soil samples are given in Table 1. The ^{232}Th activity was below detection level in most of the plant samples analyzed, though the activity was significant in soils associated to these plants. This shows that the uptake of this radionuclide by plant is negligible even though the soil on which the plant grows has significant amount of ^{232}Th. The variation of activity of ^{226}Ra shows near uniform trend in almost all plant samples. The observed ^{226}Ra activity was higher compared to ^{232}Th activity in all plants. The significant activity of ^{226}Ra in both plant and associated soil shows the higher root uptake of this radionuclide from soil. It is clear from the table that all the plants and associated soils show significant ^{40}K activity. The higher activity of ^{40}K in plants may be due to the continuous accumulation of ^{40}K through root uptake over a period of time. It is known that as an essential element of metabolism, plants take up potassium from soil in varied amounts depending upon the metabolism.

The activities of ^{232}Th, ^{226}Ra and ^{40}K in soil samples are compared with the reported values of other environs in Table 2. The average values obtained for ^{232}Th (54.7 Bq kg^{-1}) is higher in comparison with the population weighted world average value of 45 Bq kg^{-1}, the average value obtained for ^{226}Ra (64.2 Bq kg^{-1}) is approximately double the population weighted world average of 32 Bq kg^{-1}. The higher ^{232}Th and ^{226}Ra activity observed in the soil samples may be traced to the laterite type of soil prevailing in the region. The average activity of ^{40}K (384.3 Bq kg^{-1}) in the present study is less in comparison with the population weighted world average of 420 Bq kg^{-1} [5].

From the activity of naturally occurring radionuclides the radium equivalent activity was calculated. The radium equivalent activity varies in the range 143.5 Bq kg^{-1} to 189.1 Bq kg^{-1} with an average value of 172.1Bq kg^{-1}. The average radium equivalent activity in the soil samples lies much below the recommended limit of 370 Bq kg^{-1} [4].

The absorbed gamma dose rates prevailing in the study area, measured by plastic scintillometer, varies from 34.8 nGy h^{-1} to 52.2 nGy h^{-1} with an average value of 43.5 nGy h^{-1}. The radiation level follows almost a uniform pattern with dose rates close to the mean value except in some sampling stations. The present value lies between

Table 1 *The results of activity of ^{232}Th, ^{226}Ra and ^{40}K and transfer coefficients*

Sample		Activity in plant (Bq kg^{-1})			Activity in soil (Bq kg^{-1})			Transfer coefficient		
ID	Name	^{232}Th	^{226}Ra	^{40}K	^{232}Th	^{226}Ra	^{40}K	^{232}Th	^{226}Ra	^{40}K
MP1	*Putranjeeva roxburghii*	2.6 ± 0.6	5.3 ± 1.1	37.8 ± 2.9	58 ± 0.7	62.2 ± 1.2	358.2 ± 3.1	0.04	0.09	0.11
MP2	*Mamia suregia*	2.1 ± 0.9	7.8 ± 1.6	50.4 ± 4.1	41.6 ± 0.7	58.2 ± 1.2	335.3 ± 3.1	0.05	0.13	0.15
MP3	*Mesua nagassarium*	BDL	4.8 ± 1.1	35.7 ± 2.9	68.9 ± 0.8	61.4 ± 1.2	358.8 ± 3.1	–	0.08	0.10
MP4	*Saraca indica*	BDL	3.3 ± 0.7	26.4 ± 2.1	52.2 ± 0.7	58.3 ± 1.2	369.1 ± 3.2	–	0.06	0.07
MP5	*Syzygium jambolanum*	BDL	4.8 ± 1.1	39.5 ± 3.1	55.8 ± 0.8	63.7 ± 1.3	366.0 ± 3.2	–	0.07	0.11
MP6	*Garcinia indica*	BDL	3.7 ± 0.8	27.3 ± 2.2	48 ± 0.7	61.2 ± 1.3	356.7 ± 3.2	–	0.06	0.08
MP7	*Ficus benghalensis*	BDL	4.6 ± 1.1	41.2 ± 3.1	62.4 ± 0.8	64.8 ± 1.3	362.6 ± 3.2	–	0.07	0.11
MP8	*Flacartia montana*	BDL	3.7 ± 0.9	31.0 ± 2.4	59.9 ± 0.8	58.3 ± 1.2	361.5 ± 3.2	–	0.06	0.09
MP9	*Nyctanthes arbor-tristis*	BDL	7.6 ± 1.7	58.3 ± 4.7	48 ± 0.9	72.5 ± 1.5	431.0 ± 3.8	–	0.11	0.14
MP10	*Morinda citrifolia*	BDL	3.3 ± 0.7	26.1 ± 2.0	47.2 ± 0.9	70.5 ± 1.5	428.6 ± 3.8	–	0.05	0.06
MP11	*Ficus recemosa*	1.9 ± 0.7	5.4 ± 1.1	39.5 ± 3.1	60.4 ± 0.9	69.3 ± 1.5	435.3 ± 3.8	0.03	0.08	0.09
MP12	*Barringtonia acutangula*	3.3 ± 0.6	5.4 ± 1.0	32.3 ± 2.6	54.4 ± 0.9	70.3 ± 1.5	448.8 ± 3.8	0.06	0.08	0.07
Mean		0.8	5.0	37.1	54.7	64.2	384.3	0.02	0.08	0.10
Median		2.4	4.8	36.7	55.1	63.0	364.3	0.05	0.08	0.10
Range		BDL - 3.3	3.3 - 7.8	26.1 - 58.3	41.6 - 68.9	58.2 - 72.5	335.3 - 448.8	0.03 - 0.06	0.05 - 0.13	0.06 - 0.15

BDL-Below Detection Limit

the dose rate range of 18–93 nGy h^{-1} given for the world [5]. The measured dose rates were less in comparison with the all India average value of 88.5 nGy h^{-1} projected by UNSCEAR [5].

Table 2 *Comparison of 232Th, 226Ra and 40K activity in soil with other environs*

| Activity (Bq kg^{-1}) | | | | |
^{232}Th	^{226}Ra	^{40}K	Region	Reference
41.6 – 68.9	58.2 – 72.5	335.3 – 448.8	Moodabidri, Coastal	Present work
(54.7)	(64.2)	(384.3)	Karnataka	
29.8	20.1 – 62.3	61 – 316.7	Karnataka	[2]
14 – 160 (64)	7 – 81(29)	38 – 760 (400)	India	[5]
10 – 200(60)	3 – 60 (26)	40 – 800 (350)	Irland	[8]
18 – 135 (50)	35 – 228 (90)	281 – 711 (524)	Chaina	[9]
45	32	420	Popln. Wtd. world avg.	[5]

The soil-to-plant transfer coefficients (Table 1) vary in the range 0.03–0.06, 0.05–0.13 and 0.06–0.15 for ^{232}Th, ^{226}Ra and ^{40}K, respectively with corresponding average values of 0.02, 0.08 and 0.10. It is clear from table that the transfer coefficient for ^{40}K is significantly higher compared to ^{232}Th and ^{226}Ra, suggesting higher levels of uptake of ^{40}K. Similar findings were reported for Goa environs[6]. According to Argonne National Laboratory [7], the transfer coefficient values for ^{232}Th, ^{226}Ra and ^{40}K are 0.001, 0.04 and 0.3 respectively.

The soil to plant or soil to vegetation transfer factor, for a given type of plant and for a given radionuclide can vary considerably from site to site with season and time after contamination. These variations depend on such factors as the physical and chemical properties of the soil, environmental conditions and chemical form of the radionuclide in the soil. Furthermore, soil management practices such as ploughing, liming, fertilizing and irrigation can also affect the uptake of radionuclide by the vegetation. It is interesting to note that the plant *Mamia suregia* showed higher transfer coefficient for all the three radionuclides though the soil associated with *Mamia suregia* showed low activity when compared to the soil activity associated with other plants. This indicates that the uptake of radionuclides by the plant is moderately high compared to other plants. Therefore the plant can be used as bio-indicator for the future monitoring of these radionuclides contamination and for impact assessment.

4. CONCLUSION

The ^{232}Th uptake by plant is negligible when compared to the uptake of ^{226}Ra and ^{40}K by medicinal plants. The average activity concentrations for ^{232}Th and ^{226}Ra in soil in the environment of Moodabidri are higher than the population weighted world average value. The radium equivalent activity in the soil samples lie below the recommended limit. The radiation level follows almost a uniform pattern in the study area. The transfer coefficient for ^{40}K is significantly higher compared to ^{232}Th and ^{226}Ra. The transfer coefficient for a given plant and for a given radionuclide can vary considerably from site to site with season and time after contamination. The plant *Mamia suregia* can be used as bioindicator for the future monitoring of radionuclide contamination and for impact assessment.

References

1. EML procedure manual (Edited by Herbert, L.V. and Gail de Planque. 26th Edn.) *Environmental Measurement Laboratory* (1983)

2. Narayana Y, Somashekarappa H M, Karunakara N, Avadhani D N, Mahesh H M and Siddappa K *Health Phys.* **80** 24 (2001)

3. M C Abani *Methods for processing of complex gamma ray spectra using computers* (In: Refresher course in gamma ray spectrometry, June 27- July 1, BARC, India) (1994)

4. Beretka J and Mathew P J *Health Phys* **48** 87 (1985)

5. UNSCEAR United Nations Scientific Committee on the Effects of Atomic Radiation, Sources and effects of ionizing radiation. Report to the general Assembly with Scientific Annex, United Nations, New York (2000)

6. Avadhani PhD thesis of Mangalore University (2002)

7. Argonne National Laboratory (Manual for Implementing Residual Radioactive Materials Guidelines Using RESRAD, Version 5.0), ANL Argonne, IL. ANL/EAD/LD-2 (1993)

8. McAulay I R and Moran D *Radiat. Protect. Dosim.* **24** 47 (1988)

9. Ziqiang P, Yin Y and Mingquiang G *Radiat. Protect. Dosim.* **24** 29 (1988)

Measurement of Radon Exhalation Rate in Building Materials Southern Karnataka, India

C. Ningappa[1], Sannappa[2,*], Srilatha[3], and B. Venu[4]

[1]*Department of Physics, Vidya Vikas Institute of Engineering and Technology, Mysore, India*
[2]*Department of Studies in Physics, Jnana Sahyadri, Kuvempu University, Shimoga, India*
[3]*Department of Physics, Kodachadri First Grade College, Hosanagara, Shimoga, India*
[4]*Department of Physics, Jyoti Nivas College, Koramangala, Bangalore, India*
E-mail: *sannappaj@yahoo.co.in

ABSTRACT

Radon is produced in environment due to decay of radionuclides present in soil, ground water, gas deposition and building materials. The indoor concentration of radon depends upon radionuclides present in different types of building materials, construction and radon exhalation rate. Radon exhalation rates from different types of building materials collected from different places of southern Karnataka have been calculated using can technique. The radon exhalation rate in the building materials is mainly depends on ^{226}Ra present in the materials. In the present study, higher exhalation rate of radon was observed in fly ash and granite rocks of the study area. From the study, it is observed that exhalation rate of radon is higher than global and Indian average value.

Keywords: Exhalation of radon, SSNTDs, Dose, Can Technique.

Pacs No.: 92.60.Mt

1. INTRODUCTION

Radon is a radioactive gas originate from uranium decay series of natural sources has significant share in the total quantum of natural sources of exposure. Radon is produced in environment due to decay of radionuclides present in soil, ground water, gas deposition and building materials. Radon and its decay products in environment are known as carcinogens. Radon and its decay products pose significant health hazards, especially when concentrated in some enclosures such as underground mines, caves, poorly ventilated and badly designed houses. On the basis of epidemiological studies, it has been established that enhanced levels of indoor radon in dwellings can cause health hazard and may lead to serious diseases like lung cancer in human beings [1, 2]. Accurate, reliable and easy measurement of radon and radon exhalation rate from building materials are of paramount importance. The indoor concentration of radon depends upon radionuclides present in different types of building materials, construction and radon exhalation rate. The transfer rate of radon per unit area from the earth surface or any solid substance is referred as radon exhalation rate. The rate of radon exhalation from soil depends on geophysical parameters such as geology of the region, soil porosity, soil texture, soil humidity (moisture content), soil temperature and also on meteorological parameters. In this context, we select to measure radon exhalation rate in different types of building materials of Southern Karnataka state, India.

1.1 Study Area

The area of study is Bangalore rural, Bangalore urban, Ramanagara, Mysore, Tumkur and Kolar districts of southern Karnataka, India. The arhean rocks of South India are best developed in Mysore plateau. The rocks are peninsular and are widely distributed in Mysore region. These rocks are younger than peninsular gneiss. Different types of building materials have been collected in Southern Karnataka state, India for measuring radon exhalation rate.

2. METHODOLOGY

The simple and convenient method to measure the radon exhalation rate from building material samples in the laboratory is "Can Technique" using Solid State Nuclear Track Detectors [3-6].

3. RESULTS AND DISCUSSION

The measured values of radon exhalation rates from different types of building materials collected from different places of southern Karnataka that includes Mysore, Bangalore city, Bangalore rural, Tumkur and Kolar districts are summerised in Table 1. The activity of ^{226}Ra in different types of building materials varies from 25 to 178 Bq.kg^{-1} with a median of 36.5Bq.kg^{-1}. The radon exhalation rate from different types of building materials varies from 3.4 to 24.6 mBq.kg^{-1}h^{-1} with a median of 4.8 mBq. kg^{-1}h^{-1}.

Radon exhalation rate depends on geophysical parameter and radon concentrations in the construction materials. The data from the Table 1 shows that fly ash contains higher activity of radium and high radon exhalation rate. Exhalation rate also mainly depends on porosity of the building materials. Radon exhalation rate is directly proportional to the activity of ^{226}Ra present in the sample. Thus, higher exhalation rate of radon is observed in fly ash.

Slightly less exhalation rate of radon was observed in pink and gray granites when compared to fly ash, because pink and gray granites consist of slightly less activity concentration of radium than fly ash.

Minimum exhalation rate of radon was observed in soil brick, cement and cement brick. Because soil brick, cement brick and cement consists of less activity concentration of radium than other building materials [7-9].

Table 1 *Variation of 226Ra and Exhalation rate of 222Rn in differt building materials of Southern Karnataka State*

Sl. No.	Building materials	Activity of ^{226}Ra (Bq.kg^{-1})	Mass Exhal. rate of ^{222}Rn (mBqkg^{-1}h^{-1})
1	Pink granite	163	22.5
2	Gray granite	149	20.6
3	Fly ash	178	24.6
4	Mosaic	25	3.4
5	Soil brick	29	4.0
6	Fly ash brick	142	19.6
7	Cement brick	32	4.6
8	Cuddapah (Andhra)	29	3.8
9	Cement	35	4.8
10	Marble (Rajastan)	38	4.9
	Minimum	25	3.4
	Maximum	178	24.6
	Median	36.5	4.8

4. CONCLUSION

The radon exhalation rate mainly depends on 226Ra present in the materials. Higher the radium more will be the exhalation rate of radon. Exhalation of radon is different for different building materials. Higher exhalation rate of radon was observed in fly ash and granite rocks of the study area. From the study, it is observed that exhalation rate of radon is higher than global and Indian average value.

References

1. S Tokonami, T Iimoto, T Ichiji, K Fujitaka and R Kurosawa *Radiat Prot Dosim.* **63 (2)** 123 (1996)
2. UNSCEAR, *Source and Effects of Ionizing Radiation* [United Nations, New York] (2000)
3. A K Singh, D Sengupta, R Prasad *Appl. Radiat. Isot.* 51 (1999)
4. R Kumar, D Sengupta, R Prasad *Radiat. Meas.* **36** 551(2003)
5. A J Khan, R Prasad, R K Tyagi, *Nucl. Track Radiat. Meas.* **20** 609 (1992)
6. Mamta Gupta, A K Mahur, R G Sonkawade, K D Verma, Rajendra Prasad, *Indian. J of Pure and Appl. Phys.* **48**, 520 (2010)
7. M Sharaf, M Mansy, A El Sayed and E Abbas, *19th Int. Conf. on Nucl. Tracks in Solids Radiat. Meas* 491 (1999)
8. I P Farai and J A Ademola *Env. Radioact.,* **79 (2)** 119 (2005)
9. Ahmad and El – Arabi; *Eniron. Radioact.* **84** 51 (2005)

Indoor Gamma Exposure Rate Related to Radon Concentration in Dwellings Around BGML, Kolar District

Umesh Reddy[1], Vishwa Linga Prasad[2], J. Sannappa[3,*] and C. Ningappa[4]

[1]Department of Physics, Govt. First Grade College for women, Hassan, India
[2]Department of Physics, Govt. First Grade College, Sirsi India
[3]Department of Studies in Physics, Jnana Sahyadri, Kuvempu University, Shimoga, India
[4]Department of Physics, Vidya Vikas Institute of Engineering and Technology, Mysore India

E-mail: *sannappaj@yahoo.co.in

ABSTRACT

Indoor concentrations of radon and gamma exposure rate have been measured using Solid State Nuclear Track Detectors and Environmental radiation dosimeter during 2009–2010 around Bharth Gold Mine Limited (BGML), Kolar district, Karnataka state. Indoor radon concentration in the study area varies from 19.9 to 116.4 Bq.m^{-3} with arithmetic mean values of 58.7 Bq.m^{-3}. The highest activity was found in the dwellings that are close to mines. This is because of poor ventilation and high activity concentrations of radionuclides. A good correlation between indoor radon concentration and indoor gamma exposure rate has been observed in the study area.

Keywords: Radon, SSNTDs, Environmental radiation dosimeter, Gamma exposure, BGML.

Pacs No.: 92.60.Mt

1. INTRODUCTION

The knowledge of distribution of radionuclides and radiation levels in the environment is of important for assessing the radiation exposure from natural sources. Among all sources of radiation dose, inhalation of radon and its decay products contributed highest dose. Radon is a radioactive noble gas present in the atmosphere by decay of ^{226}Ra present in the environment and building materials. Exposure to radon and its decay products represents a significant fraction of the total natural radiation dose to the general public in indoor due to use of undesirable construction materials and also mining, milling and waste management of heavy minerals and sands. Gamma radiation levels mainly depend on primordial radionuclides present in earth crest, building materials and indoor radon concentration. Several studies shows that indoor radon concentration has a course correlation with the dose rate in air due to terrestrial gamma radiations [1-5].

In view of this correlation between indoor radon concentration and dose rate in air from terrestrial gamma radiation is studied in few dwellings around Kolar Gold Field (KGF) city. Bharat Gold Mine Limited (BGML) is one of the major gold mines in India and is located at 12°57′ N and 78°16′ E in Kolar district, close to Bangalore city.

2. METHODOLOGY

Radon concentrations in some dwellings around BGML were measured using Solid State Nuclear Track Detectors (SSNTDs) and indoor gamma exposure rate have been measured using Environmental Radiation Dosimeter [6-8].

3. RESULTS AND DISCUSSION

The measured values of indoor concentrations of radon and gamma exposure in the dwellings around BGML of Kolar district is summarized in Table 1. The data shows that indoor radon concentration in the study area varies from 19.9 to 116.4 Bq.m^{-3} with arithmetic mean values of 58.7 Bq.m^{-3}.

The maximum indoor concentrations of radon and its progeny have been observed in the dwellings at Champion place and Oorgaum place where the shafts are made for mining activity. This is because of burst that occurs rarely. Due to this the rocks are cracked and fissured. Through this radon enters into the buildings and environment. These two places are surrounded by huge quantity of waste materials dumped after extraction of gold. Also, the dwellings in these places are poorly ventilated [9-10]. Hence, in these places higher concentrations of radionuclides have been observed.

The depth of the mine is less compared to champion reef mine at Marikuppam. Hence slightly lower concentration of radon has been observed. Krishnavaram and Andersonpet are away from the mines and are surrounded by granite rocks. Hence slightly low concentration has been observed in these places.

The low concentration of radon has been observed at Robersonpet and BEML Nagar because these places are far away from the mines. A good correlation of 0.88 between indoor radon concentration and indoor gamma exposure rate has been observed.

Table 1 *Concentrations of radon and gamma exposure raye in the dwellings around BGML, Kolar district*

Sl. No	Location	Conc. of ^{222}Rn (Bq.m^{-3})	Gamma Exp. Rate (nGyh^{-1})
1	Robersonpet	22.8	156.6
2	Krishnavaram	63.7	278.4
3	Andersonpet	69.6	217.5
4	Marikuppam	87.1	565.5
5	Kolar city1	19.9	174.0
6	Oorgaum	104.7	748.2
7	Champion	116.4	765.6
8	Kolar city2	52.0	278.4
9	Bangarpet	57.9	287.1
10	BEML Nagar	31.6	139.2
	Minimum	19.9	139.2
	Maximum	116.4	765.6
	Median	60.8	278.4

4. CONCLUSION

This study has clearly established that fairly elevated activity of radon present in many non-uranium mines in India. The highest activity was found in the dwellings that are close to mines. This is because of poor ventilation and high activity concentrations of radionuclides. A good correlation of 0.88 between indoor radon concentration and indoor gamma exposure rate has been observed in the study area.

References

1. K Fujimoto *Health Phys.* **75** 236 (1998)
2. Pierre Ayotte, Benoit Levesque, Denis Gauvin, G Richard McGregor, Richard Martel, Suzanne Gingras, B William Walker, G Ernest Letourneau *Health Physics* **75(3)** 297(1998)
3. RC Ramola, Ganesh Prasad, Yogesh Prasad *Indoor and Built Environment* **16(1)** 83 (2007)
4. Ivan Barnet and Ivana Fojtíková *Radiat Prot Dosimetry* **130 (1)** 81(2008)
5. Makelainen I, Arvela H, Voutilainen A; *The Science of the Total Environment* **272(1)** 283 (2001)
6. Y S Mayya, K P Eappan, K S V Nambi *Radiat. Prot. Dosim.* **77(3)** 177 (1998)
7. UNSCEAR *Effects and Risks of Ionising Radiation* [General Assembly, United Nations, New YorK] (2000)
8. K S V Nambi, V N Bapat, M David, V K Sundram, C M Sunta, S D Soman *Rad. Prot. Dosim.* **24** 27(1987)
9. J Vaupotic, M.Sikovec and I Kobal *Health Physics,* **78** 559(1999)
10. J Sannappa J, M S Chandrashekara, L A Sathish, L Paramesh, P Venkataramaiah *Radiation Measurements,* **37(1)**, 55 (2003)

Radon and Thoron Levels in Few Dwellings of Bellary District

Ujjinappa[1], Nagabhusan[2], C. Ningappa[3] and J. Sannappa[4,*]

[1]*Department of Physics, Govt. First Grade College, Chamarajanagara, India*
[2]*Department of Physics, Govt. First Grade College, Hassan, India*
[3]*Vidya Vikas Institute of Engineering and Technology, Mysore, India*
[4]*Department of Studies in Physics, Jnana Sahyadri, Kuvempu University, Shimoga, India*
E-mail: **sannappaj@yahoo.co.in*

ABSTRACT

The measurement of concentrations of radon, thoron and their progeny levels in dwellings has been carried out using twin cup dosimeter based Solid State Nuclear Track Detectors in Bellary district, India. In the study area, the indoor radon and thoron concentration found to be varies from 25.5 to 72.4 $Bq.m^{-3}$ and 12.0 to 30.0 $Bq.m^{-3}$ with median values of 37.4 and 17.4 $Bq.m^{-3}$, respectively. Maximum concentrations have been observed in the dwellings having red oxide and granite floorings. Dwellings having mosaic, kadapa and marble show less concentration of radon and thoron. The dose due to radon, thoron and their progeny in the study area is less than global and Indian average values.

Keywords: Radon, Thoron, SSNTDs, Bellary, Dose, BGML.

Pacs No.: 92.60.Mt

1. INTRODUCTION

The risk of lung cancer for Workers in Uranium mines is known for a long time and is related to radon exposure. Radon, thoron and their progeny present in air contribute to nearly 50% of the average effective dose received by the human beings from the natural radiation environment [1]. Radon is a noble gas in the decay series of Uranium with a fairly long half life of 3.8 days. Being an inert gas, it can easily disperse into the atmosphere as soon as it released. Thoron is a decay product of thorium with half life of 55.6 sec. Radon and thoron in dwellings is generated from radionuclides such as ^{226}Ra, ^{235}U and ^{40}K present or existing or accumulated in soil, construction materials and water. Radon, thoron and their progeny may pose significant health hazards, especially when concentrated in some enclosures such as underground mines, caves, cellars or poorly ventilated and badly designed houses [2]. Thus measurement of concentrations of radon, thoron and their progeny levels in dwellings is of very important due to the health risk and to determine the design to of control strategies. The area of study is Bellary district of Karnataka State, India.

2. METHODOLOGY

Radon, thoron and their progeny concentrations in some dwellings Bellary district were measured using Solid State Nuclear Track Detectors (SSNTDs) [3-4].

3. RESULTS AND DISCUSSION

The measured values of indoor concentrations of radon, thoron and their progeny dwellings of Bellary district is summarized in Table 1. The data shows that indoor radon and thoron concentration in the study area varies from 25.5 to 72.4 $Bq.m^{-3}$ and 12.0 to 30.0 $Bq.m^{-3}$ with median values of 37.4 and 17.4 $Bq.m^{-3}$, respectively.

Table 1 *Concentrations of radon, thoron and their progeny level and inhalation dose due to these in the dwellings Bellary district*

Sl No	Types of flooring	No. of Dwellings	Conc. ($Bq.m^{-3}$)		Prog. conc. (mWL)		Dose ($mSv.y^{-1}$)
			^{222}Rn	^{220}Rn	^{222}Rn	^{220}Rn	
1	Mosaic flooring	4	30.1 ± 3.8	12.0 ± 2.5	3.8	0.2	0.83
2	Granite flooring	6	62.4 ± 5.6	28.1 ± 1.8	5.3	2.5	1.73
3	Red oxide	5	72.4 ± 2.8	30.0 ± 4.3	6.2	4.1	2.00
4	Tiles	6	44.8 ± 3.4	14.7 ± 3.4	2.3	3	1.23
5	Marble	4	25.5 ± 2.1	20.2 ± 2.6	1.6	0.98	0.73
6	Kadapa	3	28.2 ± 3.8	14.1 ± 1.8	1.4	0.3	0.79
	Maximum		25.5 ± 2.1	12.0 ± 2.5	1.4	0.2	0.73
	Minimum		72.4 ± 2.8	30.0 ± 4.3	6.2	4.1	2.00
	Median		37.4 ± 3.4	17.4 ± 3.2	3.0	1.7	1.03

The maximum indoor concentrations of radon, thoron and their progeny have been observed in the dwellings having red oxide floorings. Because, these dwellings are poorly ventilated and badly designed [5, 6].

Slightly less concentration of radon, thoron and their progeny have been observed in the dwellings having granite flooring. Granite contains higher concentrations of radionuclides. Concentration of radon and thoron mainly depends on primordial radionuclides such as ^{226}Ra, ^{235}U and ^{40}K.

Minimum indoor concentrations of radon, thoron and their progeny have been observed in the dwellings having marble, kadapa and mosaic. Marble, kadapa have less concentration of radionuclides and these dwellings are good ventilated. Hence less concentrations of radon, thoron and their progeny have been observed. The inhalation dose due to radon, thoron and their progeny in the study area varies from 0.73 to 2.0mSv.y^{-1} with a median value of 1.03 mSv.y^{-1}.

4. CONCLUSION

Concentrations of radon, thoron and their progeny have been measured in 28 dwellings of Bellary district. Maximum concentrations have been observed in the dwellings having red oxide and granite floorings and poor ventilation. Dwelling having mosaic, kadapa and marble shows less concentration. The dose due radon, thoron and their progeny in the study area is less than global and Indian average values.

References

1. UNSCEAR *Source and Effects of Ionizing Radiation* [United Nations, New York] (2000)
2. S Tokonami, T Iimoto, T Ichiji, K Fujitaka and R Kurosawa *Radiat Prot Dosim* **63(2)** 123 (1996)
3. UNSCEAR *Effects and Risks of Ionising Radiation* [General Assembly, United Nations, New YorK] (2000)
4. K S V Nambi, V N Bapat, M David, V K Sundram, C M Sunta, S D Soman *Rad. Prot. Dosim.* **24** 27(1987)
5. J Vaupotic, M Sikovec and I Kobal *Health Physics,* **78** 559 (1999)
6. J Sannappa, M S Chandrashekara, L A Sathish, L Paramesh, P Venkataramaiah *Radiation Measurements* **37(1)** 55 (2003)

Calculation of Ozone Production Rate for Indus-2 Synchrotron Radiation Beam Lines

M.K. Nayak[1,*], Saleem Khan[2], G. Haridas[1] and P.K. Sarkar[1]
[1]*Health Physics Division, Bhabha Atomic Research Centre, Mumbai, India*
[2]*IOAPDD, Raja Ramanna Centre for Advanced Technology, Indore, India*
E-mail: *nayak@rrcat.gov.in

ABSTRACT

Indus-2 is a synchrotron radiation sources facility in RRCAT Indore. It is having synchrotron radiation beam lines to tap synchrotron radiation (SR) in specially designed hutches. Ozone will be produced due to interaction of SR photon with air molecules present inside the hutches. Theoretical formulation for calculating the ozone concentration in the hutches was done. The ozone production rate and saturation concentration are calculated for three beam lines: BL-8, 11 & 12. The maximum saturation concentration was found to be in Beam line-11.

Keywords: Indus-2, Synchrotron radiation beam lines, Beam line hutches, Ozone.

1. INTRODUCTION

Indus-2 is a 2.5 GeV, 300 mA electron synchrotron radiation source (SRS) which has capacity to house 27 synchrotron radiation(SR) beam lines [1]. 8 beam lines are installed in the experimental hall, in specially designed hutches, of which 5 are under trial operation. When synchrotron radiation is brought out in air for experiments, through SR beam lines, it interacts with air molecules and produces ozone through the dissociation of oxygen molecules. The production of ozone in the hutch depends on, photon flux, its energy spectrum, irradiation time, path length, energy absorption, coefficient of radiolytic yield (G-value) and the hutch volume. The produced ozone gets removed mainly by two means: by chemical decomposition and ventilation. The decomposition time for ozone is about 50 minutes [2]. If no ventilation is provided the concentration of ozone within the hutch reaches a saturation value due to equilibrium between the production rate and removal rate, which is mainly due to its chemical decomposition. If we consider ventilation, then the removal of ozone is faster due to both chemical decomposition and ventilation and hence the saturation concentration will be lesser than with no ventilation.

2. SYNCHROTRON RADIATION SPECTRUM

The synchrotron spectrum from the bending magnet of Indus-2 is simulated using XOP 2.1 program [3]. The parameters used for the simulation are: magnetic radius- 5.5m, beam energy- 2.5GeV and beam current- 300 mA. The simulated spectrum in the range 10 eV – 50 keV is shown in the Fig. 1.

3. BEAM LINES & THEIR PARAMETERS CONSIDERED

The beam lines EXAFS (BL-8), EDXRD(BL-11) & XRD(BL-12) are considered for the present calculations. The parameters used in calculations are given in Table 1. For the calculation of ozone concentration, in the beam line hutches, the volumes considered are 10 m^3 & actual hutch volume.

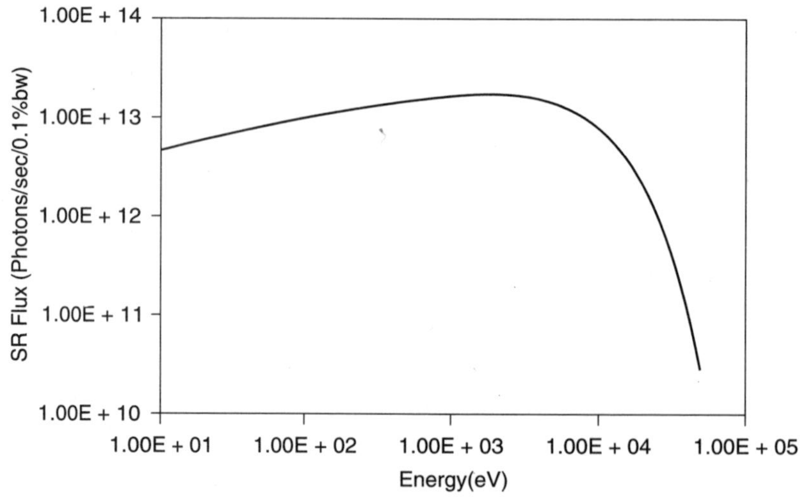

Fig. 1 *Simulated Synchrotron radiation spectrum for Indus-2*

Table 1 *Parameters of beam lines used for calculation of ozone concentration*

Beam line	Beam port (Indus–2)	Energy E (keV)	Integral flux φ (photons/s)
EXAFS (BL–8)	Dipole–4, 5^0 port	5 ± 175eV (pink)	1.07E + 13
EDXRD (BL–11)	Dipole–6, 5^0 port	5–50 (white)	8.0E + 14
XRD (BL–12)	Dipole–6, 10^0 port	5 ± 1eV (monochromatic)	2.75E + 10

4. DERIVATION OF EMPIRICAL RELATIONS

4.1 Ozone Production Rate

An empirical relation which gives the production rate of ozone when synchrotron radiation interacts with air, derived by the authors, is given below.

$$P_{03}\text{(molecules/min)} = 0.776 \times \sum_{i=1}^{n} \left[\varphi_i E_i \left(\frac{\mu_e}{\rho} \right)_i \right] Gx$$

Where,

$(\mu_e/\rho)_i$ = mass energy absorption co-efficient (cm^2/g) of ith energy bin

n = no of energy bins

E_i = Energy of SR photons (KeV) in the ith energy bin

ϕ_i = SR photon flux (photons/s) in the ith energy bin

x = path length in air (cm)

G = Radiolytic yield in air (10 ozone molecules/100 eV) [2, 4, 5]

The saturation concentration in ppm with ventilation is calculated using the formulations [2] given below.

$$C_{s,\ vent}\ (ppm) = P_{o3}\ (T_{eff}/V)$$

$$V = \text{volume of air(cc)}$$

T_{eff} is the effective ozone removal time given by

$$T_{eff} = \frac{(T_{vent} \times T_{decomp})}{(T_{vent} + T_{decomp})}$$

Where

T_{vent} = room volume divided by the exhaust volume per unit time

T_{decomp} = decomposition time (50 minute)

5. RESULTS

The results of the calculation performed on ozone saturation concentration are given in Table 2 to Table 4. Table 2 gives the energy absorption rate of SR in air, ozone production rate due to the absorbed energy and the saturation concentration for unit volume (1 m^3). Table 3 & 4 gives the saturation concentration for a volume of 10m^3 & actual hutch volume of BL–8, 11 & 12.

Table 2 *Ozone production rate & saturation concentration*

Energy absorption rate (keV/sec)	Production rate, P_{o3} (molecules/min)	Saturation concentration* (ppm) $C_{s,\ no\ vent}$
6.16E + 15	3.69E + 19	68.69

saturation concentration per unit volume ($1m^3$)

1 meter path length is assumed. SR spectrum (5 keV – 50 KeV) is sampled from the simulated spectrum for the calculations.

Table 3 *Saturation concentration with & without ventilation for BL-8, 11 & 12 for $10m^3$ hutch volume*

		Hutch volume – 10 m^3		
		Beam Line–8	Beam Line–11	Beam Line–12
Saturation concentration without Ventilation C_s (ppm)		0.30	6.87	7.8E–04
Saturation concentration with ventilation	1 Air Change Per Hour	0.17	3.75	4.26E–04
Cs, vent (ppm)	5 Air Change Per Hour	0.06	1.33	1.51E–04
10 Air Change Per Hour		0.03	0.74	8.36E–05
20 Air Change Per Hour		0.02	0.39	4.42E–05

Table 4 *Saturation concentration with & without ventilation for BL-8, 11 & 12 for actual hutch volume*

		Hutch volume – 10 m^3		
		Beam Line–8 (120.12 m^2)	Beam Line–11 (51.45 m^3)	Beam Line–12 (34.5 m^3)
Saturation concentration without Ventilation C_s (ppm)		0.025	1.34	2.3E–04
Saturation concentration with ventilation,	1 Air Change Per Hour	0.014	0.73	1.23E–04
Cs vent (ppm)	5 Air Change Per Hour	0.005	0.26	4.38E–05
	10 Air Change Per Hour	0.003	0.14	2.42E–05
	20 Air Change Per Hour	0.001	0.08	1.28E–05

From the calculation it is found that the saturation concentration is the highest for the beam line-11. Among 1the beam line considered, BL-11 is a white beam line where the integral flux at the experimental sample is the highest. The concentrations are to be measured in order to verify the calculated values for the different volumes $10m^3$ and actual hutch volume.

6. CONCLUSION

Calculation methodology for determining ozone concentration in air due to interaction of synchrotron radiation spectrum with air for Indus-2 beam line hutches is developed. The calculated saturation concentration for actual hutch volume at BL-8, 11 & 12 is 0.025 ppm, 1.34 ppm & 2.3E-04 ppm for 2.5GeV, 300mA respectively. The concentration is the highest for BL-11 and significant at BL-8, due to the reason that they are white and pink beam lines respectively. BL-12 being a monochromatic beam line shows negligible concentration even without ventilation. It is also planned to systematically measure the ozone concentration in order to validate the theoretically calculated results and suggest approximate ventilation scheme.

Acknowledgements

Authors express sincere thanks to Dr P.D. Gupta, Director, RRCAT, Shri Gurnam Singh, Head, IOAPDD for guidance and encouragement in the work. We are thankful to Dr. S.K.Deb, Head, ISUD for technical discussion during the present investigations. The authors are thankful to Dr.A.K.Sinha, ISUD, Dr. S.N.Jha, Spectroscopy Division, BARC/RRCAT, Shri Mishra and Shri K.K.Pandey, HPPD, BARC, for their help in providing data in respect of the beam lines concerned and for useful discussion of the results.

References

1. http://www.rrcat.gov.in
2. W.P.Swanson, Radiological Safety Aspects of the operation of electron linear accelerators, IAEA technical reports series no. 188, (1979)
3. www.esrf.eu/computing/scientific/
4. Guidelines for ozone mitigation at the APS (Advanced Photon Source), (May 1994)
5. W.P.Swanson, Toxic gas production at electron linear accelerators, SLAC-PUB-2470, (Feb 1980)

PIGE Analysis of Water and Soil Samples

Rajbir Kaur*, Mumtaz Oswal, Amandeep Singh, Navneet Kaur, K.P. Singh, B.R. Behera, Gulzar Singh and Ashok Kumar

Department of Physics, Punjab University, Chandigarh, India
E-mail: **rajbir.physics@gmail.com*

ABSTRACT

A direct non-destructive proton-induced gamma-ray emission (PIGE) technique for the determination of fluorine and sodium in a wide variety of water and soil samples is presented. The method is based on the nuclear reaction $^{19}F(p, p'\gamma)^{19}F$ and $^{23}Na(p, p'\gamma)^{23}Na$, respectively. The Knowledge of the elemental contents of water and soil is important because they are the most important source of minerals for both human being and animals. In the present work, the PIGE technique was used for the determination of concentrations of fluorine and sodium, respectively, in the water and soil samples, collected from areas around Chandigarh(India) and Malwa, one of the highly polluted regions of Punjab, India. PIGE proved to be an adequate, sufficiently accurate and convenient technique for the analysis of light elements in water as well as soil samples.

Keywords: PIGE, Fluorine, Sodium.

1. INTRODUCTION

Particle Induced Gamma-Ray Emission (PIGE) is an analytical technique based upon the measurement of γ–ray emission from a sample irradiated with the ions from a particle accelerator [1, 2]. In PIGE, an excited nucleus is produced by a nucleus normally by a $(p, p' \gamma)$, $(p, n \gamma)$ or $(p, a \gamma)$ reaction. The γ-ray emitted in the de-excitation is detected and the nucleus can be identified from the γ-ray energy. However, this nucleus can be a product of different nuclear reactions. It is, therefore, difficult to identify the target nucleus and furthermore to relate the detected γ-yield to the concentration of a nuclide or an element in the sample. The γ-ray yield changes rapidly with the incident particle energy. Therefore thick-target γ-ray yields have been measured and tabulated for different particle energies [3, 4, 5 and 6].

Light elements that can be determined with PIGE are often difficult to analyze with other methods. The calculation of fluorine and sodium concentrations makes use of following nuclear reactions: ^{19}F $(p, p'\gamma)$ ^{19}F and ^{23}Na $(p, p'\gamma)$ ^{23}Na and the corresponding gamma-ray energies are 197 and 440 keV, respectively.

2.   SAMPLE COLLECTION AND PREPARATION

2.1   Soil Samples

Soil samples were collected from different areas around Chandigarh and Malwa region of Punjab. The samples were dried below 60°C and subsequently ground into fine powder by using an agate mortar. The powdered samples were thoroughly mixed with high purity graphite powder in the ratio 1:1. The samples thus obtained were pressed using a hydraulic press so as to get the pellets of uniform thickness and to reduce the surface effects. These pellets were used as the targets for irradiation with proton beam.

2.2   Water Samples

Water samples were collected from the regions under study in polythene bottles which were cleaned thoroughly with water, 1 N nitric acid and then by deionised water (18 MΩ water).For sample preparation, technique similar to the one used by Aprilesi *et al.* [7] was followed. 50 ml of each of the water samples were taken in a beaker and 2.5 ml of saturated NaDDTC (sodium diethyldithio carbamate $C_5H_{10}NaS_2H_2O$) was added to precipitate the metal contents as their respective carbamates. The precipitates thus formed were collected on 25 mm diameter Nuclepore filters of pore size 0.4 µm, by vacuum filtration, using a Millipore vacuum filtration unit. A thin uniform layer was formed on the membrane that is used as the target for the PIGE measurements.

3.   EXPERIMENTAL SET-UP

A 3 MeV proton beam with a current of ~10 nA was used to bombard the samples. Proton beam was produced from Single Dee cyclotron situated at Panjab University, Chandigarh, India [8]. The beam size at the target position was 2 mm in diameter.

The target was positioned at 90° w.r.t. the beam direction and the characteristic γ-rays emitted from the samples were detected by an ORTEC HPGe detector (FWHM 1.9 keV at 1332 keV) at 135° as shown in Fig. 1. The beam current was integrated in the sample (for thick targets) and at a Faraday cup behind the target (for thin targets).

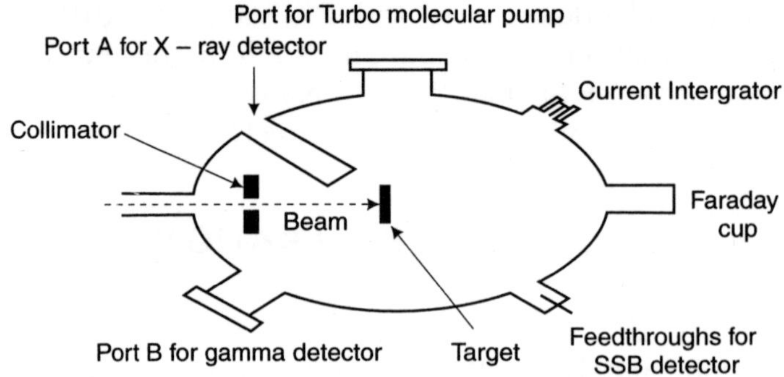

Fig. 1 *Schematic diagram of PIXE-PIGE Chamber.*

4. DATA ANALYSIS

4.1 PIGE Analysis of Soil Samples

For the quantitative PIGE analysis of soil samples, the unknown concentration for the specific element a in the analyzed sample is calculated using the following equation [9,10]:

$$C_{samp,\ a} = \frac{C_{ref,\ a}\ Y_{samp}(E_0)\ S_{samp}(E_0)}{Y_{ref}(E_0)\ S_{ref}(E_0)}$$

Where Y_{samp} and Y_{ref} are the yields of the measured γ–ray for the sample and reference material, respectively, at proton energy E_0. S_{samp} and S_{ref} are the stopping powers calculated for the sample and the reference material, respectively, at proton energy E_0.

4.2 PIGE Analysis of Water Samples

For the analysis of water samples, the gamma yield Y emitted by an isotope i of an element e within thin target bombarded by a proton beam of energy E is calculated using the following equation [11]:

$$Y(E_n) = \varepsilon_{abs}(E_Y) \cdot N_P \cdot \sigma(E) \cdot f_m \cdot f_i \cdot N_{av} \cdot A^{-1} \cdot \Gamma$$

where $\varepsilon_{abs}(E_\gamma)$ is the absolute efficiency of the detection system at the emitted energy E_γ, N_P the number of incident protons and $\sigma(E)$ is the nuclear reaction cross-section at the incident proton energy E. The parameters f_m, f_i, N_{av} and A^{-1} represent the mass fraction of the element e, the abundance of the isotope i related to the gamma emission, the Avogadro's number and the inverse of the atomic mass of the element e, respectively. Here Γ is the thin sample thickness and measured in $\mu g/cm^2$.

5. RESULTS AND DISCUSSION

Figures 2 and 3 show the spectrum of a soil sample and water sample, respectively collected from one of the areas under study. Table 1 shows the experimental results for the concentration of *F* and Na in the soil and water samples of these regions.

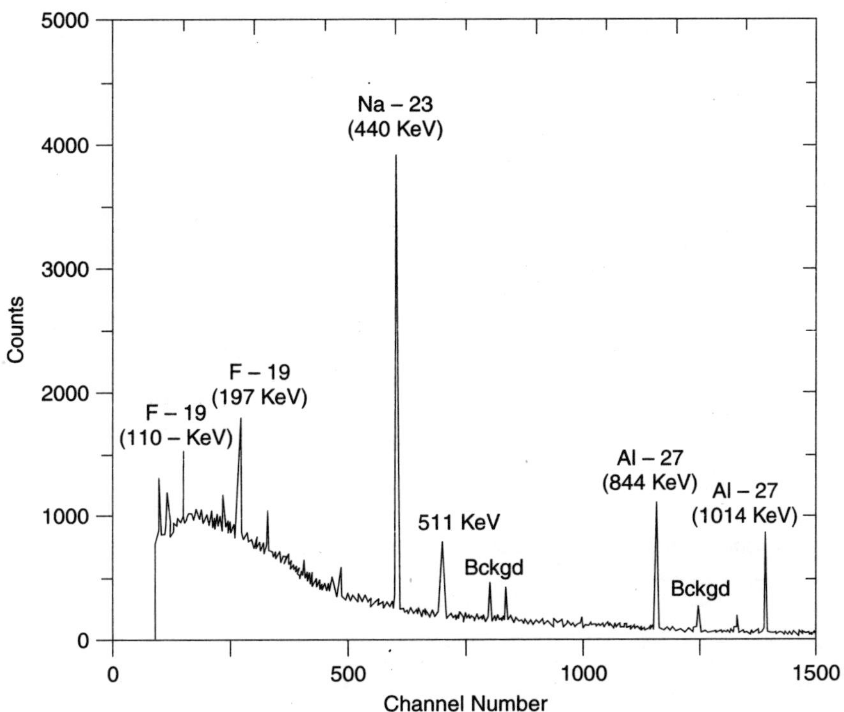

Fig. 2 *PIGE spectrum of a soil sample.*

6. CONCLUSION

The *F* content of the 13 soil samples collected from this region lies in the range 270-900 ppm whereas the standard content in the soil is 480 ppm. Fluorine has attracted much attention in recent years because of the apparent role it plays in health of human beings. Fluoride ions are readily absorbed by plants, especially from the more acid types of soils, but in any appreciable concentration they are highly toxic. Regional bedrocks are a primary factor controlling the spatial distribution of soil fluoride concentrations. Phosphate fertilizers are the major source of fluoride contamination of soil and water.

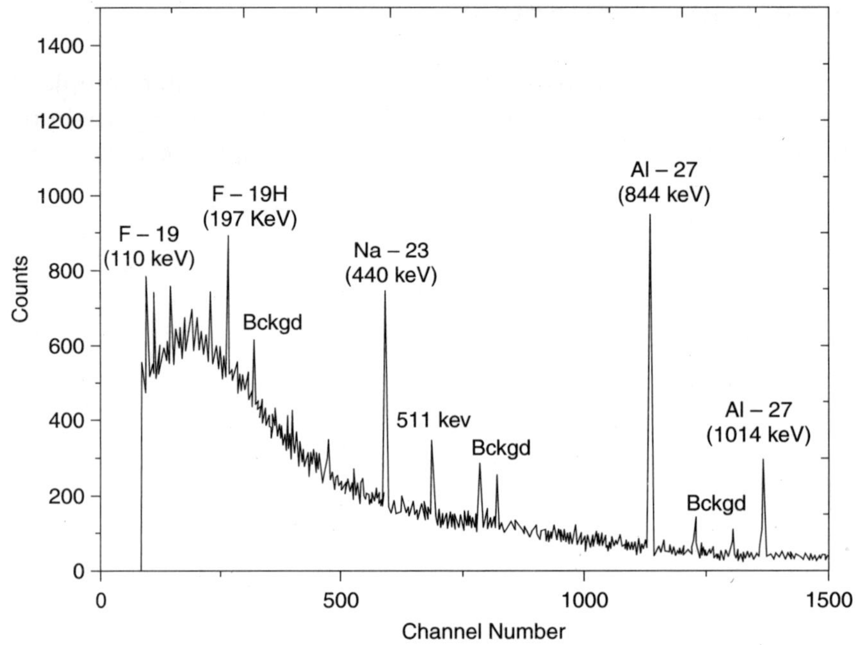

Fig. 3 *PIGE spectrum of a water sample.*

Table 1 *Experimental results for elemental concentrations.*

Sample No. & Name	Soil (ppm)		Water (mg/L)	
	F-19	Na-23	F-19	Na-23
1 KARANPUR	869 ± 2	12397 ± 124	*BDL	19 ± 2
2 PINJORE	845 ± 20	10076 ± 105	BDL	12 ± 2
3 MIRZAPUR DAM	657 ± 28	7445 ± 140	0.04	24 ± 3
4 INDO GLOBAL	587 ± 17	5956 ± 80	BDL	20 ± 3
5 BADDI INUDUS.	375 ± 16	5851 ± 90	BDL	30 ± 3
6 SESWAN DAM	535 ± 21	5857 ± 103	0.07	36 ± 3
7 HARIPUR	352 ± 17	6950 ± 111	BDL	20 ± 2
8 HMT	692 ± 22	5112 ± 86	BDL	13 ± 2
9 KIRATPUR	576 ± 20	5649 ± 69	BDL	BDL
10 BAHADURPUR	365 ± 2	10560 ± 23	BDL	15 ± 3
11 BHUCHOMANDI	300 ± 1	10540 ± 37	BDL	9 ± 1
12 BAJAKHANA	363 ± 3	10510 ± 22	BDL	15 ± 2
13 HARRAIPUR	275 ± 1	10254 ± 80	BDL	6 ± 1

*BDL is Below Detection Limit

The *F* content in water samples is either less than the optimum ranges of 0.5-1 mg/L or Below Detection Limit. Contamination of water resources with fluoride beyond acceptable limits is a health problem in many areas of Punjab. The results indicate that the fluoride level in groundwater is, in general, lower than the maximum permissible limit of 1mg/L.

The Na content of the 13 soil samples lies in the range 5000-13000ppm whereas the optimum Na content for the soil is 6300 ppm. Soil pollution in this region of Punjab may be due to pesticides and fertilizers as well as corrosion. Soils with high levels of sodium are called Sodic soils, which may affect plant growth by specific toxicity to sodium sensitive plants and nutrient deficiencies or imbalances.

The Na content in water lies in the range 5-40 mg/L whereas the optimum value is 20mg/L. High concentrations of sodium in water bodies mean high total minerals and tend to increase the corrosive action of water.

In conclusion, proton induced nuclear reaction based technique is more suitable for analyzing light elements like *F* and Na, which are often difficult to determine by other analytical techniques. PIGE proved to be an adequate, sufficiently accurate and convenient technique for the analysis of light elements in water as well as soil samples.

References

1. J.R.Bird, J.S.Williams, Ion Beams for Material Analysis, Academic Press, Australia, 1989, pp.719.
2. B. Borderie. *Nucl. Instr. Meth.* **175** (1980), p. 465.
3. G. Demortier. *J. Radioanal. Chem.* **45** (1978), p. 459.
4. M.J. Kenny, J.R. Bird and E. Clayton. *Nucl. Instr. Meth.* **168** (1980), p. 115.
5. A. Anttila, R. Hänninen and J. Räisänen. *J. Radioanal. Chem.* **62** (1981), p. 293.
6. J. Räisänen and R. Hänninen. *Nucl. Instr. Meth.* **205** (1983), p. 259.
7. G.Aprilesi, R.Cecchi, G.Ghermandi, G.Magnoni, P.Santangelo. *Nucl. Instr. and Meth. B* 3(1984)158.
8. N.K. Puri, P. Balouria, I.M. Govil, B.P. Mohanty and M.L. Garg, *Int. J.PIXE* **16** (1 and 2) (2006), pp. 7–20.
9. K.Nomita Devi, H.Nanda Kumar Sarma, Sanjiv Kumar. *Nucl. Instr. and Meth. B* 266 (2008)1605.
10. M. Mosbah, N. Metrich and P. Massiot. *Nucl. Instr. and Meth. B* **58** (1991), p. 227.
11. R. Mateus, A.P. Jesus and J.P. Ribeiro, *Nucl. Instr. and Meth. B* **229** (2005), p. 302.

Study of Distribution of ^{226}Ra, ^{232}Th and ^{40}K in Soil Samples of Chickmagalur Area and their Dose Estimations

Jayasheelan A.[1] and Manjunatha S.[2*]

^{1}*Department of Physics, GFGC, Sira-Karnataka, India*
^{2}*Department of Physics, PESIT & M, Shimoga-Karnataka, India*
E-mail: *manjunatha_s@yahoo.com*

ABSTRACT

In the present study, activity concentrations of ^{226}Ra, ^{232}Th and ^{40}K in soils are measured which were used to estimate the distribution of natural and artificial radioactivity in the environment of Chikmagalur. The study has three regions (i) U-QPC horizon, (ii) area surrounding the U-QPC and (iii) Mundre QPC region. The annual effective dose assuming 20% occupancy factor in region (i), due to ^{226}Ra, ^{232}Th and ^{40}K were 7.58×10^{-5} Sv, 8.01×10^{-5} Sv and 3.06×10^{-5} Sv with an average dose of 26.26×10^{-5} Sv due to all the three elements, in region (ii), were, 1.1×10^{-5} Sv, 2.42×10^{-5} Sv and 1.5×10^{-5} Sv with an average dose of 9.3×10^{-5} Sv, in region (iii), were, 1.08×10^{-5} Sv, 2.04×10^{-5} Sv and 1.51×10^{-5} Sv with an average dose of 8.86×10^{-5} Sv respectively. The concentration of radionuclide at U-QPC horizon is found to be higher in comparison with its surrounding areas and as well as world average [1-4].

Keywords: U-QPC, ^{226}Ra, ^{232}Th, ^{40}K, Soil, Activity concentration, Dose estimation.

Pase No: 89.60.-K, 91.62.Rt

1. INTRODUCTION

Chickmagalur is part of the bababudan belt and enriched with geological important mineral kingdom. The Uraniferous Quartz Pebble Conglomerate (UQPC) horizon lies near Chickmagalur town and stretches nearly 34 km long. UQPC are rare of its kind of sedimentary formation in the world and it is associated with relatively higher concentration of uranium and thorium and its daughters. The study of distribution radionuclide concentration in soil is essentially important to form a base line data and to take necessary precautions by the living population in the area. In the present study, concentrations of ^{226}Ra, ^{232}Th, and ^{40}K in soil were measured which was used to estimate the distribution of natural radioactivity in the environment of Chikmagalur.

2. MATERIALS AND METHODS

In present study, a total of 116 soil samples from 29 different locations around Chickmagalur were collected according to the standard procedure. The concentrations of ^{226}Ra, ^{232}Th, and ^{40}K in soil were measured using HPGe detector and estimated the distribution of natural and artificial radioactivity in the environment of Chikmagalur. The calibration of the instrument was done using recommended IAEA standard source of the same geometry and composition as the sample. To ensure better accuracy, some of the samples were analysed using similar system at Indira Gandhi Centre for Atomic Research (IGCAR), and the results obtained were same with negligible deviations. HPGe gamma ray spectrometer (GMX - 10190 – P, Resolution of 1.75 keV (FWHM) at 1.33 MeV and 17% Relative efficiency at 1.33 MeV) in the Department of Physics, University of Mysore, was used for the measurement of the activities of the ^{226}Ra , ^{232}Th, and ^{40}K in soil samples. Gamma spectra of the samples were obtained over an average period of 25,000 seconds. The calibration of the counting system was made with a standard source from Environmental Survey Laboratory, Kalpakkam. Some of the samples were analyzed using similar system in the IGCAR.

3. CALCULATION OF ACTIVITY AND ANNUAL EFFECTIVE DOSE

The activity concentrations of all the three elements were calculated by using the count rates and the corresponding dose calculations were made using the formula [5] $D = 0.462\ A_{Ra} + 0.604\ A_{Th} + 0.0417\ A_K$

Where D is the dose rate in nGy h-1 and A_{Ra}, A_{Th} and A_K are the activity concentrations of radium, thorium and potassium respectively in Bqkg^{-1}.The annual effective dose in Sv is calculated using the formula [6] for 20% occupancy dose (Sv) = $D \times 24 \times 365 \times 0.7 \times 0.2$

4. RESULTS AND DISCUSSION

The thorium/radium ratio was calculated for all the three regions, which can be used as an indicator of the relative occurrence of uranium and thorium [7]. The variation of activity concentration of ^{226}Ra, ^{232}Th and ^{40}K with different locations are given in Table 1. The annual effective dose in U-QPC region, due to ^{226}Ra, ^{232}Th and ^{40}K were

332   *Jayasheelan A. and Manjunatha S.*

7.58 × 10^{-5} Sv, 8.01 × 10^{-5} Sv and 3.06 × 10^{-5} Sv with an average dose of 26.26 × 10^{-5} Sv (Cosmic ray component of 30 nGy.h^{-1} included) due to all the three elements. In the area surrounding U-QPC, these were 1.1 × 10^{-5} Sv, 2.42 × 10^{-5} Sv, in the Mundre U-QPC region, the doses were 1.08 × 10^{-5} Sv, 2.04 × 10^{-5} Sv and 1.51 × 10^{-5} Sv with an average dose of 9.3 × 10^{-5} Sv due to the three elements with an average dose of 8.86 × 10^{-5} Sv.

5. CONCLUSION

The activity concentration and hence the gamma dose rates in the regions (i) U-QPC horizon, (ii) surroundings of U-QPC and (iii) Mundre QPC region were studied, they were found to be higher in U-QPC region in comparison with its surrounding areas and as well as world average [1-4,8].

Table 1 *BDL – Below Detection Limit *(plus cosmic ray component 30nGyh^{-1} [2])*

Location		Activity in Bqkg^{-1}			$\frac{^{32}Th}{^{226}Ra}$	Mean dose in nGyhr^{-1}		Total dose* in nGyhr^{-1}
		^{226}Ra	^{232}Th	^{40}K				
U-QPC	Min	BDL	25.1	171.0	0.35	^{226}Ra	62.2	
horizon	Max	509.9	341.5	1283.2	0.83	^{232}Th	65.8	215.4
	Mean	177.4	99.4	585.1	0.57	^{40}K	25.2	
Surroundings	Min	8.50	12.9	BDL	0.60	^{226}Ra	8.8	
of U-QPC	Max	47.7	43.5	20.8	2.53	^{232}Th	18.8	72.0
	Mean	20.6	28.4	276.2	1.40	^{40}K	11.9	
Mundre region	Min	8.6	12.9	BDL	0.42	^{226}Ra	8.34	
	Max	31.0	55.5	645.0	2.84	^{232}Th	15.8	68.5
	Mean	19.5	23.8	271.6	1.20	^{40}K	11.7	

References

1. United Nations Scientific Committee on the effects of Atomic Radiation (UNSCEAR), Effects and risks of ionising radiation, Report to the General Assembly, United Nations, Newyork, 2000.
2. Nambi K S V, Bapat V N David M, Env. Rad. **24** (1987) 31,155 (2003).
3. IAEA/RCA, Health Physics Division, BARC, Kalpakkam, India, 1989, pp 85-92.
4. Ramola R C & Negi M S, Einviron, Pollution, **22** (2004) 628.
5. Ramoloa R C, Prasad G & Prasad Y, Indor & Built Environ, **16** (2007) 83
6. S Dragovic et al Radiation Protection Dosimetry (2006) Vol 121 No. 3
7. Manjunatha S, Ph D Thesis, Univ. of Mysore, 1999.
8. Sonkawade R G, Kant k 7 Muralithar, Atm Environ, **42** (2008) 2254.

Committees

International Advisory Committee

S S Bajaj	AERB (India)
S Banerjee	AEC & DAE (India)
R K Bhandari	VECC (India)
S K Brahmachari	CSIR (India)
S Das	Calcutta University (India)
A Esposito	INFN, Italy
P D Gupta	RRCAT / BARC (India)
S Kailas	BARC (India)
H S Kushwaha	BARC (India)
Y S Mayya	ECIL (India)
T Nakamura	Tohoku University (Japan)
B Raj	IGCAR (India)
P Rama Rao	BRNS (India)
T Ramasami	DST (India)
M K Sanyal	SINP (India)
R K Sinha	BARC (India)
T Shibata	JPARC (Japan)
L Tomasino	ex-APAT (Italy)

Organising Committee

T. Bandyopadhyay (BARC/VECC)	
P Banerjee (SINP)	
S Basak (SINP)	Chairperson
S Basu (SINP)	Treasurer
S Bhattacharya (SINP)	
A De (SINP)	
D Ghose (SINP)	
M Nandy (SINP)	Convener
M Sarkar (SINP)	
P K Sarkar (BARC)	

<h1 style="text-align:right">Index</h1>